MOLECULAR BASIS OF TRANSITIONS AND RELAXATIONS

MIDLAND MACROMOLECULAR MONOGRAPHS

A series of monographs based on special symposia held at the Midland Macromolecular Institute.

Editor: Hans-Georg Elias

Volume 1   TRENDS IN MACROMOLECULAR SCIENCE
Edited by Hans-Georg Elias

Volume 2   ORDER IN POLYMER SOLUTIONS
Edited by Karel Šolc

Volume 3   POLYMERIZATION OF ORGANIZED SYSTEMS
Edited by Hans-Georg Elias

Volume 4   MOLECULAR BASIS OF TRANSITIONS AND RELAXATIONS
Edited by Dale J. Meier

Other volumes in preparation

ISSN 0141-0342

MOLECULAR BASIS OF TRANSITIONS AND RELAXATIONS

Edited by

DALE J. MEIER

Midland Macromolecular Institute

GORDON AND BREACH SCIENCE PUBLISHERS

London   New York   Paris

Editorial office for the United Kingdom

Gordon and Breach Science Publishers Ltd.
42 William IV Street
London WC2N 4DE

Editorial office for France

Gordon & Breach
7—9 rue Emile Dubois
Paris 75014

Library of Congress Cataloging in Publication Data
Main entry under title:
Molecular basis of transitions and relaxations.
    (Midland macromolecular monographs; v. 4)
    Symposium held Feb. 6, 1975 in honor of Raymond
Boyer, and sponsored by the Midland Macromolecular
Institute and the Dow Chemical Co.
    Includes bibliographical references and index.
    1. Polymers and polymerization — Congresses.
2. Mechanical chemistry — Congresses.   3. Boyer,
Raymond F.   I. Boyer, Raymond F.   II. Meier, Dale J.
III. Midland Macromolecular Institute.   IV. Dow
Chemical Company.   V. Series.
QD38L.M64   547'.84   77-76459
ISBN 0-677-11240-8

Contents

Dedication

TO RAYMOND F. BOYER ON HIS 65th BIRTHDAY

On February 6th, 1975, Raymond Boyer celebrated his 65th birthday. To honor him on this occasion, it seemed appropriate that a symposium be organized and co-sponsored by The Midland Macromolecular Institute (MMI) and The Dow Chemical Company. At the time of the symposium, he was a member of the Board of Directors of the Midland Foundation for Advanced Research (MFAR), the parent Foundation of MMI, and as a past advisor to MFAR he had played a major role in the setting-up of MMI. His career at The Dow Chemical Company was spectacular, for he was able to combine the top scientific position as a Science Fellow (his was the first such appointment at Dow) with a senior research management position as Assistant Director of Corporate Research for Polymers. The theme for the symposium to honor Ray Boyer was chosen to be molecular transitions, a field to which he had made (and continues to make) major scientific contributions. This volume represents the collection of papers delivered during the three day symposium, February 5–7, 1975.

The symposium was organized to broadly cover the selected field, with invited speakers asked to present overviews of their topics, ranging from theory to mechanical and dielectric relaxational spectroscopy to relationships between molecular transitions and physical properties. The sympoisum was originally planned as a two-day event, but the response of the many friends of Ray Boyer who wished to present contributed papers in his honor was such that the event was extended to three days, and even then the sessions were long.

A biographical and anecdotal account of Ray Boyer is given in the following pages so it is unnecessary to recount here the many honors and awards he has received for his work. However, a biography cannot adequately convey the impact and influence that he has had on others. Much of the work reported in this volume has been inspired by him, both directly and indirectly. His direct influence is shown by the fact that he is coauthor of five papers in this volume, and three of the five are coauthored with individuals from other organizations.

My thanks go to the speakers at the symposium, who, without exception, presented definitive accounts of their subjects. Also to be thanked are the session chairmen (Drs. J. D. Hoffman, E. Baer, W. H. Stockmayer, J. K. Rieke and T. Alfrey, Jr.) who managed to allow the lively and prolonged discussions and still maintain a reasonable schedule. The Symposium Secretary, Mrs. Julia T. Lee, deserves a special commendation, for it was she who so efficiently took care of the endless details concerning transportation, accommodations, meeting arrangements, meals, and so on. Finally, we must thank Ray Boyer for providing the

excuse and incentive to bring so many of his friends and collaborators together for the three days at what many regard as the most stimulating and delightful symposium they had ever attended.

We at MMI welcome Ray Boyer now as a colleague and join his many friends around the world in wishing him many more years of creative scientific work.

Midland, Michigan   D. J. MEIER

Biographical sketch of Raymond F. Boyer

Born: February 6, 1910 — Denver, Colorado.

Education: BS in Physics (1933); MS in Physics (1935); Hon. D Sc. (1955); all from Case Institute of Technology (now Case Western Reserve University).

Employment: At the Dow Chemical Company continuously since 1935, mostly in research and research administration in the field of polymers. Retired Dow February 28, 1975; joined MMI March 1, 1975 as Research Affiliate.

Publications: 90 papers, 20 U.S. patents, mostly in the area of polymers.

Fields of Specialization: Glass temperature, multiple transitions, molecular weights and molecular weight distributions, physical property—chemical structure correlations.

Honorary Positions: Chairman — Gordon Research Conference on Polymers, 1955.

Chairman — High Polymer Division APS, 1953.

Chairman — High Polymer Division ACS, 1956.

Chairman — Ad Hoc Committee on Characterization of Polymers for Materials Advisory Board, National Academy of Science 1965—66.

Program Chairman — ACS Biennial Polymer Symposium, 1962, 1972.

Member — Macromolecular Committee of IUPAC, starting 1972.

Member and Sometimes Chairman, Advisory Committee to Polymer Division, National Bureau of Standards, over past ten years.

Member — Committee on Science of Materials (COSMAT), National Academy of Science/National Academy of Engineering, 1971—72.

Member — Advisory Panel to Polymer Departments at Princeton University, University of Massachusetts and University of Detroit.

Member — Ad Hoc Advisory Panel to National Science Foundation on the Funding of Polymer Science.

Special Invitational Lectures:

1) IUPAC Polymer Group, Montreal, 1961.

2) Gordon Research Conferences on Polymers and Coatings.

3) Moreton Hampstead Polymer Conferences.

4) University Polymer Groups
   Massachusetts Institute of Technology (4)
   Case Western Reserve University
   University of Massachusetts
   Princeton University (3)
   University of Arizona
   University of Southern California
   Wayne State University (many times)
   Lowell Technological Institute
   University of Utah (3rd time in 1974)
   Florida State University
   University of Clausthal, Germany
   University of Manchester, England (4)

ETH, Zurich
University of Mainz, Germany
University of Marburg, Germany
Moscow Lomonosov University, USSR
University of Birmingham, England
Technical University, Łodz, Poland
Clausthal Technical University
Lehigh University
Arizona State University
Saginaw Valley State College
University of Michigan
University of Strathclyde
Herriot-Watt University
University of Liverpool
Imperial College, London
University of Manchester
Institute of Science and Technology, England

5) National Bureau of Standards (4).

6) Army Technical Center, Natick, Massachusetts.

7) Third International Conference on Thermal Analysis, Davos, 1971, Main Lecture.

8) Institute of High Molecular Compounds, Leningrad, USSR.

9) Georgian Academy of Sciences, Tbilisi, USSR.

10) Invited lecturer for 6 to 8 sessions, Course in Special Topics in Polymer Physics, Case Western Reserve University, Spring term, 1974.

11) Great Lakes Polymer Conferences, Detroit (3 times).

Awards: Society of Plastics Engineers, International Award in Polymer Science and Engineering, 1968.

American Chemical Society, Borden Award in Chemistry of Plastics and Coatings, 1970.

Plastics Institute (Great Britain): Biennial Swinburne Award, November 30, 1972.

July 8–22, 1972, Guest of Soviet Academy of Sciences.

On September 7, 1972, R. F. Boyer was appointed by the Board of Directors and Research Management of The Dow Chemical Company to be its first Research Fellow, whose title represents the top of the research professional ladder in Dow.

September 16–23, 1973, Guest of Polish Academy of Sciences. Visiting Professor, Case Western Reserve University, 1974. September 1974, Invited Symposium Lecturer, IUPAC Meeting on Macromolecules, Madrid. Senior Visiting Fellow, Manchester University, Fall, 1975.

Hobbies: Theories of Dieting and Health
Causes of Coal Mine Explosions
Dynamics of Invention
Sailing

Anecdotal account[†]

My interest in multiple mechanical transitions and relaxations was inspired by a 1960 Gordon Polymer Conference lecture by the late Dr. Karl Wolf, then of BASF. He described an extensive body of torsion pendulum data by himself and his colleagues (Heinze, Schmieder and Schnell) on amorphous polymers and copolymers, crystalline polymers, and vulcanized elastomers. His numerous plots of dynamic loss peaks at 1 Hz as a function of temperature permitted me to make mental comparisons with quasi-static data already familiar to me from thermal expansion, heat capacity and other methods. Moreover, it seemed clear to me that if his data were plotted on a relative temperature scale, T/Tg, such that Tg was the reducing parameter, many of his loss curves would coincide approximately. These concepts were written up in a privately circulated report which was distributed to interested persons prior to the 1961 Gordon Polymer Conference where discussions were held with some of the recipients. Moreover, this subject was treated in some detail at my 1961 IUPAC main lecture in Montreal. Dr. Gordon Kline invited me to repeat this lecture before his National Bureau of Standards group in October 1961. Dr. Norman Bekkedahl, who was in the audience, proposed to the Editorial Board of Rubber Chemistry and Technology that I prepare a review paper on the subject. This paper subsequently appeared in 1963, greatly expanded from the original 1961 version.[1] I have sent out over 500 reprints of this article, it was on several occasions reprinted in lots of 100 copies on one college campus; and it has been translated into Chinese. Requests for reprints are still received.

This topic was discussed by me at the 1963 Gordon Polymer Conference under the chairmanship of Professor M. Szwarc, with Dr. Turner Alfrey as discussion leader. A question by Dr. J. J. Hermans gave rise to a long and heated discussion: "Why were there discrete, reasonably sharp loss peaks instead of one single broad peak?" This question was certainly not answered on that occasion and only became clarified in part on subsequent occasions as various scientists proposed specific loss mechanisms to explain specific loss peaks in a given polymer or classes of loss peaks common to a variety of polymers.

Whereas the 1963 review was concerned primarily with amorphous polymers and copolymers, the intervening years have been spent in preoccupation with problems presented by highly crystalline polymers such as PE, PEO and POM. This work has been summarized in my 1972 Swinburne Award lecture[2] and in even more completely my 1974 IUPAC Madrid lecture.[3] Of course, the monumental book of McCrum, Read and Williams[4] appearing around 1967 greatly simplified all of our subsequent studies.

[†] This is part of the anecdotal account I gave at my MMI Symposium Banquet on February 6, 1975.

References

1. R. F. Boyer, *Rub. Chem. Tech.* **36**, 1303−1421 (1963).
2. "Multiple Transitions in Semi-Crystalline Polymers," Seventh Swinburne Award address, in five parts, available as a reprint from the author or from the Plastics Institute, 11 Hobart Place, London SW1W 0HL, England. Part I appeared in *Plastics and Polymers* **41**, 15−21 (1973); parts II and III, *ibid.,* pages 71−78.
3. R. F. Boyer, *J. Poly. Sci.* **C-50**, 189−242 (1950).
4. N. G. McCrum, B. E. Read and G. Williams, *Anelastic and Dielectric Effects in Polymeric Solids,* Wiley, New York, 1967.

Transitions and Relaxations in Polymers: Phenomena and Methods

G. ALLEN,

Science Research Council, State House, High Holborn, London, United Kingdom.

In this review paper the dilatometric method is used to identify transition phenomena observed in polymeric materials. Two classes of polymer, amorphous and crystallisable, are thus defined and the nature of the glass and crystallisation transitions documented. On closer inspection subsidiary transitions of a "second" order nature are also observed in amorphous polymers; partially crystalline polymers show subsidiary first and "second" order transitions.

The molecular interpretation is greatly facilitated by the use of relaxation methods since different methods have different sensitivities towards particular motions. The wide frequency range of relaxation methods is also important. The "second" order transitions are characteristically activated processes. Experimental data and molecular models can be correlated also through time correlation functions. Thus the molecular basis of transitions is established.

In recent years photon correlation spectroscopy, Brillouin spectra and neutron scattering spectroscopy have begun to contribute to our understanding of the molecular motions of transitions and relaxations in polymeric materials.

This introductory paper can give no more than a cursory survey of this enormous topic, leaving others to delve into detailed analyses of theories or specific techniques or individual materials. Nevertheless I hope to reflect something of the historical development, to note changes in approach and even to remind you of some techniques which are not actually covered in the succeeding papers.

It is appropriate to note that there *are* papers[1-3] in the literature concerned with transition phenomena in polymeric systems which predate the distinguished and stimulating career of R. F. Boyer. In the reign of King George III John Gough[1] observed in a letter published in the Proceedings of the Manchester Literary and Philosophical Society:

if a thong of Caoutchouc be stretched in water *warmer* than itself it retains its elasticity . . . If the experiment be made in water *colder* than itself it loses part of its retractile power, being unable to recover its former figure. But let the thong be placed in hot water while it remains extended for want of spring and the heat will immediately make it contract briskly.

1

Gough had observed crystallisation in natural rubber. No doubt your forebears in North America were also witnessing the same transition in the properties of materials made from raw natural rubber in the winters of the same period.

The crystallisation of natural rubbers compounded in various forms has been well documented and in 1934 Ferri[4] published his much presented results on the glass transition of natural rubber (Figure 1) observed by measuring stress at constant length under conditions where the material did not crystallise.

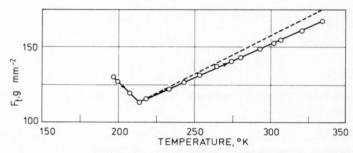

FIGURE 1   Stress F_t required to maintain a vulcanised natural rubber sample at constant length (350% extension). The broken line represents a correction for thermal expansion.

1   DILATOMETRIC STUDIES

Nevertheless we will commence our survey of the phenomena by concentrating on possibly the most simple technique, combining adequate sensitivity with good resolution, and certainly the one favoured by Boyer in his early contribution to the field.[5]

Dilatometric studies, carried out at reasonably low rates of cooling with polymers initially in their rubbery state immediately subdivide the materials into those which crystallise and those which don't. If we take natural rubber as typical of the crystallisable materials and Boyer's beloved atactic polystyrene as a non-crystallisable polymer, Figures 2 and 3 show the salient features observed.

Crystallisation in the natural rubber[6] sample is accompanied by a contraction in volume in the region of $0°C$, not dissimilar to that observed in a simple liquid, though somewhat more diffuse and rather smaller in magnitude. It does have the semblance of a first order thermodynamic transition. The rubber becomes opaque tough leather. There is no such transition in polystyrene; instead there is a transition embodying a change in slope of the V vs. T plot in the region of $100°C$ which corresponds to a change in physical state from a sluggish rubber to a transparent brittle glass. This transition shows features typical of a second order thermodynamic transition. Note however that a similar transition is observed in the

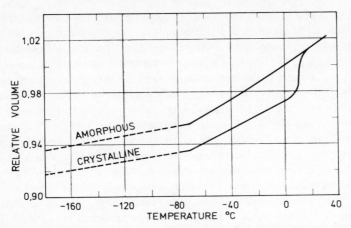

FIGURE 2   Changes in volume of non-crystalline and partially crystalline natural rubber as a function of temperature.

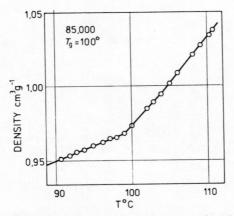

FIGURE 3   Changes in volume of atactic polystyrene as a function of temperature.

crystallised natural rubber at about −70°C. Now the natural rubber is transformed into a hard brittle material. The molecular explanation of these phenomena is universally accepted to be associated with the loss of motional degrees of freedom of the individual polymer molecules. In the rubber each molecule is rapidly wriggling through a whole set of conformations available to it by means of internal rotations about σ-bonds in the backbone of the molecule. As the matrix of entangled wriggling chains cools long range molecular motions slow down, and in the case of molecules having long sequences of repeat units in steroregular arrays nucleation, followed by crystal growth, occurs on a time scale which might vary from fractions of a second to days according to the chemical nature of the polymer. An unstrained sample in a dilatometer will nucleate sporadically throughout the matrix and randomly oriented crystallites of various sizes will develop. In a high molecular weight polymer a given polymer molecule may pass through several crystallites and in the intervening amorphous rubbery regions be hopelessly entangled with other chains which may also have sequences locked in neighbouring crystallites. Thus crystallisation of the materials tends to be far from complete and may stop in dynamic equilibrium with, say, 50% of the material still present as a rubber. The relatively small volume contraction noted above arises because only a limited amount of crystallisation occurs. Unit cell measurements on crystalline polymers[7] show that the contraction to be expected for complete crystallisation is comparable in magnitude to that observed in the crystallisation of simple liquids. There is however a more pronounced difference for polymeric materials in the temperature of crystallisation and melting. For natural rubber[8] this relationship is shown in Figure 4.

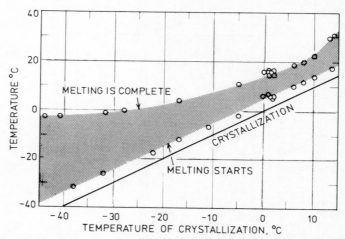

FIGURE 4   Relation between the temperature of crystallisation and the melting range for partially crystalline natural rubber.

Unlike crystallisation the transition observed in a noncrystallising polymer does not involve the creation of a new phase. On a molecular basis it corresponds to the freezing in of the tangled array of randomly coiled chains, each chain in one of the millions of conformations available to it. The long-range wriggling motion has disappeared in the glass. The rubber has a coefficient of thermal expansion ca 6×10^{-4} K^{-1} and the glass ca 2×10^{-4} K^{-1}. The transition in slope of the V, T plot is gradual rather than sharp and of course no discontinuity in volume is involved.

The second transition observed in the crystallised rubber at $-70°C$ (Figure 2) is also a glass transition. It corresponds to the amorphous regions of the partially crystalline polymer undergoing its rubber to glass transition and below this temperature the crystallites are embedded in a glassy rather than a rubbery matrix. Thus the transitions in the two classes of polymer are explained. It is important to note here that for linear polymers both T_m and T_g measured as a function of molecular weight, rapidly reach limiting constant values. These values are characteristic of a particular material. In the rest of this paper, where transition temperatures are quoted, the values will refer to the high molecular weight limit. Note also that in the volume—temperature curve of polystyrene a transition is observed apparently in the melt (i.e. rubber). We shall return to this in Section 2.3.

1.1 Thermodynamic effects

The phenomenology of the transitions can be explored by conventional thermodynamic investigations. The crystallising/melting transition shows all the phenomena expected of a first order liquid/crystal transition. The melting point, T_m, though more difficult to define because of the diffuse nature of the transition, obeys the Clausius—Clapeyron equation:

$$\frac{dT_m}{dp} = \frac{T\Delta V_m}{\Delta H_m}$$

Indeed, in a crosslinked polymer, advantage can be taken of the fact that the system supports a stress, to measure T_m at different levels of constant stress and it is found that the corresponding equation is obeyed at constant pressure:

$$\frac{df}{dT_m} = \frac{f}{T} - \frac{\Delta H}{\Delta L}$$

where ΔL = difference in length of sample when amorphous and when completely crystalline. The glass transition shows superficially the behaviour expected for a second order transition:

$$\frac{dT_g}{dp} = \frac{TV\Delta\alpha}{\Delta C_p}$$

where $\Delta\alpha$ is the difference in coefficient of thermal expansion of the rubber and glass and ΔC_p is the difference in heat capacity of the rubber and glass. In the case of polystyrene for example $dT_g/dp \sim 3100$ kN^{-1}m^2 and $(TV\Delta\alpha/\Delta C_p) = 3600$ kN^{-1}m^2. There are however subtle factors which obtain in the use of these equations in practice,[9] relating to whether the glasses formed in successive experiments have higher than normal densities and whether they have normal enthalpies. Since

we are principally interested in molecular motion we shall not enlarge further on
thermodynamic studies.

1.2   Rates of cooling

Studies of both types of transition at different rates of cooling are informative.
For many crystallisable polymers it is possible to cool sufficiently rapidly to
eliminate crystallisation entirely (e.g. natural rubber, isotactic polystyrene). At
slower rates, the degree of crystallisation is affected and also the melting point of
the sample produced, since most polymers have an optimum rate of crystallisation roughly midway between T_g and T_m as illustrated in Figure 5. The observations are consistent with conventional explanations in terms of the requirements
for nucleation and crystal growth.

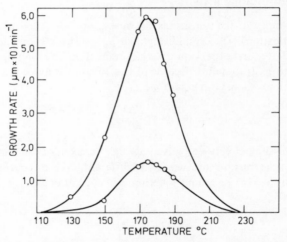

FIGURE 5   Rate of crystallisation, measured as rate of spherulitic growth, as a function of
temperature for isotactic polystyrene.

The effect of rate of cooling on the glass transition is simple. The observed
value of T_g is raised with increasing rate of cooling. The effect is characteristic
of a rate phenomenon: the value of the measured parameter being simply related to
the time-scale of the experiment since as the wriggling motion of the chains slows
down the matrix appears to be a glass when the time-scale of molecular conformational rearrangements is longer than the time-scale on which the physical
property being influenced is observed. The energy barrier involved in the rate
process is the hindrance to segmental motion in the polymer chain due to interchain interactions and barriers to internal rotation.

1.3 Free volume

The statistical concept of "free volume" was introduced by Eyring and others in the mid-1930's in an attempt to explain the thermodynamic properties of simple liquids using statistical mechanics to calculate the conformational entropy of an assembly of molecules plus holes. This concept carried over into the molecular understanding of melt flow in polymers — free volume providing "room for manoeuvre" for the activated wriggling segments of polymer molecules. The observation that high rates of cooling lead to glasses of lower densities was explained therefore in terms of a larger fraction of free volume being trapped in the glass matrix. Of course it implies that different rates of cooling (and variations in other combinations of external variables) produce from the same melt glasses of different structure. There is a considerable literature[10] on free volume treatments of the physical properties of glasses. Because the properties of the glass depend on its history, apparently similar materials give detectably different properties and caution is needed in testing thermodynamic identities as noted in Section 1.1. However the elegant work of Petrie[11] has shown that with sufficient care and patience glasses from a given melt can be transformed to a reproducible state by annealing at temperatures a few degrees below T_g. This work emphasises the importance of applying thermodynamic analyses only in circumstances in which the glasses compared *are* identical.

In a sense with the Gibbs and di Marzio[12] statistical theory of glass formation, the free volume analysis comes full circle. In this theory the glass temperature is related to a temperature T' at which the conformational entropy of the assembly of polymer chains on its lattice vanishes. The lattice includes a certain fraction of vacant sites, i.e. free volume.

1.4 Subsidiary transitions

More careful examination of volume–temperature curves, especially below T_g, shows that subsidiary first order and second order transitions occur calling for some detailed modification of molecular explanations and implying that molecular motion, other than the normal modes of vibration of the molecules persists at lower temperatures.

Crystallised polymers often show subsidiary first order transitions at low temperatures. X-ray analyses show that generally these involve a change in crystal structure and sometimes, in fibres, loss in long range order in one direction. Corradini[13] has reported a broadening in the x-ray diffraction pattern in for example poly (*trans* 1:4 butadiene) at 106°C, where a first order transition occurs 38°C below T_m. From estimates of ΔH_m and ΔS_m at T_m and ΔH_T and ΔS_T the degrees of conformational freedom released in these subsidiary transitions can be assessed. What is quite clear is that melting and recrystallisation occurs on a major scale at these transitions. Obviously such phenomena occur more readily if the amorphous residue in the polymer sample is in its rubbery state, i.e. if the temperature of the subsidiary transition is above T_g. In addition second order transitions are observed in the crystalline phase over a very wide temperature range.

Uncrystallised polymers often have subsidiary transitions similar in character to glass transitions below T_g. However, they correspond to very small changes in the coefficient of thermal expansion (e.g. $2 \rightarrow 1.5 \times 10^{-4}$ K^{-1}) and are often difficult to locate precisely by dilatometry. The amorphous regions of crystalline polymers probably show similar transitions but they are even more difficult to locate, especially at high degrees of crystallisation. Figure 6 shows dilatometric results for poly(methyl methacrylate) where two subsidiary transitions at $\sim 7°C$ and $\sim 60°C$ can be clearly defined. It also shows the effect of rate of cooling on the transitions.

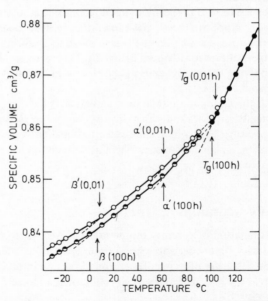

FIGURE 6   Changes in volume of non-crystalline (atactic) poly(methylmethacrylate) at two different rates of heating. Three transitions are observed at respectively 104°C (T_g), 60°C and 7°C (Wittman and Kovacs).

To establish the molecular origins of these subsidiary transitions we need to turn to other techniques. We could list the results of calorimetry or measurements of other bulk static properties which would corroborate our observations, but we need spectroscopic techniques which will act as probes for the dynamic behaviour of these materials in order to provide a basis for the molecular understanding for the phenomena. Thus relaxation methods are important because they cover a very wide frequency range (20 decades) and also show selective sensitivity to different transitions.

2 RELAXATION METHODS

All relaxation methods for studying transitions have one common feature; an external field is applied to the materials under investigation and the response of the material is measured. The field couples weakly to the specimen so that the natural dynamic behaviour of the material is not seriously perturbed. The response may be observed in principle as a function of the frequency variation of the external field (continuous wave experiments) or of the time variation of the field (pulse methods in which the decay in response to an applied pulse is monitored). The data from the two methods can be interconverted by use of the appropriate Fourier transforms. In all experiments of this kind the operative physical process is the conversion of energy acquired by the sample into heat.

A sinusoidal field of period $f = f_0 \sin \omega t$ induces a strain in the sample

$$\sigma = \sigma_0 \sin(\omega t + \delta)$$

If the response of the sample is much faster than the period of the field then the applied field and strain remain in phase ($\delta = 0$) but if the response of the system is sluggish a phase lag develops. If the response of the system is characterised by a relaxation time τ then when $\omega < \tau^{-1}$, δ is small, but δ increases to a maximum value when $\tau \sim \omega$ and decreases again when $\omega > \tau^{-1}$. The ratio of the amplitude of the applied field to the resulting stress is a simple quantity X when the response of the system is very fast. In the region where response is sluggish X becomes a complex quantity which can be resolved into an in phase (X') and an out of phase (X'') component.

$$X^* = X' + iX''$$

and

$$X' = \frac{X}{1 + \omega^2 \tau^2} \qquad X'' = \frac{X \omega \tau}{1 + \omega^2 \tau^2}$$

The behaviour of the system over a range of frequencies is shown in Figure 7(a). The maximum value of X'' indicates the frequency at which the response of the sample is of the order of the time scale of the experiment. In a polymer sample τ will be expected to decrease as the temperature of the sample is lowered on the general grounds that molecular motion is slowed down. Thus if we inspect a sample at constant frequency we can expect to pass through the temperature

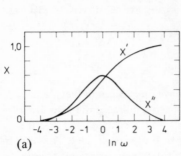

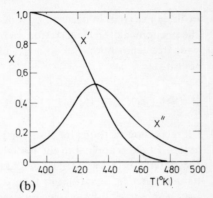

FIGURE 7   (a) X′ and X″ as a function of frequency in a dynamic experiment carried out at constant temperature, for a system characterised by a single relaxation time $\tau \simeq 1$ s. (b) X′ and X″ as a function of temperature at constant frequency for the same system.

interval where $\omega \sim \tau$ and hence curves of the kind depicted in Figure 7(b) are obtained. The maximum in X'' is displaced to higher temperature as the frequency is raised.

When a pulse is applied to the sample then τ simply determines the decay in time of strain induced in the sample

$$\sigma(t) = \sigma_0 \exp(-t/\tau),$$

where

$$\sigma_0 = Xf_0$$

2.1   Relaxation techniques

Three main groups of relaxation studies are practised on polymerised materials

   i) mechanical and acoustical

  ii) dielectrical

 iii) NMR.

In addition dynamic birefringence, electron spin labelling and UV depolarisation are used less often; in the case of the latter two techniques particularly where very fast relaxation processes are involved. This symposium concentrates on the three main methods.

Dynamic mechanical techniques including torsion pendula, forced oscillations and resonance methods combined with acoustical techniques allow the measurement of the complex shear modulus of amorphous and crystalline polymers $(G^* = G' + iG'')$ over a wide range of frequency and temperature. Step-function experiments in the form of stress relaxation and creep can also be used to extend the low frequency range. The interpretation of the results for amorphous polymers is more straightforward than for partially crystalline samples because the latter have morphologies in which the modulus of the amorphous regions can differ by several orders of magnitude from that of the crystalline regions and it is not clear that the applied stress is transmitted uniformly through the bulk of the specimen. Mechanical and acoustical excitation create coherent phonons at the frequency of the experiment which can excite molecular mechanisms in the material. The excitation decays either by emission of phonons coherent with the radiation or by emission of phonons of different phase leading to the dissipation of energy in the form of heat. A typical set of data[14] for atactic poly(methylmethacrylate) are presented at 1 Hz in Figure 8(a), in the form of tan δ as f(T). Note that several relaxation phenomena are observed and that two are located at temperatures close to transitions identified by dilatometry in Figure 6. As the frequency of the experiment is raised all the loss peaks are displaced to higher temperatures, though at different rates. Thus all the relaxations show the characteristic behaviour of activated processes.

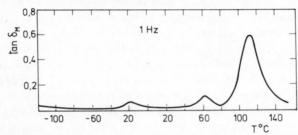

FIGURE 8   Relaxation curves for atactic poly(methylmethacrylate).
(a) Dynamic mechanical loss (1 Hz)

Dielectrical techniques are also available over a very wide range of frequency and temperature though they are restricted to polar materials. The complex permittivity $\epsilon^* = \epsilon' + i\epsilon''$ can be measured by continuous wave methods in the range 10^9–10 Hz for polymers and voltage-step response methods operating in the time domain give the Fourier transform of ϵ^* in the range 1–10^{-5} Hz. Dielectric relaxation can be explicitly analysed in terms of the time dependence of the orientation of molecular dipoles in the sample and there is not an extreme difference between the electrical properties of the two phases of a partially crystalline polymer which obtains in the mechanical response. Figure 8(b) shows a dielectric loss spectrum for[15] atactic poly(methylmethacrylate) at 20 Hz. Compared with the corresponding mechanical loss spectrum (8(a)) the relaxation processes lie at higher temperatures because of the higher frequency of the experiment. Note also that the relative intensities of the transitions are different in the two experiments; this is important in the analysis of the molecular basis of the phenomena.

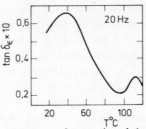

FIGURE 8   Relaxation curves for atactic poly(methylmethacrylate).
(b) Dielectrical loss (20 Hz)

NMR relaxation studies usually use ^{1}H or ^{13}C nuclei. The simplest proton relaxation studies are continuous wave measurements of the NMR line width of the polymer as a function of temperature. This is in essence a measure of the spin—spin relaxation time T_2 which decreases as molecular motions of groups containing protons are effectively frozen out because the dipolar coupling between nuclei increases as reorientational motions slow down. Pulse sequences allow T_1 the spin—lattice relaxation time and T_2 to be measured and now relaxation in the rotating frame, characterised by the relaxation time $T_{1\rho}$ can also be measured. T_1 is usually measured in the frequency range 10—200 MHz, since it observes the rate of dissipation of the nuclear spin energy into the lattice the results are comparable to high frequency dielectrical or mechanical measurements though minima in T_1 (rather than maxima in the relaxation rate $\propto 1/T_1$) are plotted. Figure 8(c) shows results for atactic poly(methylmethacrylate);[16] again the relaxation intensities differ from those displayed in 8(a) and 8(b) but, taking into account the higher frequency they are located at corresponding temperatures. The spin—spin relaxation time, as exemplified by the line widths measured on poly(methylmethacrylate), shows a similar set of relaxations. Since the line widths correspond to 10^4–10^5 Hz, the measurements are effectively comparable to this lower range of frequencies. $T_{1\rho}$ measurements, made in the r.f. field H_1, also correspond to frequencies of about 10^4 Hz. The understanding of NMR relaxation effects is well established in terms of the changes in bulk magnetism of the sample and the molecular understanding of these changes in terms of spin relaxation phenomena associated with groups of nuclei. This makes the technique a particularly powerful one for probing the molecular origins of the relaxation phenomena especially when it is realised that selective substitution of ^{1}H by ^{2}H in individual chemical groups can reduce the relaxation strength by over an order of magnitude because the magnetogyric ratio $\gamma^2 H/\gamma^1 H = 0.14$.

^{13}C NMR measurements are extending the study of relaxation phenomena in two important ways. Because of the larger chemical shift and relatively narrow line-width it is possible to measure T_1 and T_2 values for chemically different C atoms in rubbers, whereas in ^{1}H NMR the line widths even in liquids and rubbers usually only allow the overall relaxation of the specimen to be monitored. Also

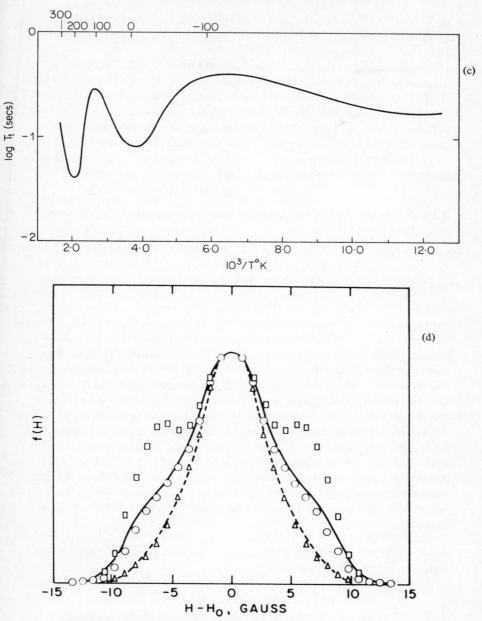

FIGURE 8   Relaxation curves for atactic poly(methylmethacrylate).
(c) ^{1}H spin-lattice relaxation time (30 MHz)
(d) ^{1}H NMR line width (△) 300°K, (○) 77°K, (□) rigid (calc)

Dr. Schaefer will report important new developments in the use of high H_1 fields to observe spectra of solids and no doubt produce information on poly(methylmethacrylate) complementary to that given above.

The correlation of the results from mechanical, acoustical, dielectrical and NMR measurements over a wide range of frequencies, of course, enables us to identify the subsidiary relaxations observed below the glass temperature with specific molecular motions of the polymer chains both in the amorphous state and also in the case of crystalline polymers related relaxation processes can be identified in the crystalline regions. The next problem is to determine the relationship between the macroscopic relaxation time identified in the dynamic equations given above and the correlation times of the various molecular motions.

2.2   Molecular motion as an activation process and time correlation functions

From the macroscopic experiments the maxima in the mechanical and dielectrical loss plots can be used to identify a characteristic relaxation time associated with macroscopic experimental effects.

$$\tau = (2\pi f_m)^{-1}$$

where f_m is the frequency of maximum loss. Similarly relaxation times can be obtained. The minima in the spin relaxation time plots can be used to identify the relaxation time in the magnetic experiment. The correlation times τ_c for the molecular phenomena involved, particularly in the case of mechanical and dielectric experiments, differ by only a relatively unimportant numerical factor from τ. The minima in T_1 or $T_{1\rho}$ plots provide values of τ_c. The results of measuring τ_c over a wide range of temperature using the three techniques are usually collated in the form of a transition map which plots τ_c or the correlation frequency $\nu_c \equiv 1/2\pi\tau_c$ as a function of T^{-1}. The map for atactic poly(methylmethacrylate)[17] is shown in Figure 9 for a glass transition loss which occurs at the highest temperature and the two subsidiary losses. It is interesting to note that all the three relaxation processes seem to converge at high temperatures, indicating that in the rubber state all the motions contribute simultaneously to an overall complex relaxation process.

These transition maps are analysed in terms of Eyring's theory of rate processes, an activation enthalpy ΔH^* and an activation volume ΔV^* for which transition can be defined thus

$$\Delta H^* = R\left(\frac{\delta \ln \tau_c}{\delta(T^{-1})}\right)_p$$

and

$$\Delta V^* = RT\left(\frac{\delta \ln \tau_c}{\delta p}\right)_T$$

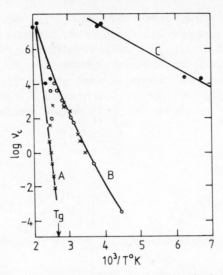

FIGURE 9   Transition map for atactic poly(methylmethacrylate): mechanical (X), dielectrical (○), T_1 and $T_{1\rho}$ (●).
A: Main chain motion   B: Carboxylate group rotation   C: α Methyl group rotation.

The activation energy at constant volume is given by

$$\Delta E_V = \Delta H^* - p_i \Delta V^*,$$

where $p_i = \left(\dfrac{\delta U}{\delta V}\right)_T$ is the internal pressure.

The magnitude of ΔH^* and sometimes ΔV^* can be used intuitively to identify the nature of the molecular processes. For example, for main chain motions
of the type that occur in the glass transition region one would expect large values
of ΔH^* and correspondingly large values of ΔV^*. For more localised motions
such as side group motions, one would expect both parameters to be small. This
kind of analysis is often useful for setting up a quantitative model for more
detailed investigations.

Another aspect of the analysis of the loss spectra is that the maximum and
minimum only define average values of τ_c and τ. If one attempts to fit the curves
of, for example, $G'\epsilon'$ with the macroscopic equations given above it is found that
a single relaxation time is not sufficient and one has to deal with distributions of
relaxation times. That is to say that the relaxations are more diffuse than the
simple equations would predict.

For many of the complex motions, especially those observed in the glass
transition region, it is very difficult to predict a priori suitable distribution
functions.

In the last decade another type of analysis of the loss spectra has been favoured involving the use of time relaxation functions which avoids the need to extract an average relaxation from the spectra. The correlation function is determined by the actual shape of the spectrum, for example the dielectric loss spectrum of a polymer measures the rate of absorption of energy from the magnetic field. The shape of the out of phase component of the spectrum ϵ'' is governed by the Fourier transform of the macroscopic dipole moment time correlation function

$$G(t) = \frac{\langle \mathbf{M}(o) \cdot \mathbf{M}(t) \rangle}{\langle \mathbf{M}(o) \cdot \mathbf{M}(o) \rangle},$$

i.e.

$$\epsilon''(\omega) \cong (\epsilon_0 - \epsilon_\infty) \int_0^\infty -\left[\frac{dG(t)}{dt} \right] \sin \omega t \, dt.$$

It is a property of Fourier transforms that they may be inverted, and it follows that

$$G(t) = \frac{2}{\pi} \int_0^\infty \left[\frac{\epsilon''(\omega)}{\epsilon_0 - \epsilon_\infty} \right] \cos \omega t \, \frac{d\omega}{\omega}.$$

Thus, if the spectrum $\epsilon''(\omega)$ is measured over a sufficiently wide range of frequency, $G(t)$ may be obtained using this equation. Conversely, if the form of $G(t)$ was already known, either through computation from a theoretical model or through direct experimental measurement in the time-domain, then the frequency-domain spectrum can be predicted.

In the case of NMR experiments a nuclear spin time correlation function is obtained, but for mechanical and acoustical measurements we still don't really know how to set up the correlation function theoretically. In order to relate the correlation functions obtained experimentally to molecular phenomena one then requires a model based in time dependent statistical mechanics of the molecular dynamics of the system which will generate theoretical correlation functions to be compared with those obtained experimentally. Thus we have experimental methods (which measure the spectrum directly or the Fourier transform of the spectrum) from which time correlation functions can be calculated and ultimately related to the molecular dynamics of the material by comparison with theoretical forms of these correlation functions. In a recent paper,[18] for example, the dielectric relaxation and visco-elastic retardation in a series of liquids show that the same empirical correlation functions could give a reasonably good representation of both forms of relaxation. The implication is that since we are reasonably sure that the dielectric relaxation is essentially a reorientation of the polar molecules in the force field of surrounding molecules, then the orientation part of the visco-elastic process must also be associated with local reorientation adjustments.

2.3 Molecular basis of transitions and relaxations

The current picture of the origins of the various transitions observed in polymeric systems can be grouped under the following headings:

  i) Main chain motion in amorphous materials,

  ii) Main chain motion in crystals,

 iii) Local main chain motions in amorphous materials and crystals involving 2—4 main chain atoms,

 iv) Side group motions in amorphous materials,

  v) Defect and end group motions in crystals.

There is little doubt that the primary main chain motion in amorphous polymers is identified with the glass transition which marks the on-set of the long range main chain wriggling motions. The rigid glass becomes rubber and the main chain motion becomes more and more active as the temperature is raised. The amorphous regions in partially crystallised materials undergo similar changes. In the amorphous state these long range wriggling motions are inhibited by the physical entanglements of the chains and by the chemical crosslinks in the case of crosslinked rubbers. However in certain polymers overall rotation of the polymer molecule is observed at suitably low molecular weights. An elegant example is the study[19] of polypropylene oxide by Stockmayer and his colleagues; the dielectric loss has two peaks at molecular weights of the order of 10^3. The smaller peak, which represents the $k = 2$ Rouse normal mode of the molecules, disappears into the larger peak representing the wriggling motion of the chain, at high molecular weights. At 30°C for polypropylene oxide fluid of molecular weight 2000, $\tau_c \simeq 10^{-4}$ s.

Main chain motion in crystals is a hindered rotation about the main chain axis. All of the chains in the crystal region can be involved. It could well be that this type of motion is responsible for the on-set of subsidiary polymorphic transitions referred to in Section 1.

When it is possible to study relaxation behaviour as a function of percentage crystallinity α_{cryst}, observed processes which increase in intensity as α_{cryst} increases are clearly identified as belonging to the crystal. Detailed morphological information which is now available enables the effect on relaxation behaviour of varying lamellar thickness and defect type and concentration to be evaluated.

On the basis of current morphological information, the following types of crystalline process are possible:

a) cooperative motions of the molecular chains along the length of the crystallite — these relaxations will be related to the lamellar thickness,

b) motions associated with defects such as end groups incorporated in the crystal, or mobile imperfections in the form of kinks which diffuse along the chain axis, which could also involve chain movement,

c) motions associated with morphological details, such as chain folds, or of large morphological features — such as interlammelar shear or interfibrillar shear

d) motions of restricted chain lengths — such as local motions, also postulated to exist in the glassy amorphous state.

In evaluating these models it must be remembered that different relaxation techniques will respond with different sensitivity to the various types of molecular motion. Thus for example, dielectric loss peaks are observed for polyethylene, and these are attributed to the presence of polar groups formed by oxidation at the chain ends. The relaxation behaviour for crystals of polyethylene and of n-alkanes have been analysed[20] by Hoffman *et al.* in terms of simple site models and no doubt the paper by McCrum[21] will refer to the work in detail.

Studies of amorphous polymers without side groups have revealed that there are small subsidiary relaxation processes at low temperatures; for example[22] polyalkylene ethers show a series of relaxation peaks in the low frequency in the neighbourhood of $-100°C$. The location of the peak is related to the number of methylene groups in the particular alkylene oxide. Such relaxations are usually attributed to localised motion of restricted segments of the main chain, though in fact it is very difficult to give a unique assignment to the motion envisaged. Similarly secondary main chain motions in crystals have been noted in (d) above.

Side group motion leading to relaxation effects can be identified with much greater certainty simply because the dielectric and NMR techniques have proved a powerful combination in determining the manifestations of these motions. There are many examples of methyl group rotations being specifically identified by the effects of deuteration on the NMR T_1 or $T_{1\rho}$ relaxation time curves. Furthermore the strength of relaxations of polar groups is a useful feature in assigning loss in dielectric spectra. If we now refer back to Figure 8 it will be noted that whereas the glass transition relaxation is the dominant peak in the mechanical spectrum, the first subsidiary peak at $60°C$ dominates the dielectric spectrum and therefore this peak is assigned to the relaxation of the methylacrylate side group. It will also be seen that the T_1/T plot shows a second relaxation phenomenon at a somewhat lower temperature. This can be identified as the αCH_3 relaxation by deuteration which reduces the strength of these relaxation processes by a factor of approximately 20. Lastly, it should be possible to locate the other methyl group motion from the minimum in T_1 which develops at very low temperatures.

The successive broadening in the NMR line width in poly(methylmethacrylate) arises from freezing in of (i) main chain motion at $\sim 110°C$, (ii) αCH_3 side groups at $\sim -20°C$ and (iii) OCH_3 groups at the lowest temperature. Thus we can build up a qualitative picture of molecular motions which generate transitions in physical properties.

Although relaxations corresponding to these side group processes are normally thought of as predominantly intra-molecular, the recent work by Petrie[11] on polystyrene shows that caution is still necessary. There is in polystyrene a subsidiary transition in the neighbourhood of 60°C. Petrie has shown that by annealing the sample for approximately 70 hours at 92°C this transition disappears completely, but when the sample has melted and the glass re-formed under normal conditions on cooling, this subsidiary relaxation re-appears. The experimental evidence is shown in Figure 10. At first sight the relaxation would be assigned to a local motion or to cooperative motion of the phenyl groups, but the evidence in Figure 10 suggests very strongly that the local structure of the glass probably dominates the molecular origin of this process. It is quite likely that relaxation processes in other polymeric glasses may well show intermolecular contributions when careful experiments of this kind are carried out.

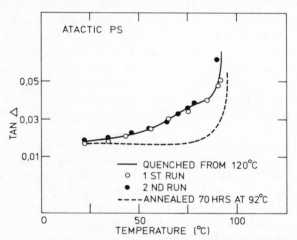

FIGURE 10   Mechanical loss in atactic polystyrene measured first (o) after quenching from 120°C to room temperature, then after annealing for 70 h at 92°C, finally (•) after melting and quenching from 120°C again.

One question that still has to be resolved is the status of the transition in polystyrene melt at $\sim 160°C$. It is depicted in Figure 2 and has been termed a liquid–liquid transition by Boyer. As yet it is insufficiently documented and certainly not fully understood. Boyer and Gilham will comment on this transition in a later paper.

In Section 2.2 we mentioned the use of activation enthalpy as a method of identifying physical processes involved in relaxations. The activation energies involved in main chain motions in the glass transition region are very large, often in the region of 20–100 kcal mol^{-1}. Some of the main chain motions postulated in crystal lattices have characteristically high activation energy in the region of 30–60 kcal mol^{-1}. The localised main chain motions tend to have smaller activation energies, usually in the neighbourhood of 10 kcal mol^{-1}. Side group motions and secondary groups such as methyl groups are generally in the region of 2–4 kcal mol^{-1} except where the group is severely sterically hindered. Some of the large polar groups have correspondingly larger activation energies in the region of 10–20 kcal mol^{-1}. In general however one must approach the identification of molecular motions in terms of $\Delta H^* \Delta V^*$ with caution.

3   SCATTERING PHENOMENA

Judging from the titles of the papers, very little will be said in this symposium about the use of scattering techniques to study molecular motions. Nevertheless these methods are beginning to add to our knowledge of molecular motions. Use of these methods immediately presents a difficulty in terms of analysis because one is now concerned with a space–time correlation function rather than with the somewhat simpler time correlation function. A direct consequence of molecular motions in polymeric material is that radiation is scattered from the materials in a quasi-elastic function. Doppler shifts from the frequency of the incident radiation are generated by microscopic fluctuations in density or dielectric permittivity arising from the collected thermal motions of the molecules in the material.

For example,[23] the Doppler broadening of Rayleigh scattered light from macromolecular systems especially in solution has enabled the measurement of the centre of mass diffusion constant, i.e. corresponding to the Rouse mode $k = 0$, for the macromolecule. For molecules of exceptionally high molecular weight ($> 10^7$) it has been possible[24] to observe the $k = 2$ Rouse mode of the polymer chain in the form of departures from linearity of the plot of Doppler broadening vs. (momentum transfer)^2.

There are preliminary reports of observation of Doppler broadening effects in Rayleigh light scattered from concentrated gels and rubbers but a detailed molecular interpretation of the origin of the phenomenon has not been given.

Brillouin scattering measurements are also being used to observe very low frequency phonons in amorphous polymers. Patterson,[25] for example, has used the temperature dependence of the separation of the Brillouin peaks to locate the glass transition in carefully annealed poly(methylmethacrylate). It is noteworthy that prolonged annealing gave $T_g \sim 88°C$, some 12°C below the value which would be obtained in a dilatometric study carried out at a low rate of cooling. However the technique has yet to make a substantial contribution to the molecular interpretation of relaxation phenomena.

Quasi-elastic neutron scattering studies are now being made on rubbers and gels. They are of interest because the large mass of the neutron enables much larger momentum transfers in the scattering process to be studied. Measurements at large momentum transfers on rubbers suggest that the local segmental wriggling process is being observed, and the Doppler broadening is proportional[26] to (momentum transfer).[2] At intermediate values of momentum transfer[27] the exponent approaches 4 and theoretical analysis[28] suggests that it may reach a value of 8 before returning at very low momentum transfers to the limit of 2 as observed in Rayleigh light scattering. Some preliminary measurements on concentrated solutions[29] have also been made; the data obtained at intermediate values of momentum transfer (i.e. $q = 0.1 - 0.3$) are compatible with the Rouse model.

We can now expect quasi-elastic scattering to contribute to the molecular interpretation of molecular motion in rubbers, gels and solution. Finally however it is worth noting that in recent years inelastic neutron scattering has made a contribution to the study of molecular motion in crystals and glasses. The accordion and torsional intra-molecular acoustic modes of the polyethylene chain in oriented crystals which are inaccessible in infra red and Raman spectroscopy have been observed by King[30] using inelastic neutron scattering. White[31] is extending this work to a careful documentation of the lattice modes of the crystal which include overall chain torsion. Work has also begun on poly(tetrafluoroethylene) and poly(oxymethylene). The torsional frequencies of methyl and phenyl side groups have also been observed by inelastic neutron scattering and thus barriers hindering internal rotation of these groups can be computed and compared with activation energies obtained from relaxation processes.

Thus scattering techniques are providing information, often at a molecular level, to aid the understanding of transitions and relaxations observed by other methods in polymeric materials.

Postscript

This review has attempted to survey the field in which Boyer has had a major influence. He has helped and encouraged experiments to establish the phenomena, the development of new methods and the qualitative understanding of the results. The major challenge confronting theoreticians in polymer science, as Stockmayer's paper demonstrates, is the quantitative description of the dynamics of macromolecules.

References

1. J. Gough, *Proc. Lit. and Phil. Soc.* Manchester, (1805).
2. K. Ueberreiter, *Z. Phys. Chem.* **B45**, 361, (1940).
3. T. Alfrey, H. Mark and G. Goldfinger, *J. Appl. Phys.* **14**, 700, (1943).

4. K. H. Meyer and C. Ferri, *Rubber Chem. Tech.* 8, 319, (1935).
5. R. F. Boyer and R. S. Spencer, *J. Appl. Phys.* 15, 398, (1944).
   R. F. Boyer and R. S. Spencer, *J. Appl. Phys.* 16, 594, (1945).
6. N. Bekkedahl, *J. Res. Natl. Bur. Stand.* 13, 411, (1934).
7. See for example R. L. Miller, in *Polymer Handbook,* edited by J. Brandrup and E. H. Immergut (Interscience, New York, 1966) III–1.
8. N. Bekkedahl and L. A. Wood, *J. Chem. Phys.* 9, 193, (1941).
9. G. Gee, *Contemporary Physics* 11, 313, (1970).
10. See for example *Physics of Glassy Polymers,* edited by R. N. Howard (Applied Science, London, 1973).
11. S. E. B. Petrie, *A.C.S. Polymer Preprints* 15/2, 336, (1974).
12. J. H. Gibbs and E. A. di Marzio, *J. Chem. Phys.* 28, 373, (1958).
13. G. Natta and P. Corradini, *Nuovo Cimento, Suppl.* 15, 9, (1960).
14. F. Vernon, Ph.D. Thesis, Manchester University, 1965.
15. N. G. McCrum, B. E. Read and G. Williams, *Anelastic and Dielectric Effects in Polymeric Solids* (J. Wiley, New York, 1967) p. 245.
16. T. Kawai, *Phys. Soc. Japan* 16, 1220, (1961).
17. G. Allen, in *Molecular Spectroscopy* (Inst. of Petroleum, 1969) p. 410.
18. M. F. Shears, G. Williams, A. J. Barlow and J. Lamb, *J. Chem. Soc., Faraday Trans. II,* 70, 1783, (1974).
19. W. H. Stockmayer, *Pure Appl. Chem.* 15, 539, (1967).
20. J. D. Hoffman, G. Williams and E. Passaglia, *J. Polymer Sci. C* 14, 173, (1966).
21. G. McCrum, Polyethylene: detailed interpretation of mechanical relaxation in a crystalline polymer, this volume, pp. 167–191. Gordon and Breach, New York (1978).
22. N. G. McCrum, B. E. Read and G. Williams, *Anelastic and Dielectric Effects in Polymeric Solids* (J. Wiley, New York, 1967).
23. T. A. King, A. Knox and J. D. G. McAdam, *Polymer,* 14, 1, 151, (1973).
24. J. G. McAdam and T. A. King, *Chem. Phys. Letters,* 6, 109, (1974).
25. G. D. Patterson, *A.C.S. Polymer Preprints,* 15/2, 14, (1974).
26. G. Allen, J. S. Higgins and C. J. Wright, *J. Chem. Soc., Faraday Trans. II* 70, 348, (1974).
27. G. Allen, R. E. Ghosh, A. Heidemann, J. S. Higgins and W. S. Howells, *Chem. Phys. Lett.* 27, 308, (1974).
28. M. Warner, unpublished work.
29. G. Allen, J. P. Cotton, B. Farnoux, R. E. Ghosh, J. S. Higgins and G. Jannink, unpublished work at ILL, Grenoble.
30. W. J. Meyers, G. C. Summerfield and J. S. King, *J. Chem. Phys.* 44, 185, (1965).
31. J. W. White and J. F. Twistleton, *Neutron Inelastic Scattering* (Vienna IAEA, 1972) p. 301.

DISCUSSION

S. L. Aggarwal (*General Tire and Rubber Co.*): It seems to me that not sufficient attention has been paid to broadening of transition peaks, especially in copolymers where the environment in which the molecular motion occurs has a pronounced effect on the breadth of the transition peak. Do you think that the models that relate molecular motions to transition behavior are adequate to account for the breadth of transition peaks?

G. Allen: The problem of presenting theoretical models which match the observed breadths of transition peaks is not trivial. In simple phenomenological models this can be done by adjusting the distribution of relaxation times but of course that is not necessarily an objective test. On the other hand the use of correlation functions automatically fits the peaks so one takes the experimental observations into account in selecting a model giving a correlation function in reasonable accord with that derived from experimental data. On the whole, I am inclined to agree with you that insufficient attention has been paid to the broadening of transition peaks and at the present time there are very few models capable of dealing objectively with this situation.

W. H. Stockmayer (*Dartmouth College*): Would you care to elaborate on your recent observations of phenyl group torsional motion in polystyrene? It should be much freer in some copolymers, as shown, for example, by Gimok's NMR relaxation studies of styrene-butadiene copolymers.

G. Allen: Dr. C. J. Wright and I have observed a phenyl group torsion in poly-styrene at 58 cm^{-1}. We are now satisfied that this is the correct assignment in bulk polystyrene and in some of its poly substituted derivatives. However, al-though we have asked Dr. John Ebdon of the University of Lancaster to supply an alternating copolymer, we have not yet been able to run its spectrum. We have in mind doing this for just the reason that you mention in your question.

M. Litt (*Case Western Reserve University*): In an amorphous polymer, or the amorphous regions of a crystalline polymer, the chain has many conformations. Also, the local environment of any group is far from constant. Would not both of these effects broaden any relaxation, except possibly that of methyl rotation? Then one would expect that in polymers even single relaxation process could have a distribution of relaxation times. Would you please comment on this.

G. Allen: Yes, one would expect local environmental effects to broaden relaxa-tions, though it is not inconceivable that situations might occur when they might sharpen them up. I agree that one could imagine a situation where a simple re-laxation phenomena characterised by a single relaxation process might have a series of relaxation times relating to different local chain conformations. In the case of methylmethacrylate however the α-methyl group rotation is not so much broadened as shifted in the temperature frequency flow by a change in local interactions arising from a change in stereo-regularity going from iso- to syndio-tactic, but of course in each of the stereo isomers there is still a range of con-formational isomers in the amorphous state, so that no doubt each of these losses is broadened.

E. Baer (*Case Western Reserve University*): My question is concerned with the "so-called" γ-relaxation in polystyrene. Recent results have indicated that this dispersion maximum is caused by mineral oil and possibly other additives (Armeniades, Rieke, and Baer, *J. Appl. Polymer Sci.* 14, 2653 (1970)). Have you probed this question recently?

G. Allen: No; your observation is very interesting but it is one that we have not followed up.

Motions in Low Molecular Weight Fluids and Glass Forming Liquids

JOHN LAMB

Department of Electronics and Electrical Engineering, The University, Glasgow, G12 8QQ, Scotland.

A review is given of the viscoelastic behaviour of supercooled liquids and undiluted polymers, as determined by experiments using alternating shear or creep recovery. Results for supercooled liquids and polymer melts are analysed in terms of retardational compliance.

Through comparisons with results of dielectric studies, it is argued that the observed viscoelastic retardational process is closely related to the corresponding process of dielectric relaxation in polar liquids.

Since this latter is essentially a reorientation of the molecule in the force field of surrounding molecules, then the retardational process must likewise be one of local reorientational adjustment. The time constant of this is controlled by the availability of sufficient local free volume through the diffusion of defects.

With increasing molecular weight of the liquid, the resulting internal flexibility of the molecule causes a broadening of the relaxation region. As the molecule becomes progressively longer a separate process is observed, due to local segmental motion.

The behaviour at high frequencies of long chain polymer melts is similar to that of relatively low molecular weight liquids of the same polymer type.

1. INTRODUCTION

Historically, the approach which my colleagues and I have followed in our studies of viscoelastic properties of liquids stemmed initially from an interest in the behaviour of lubricants. In the natural sequence of events, therefore, our first concern was with molecules which, though often complex in a strictly chemical sense, were of relatively low molecular weight and might perhaps be regarded as simple liquids from the standpoint of a polymer scientist. We were thus led by interest and our curiosity into that region of ill-defined molecular behaviour between what are conventionally regarded as non-polymeric liquids and recognisable longer chain polymers. I refer particularly to the region in the conventional plot of log (viscosity) against log (molecular weight) below the accepted break point at the critical molecular weight M_c; or, more precisely, to the plot which results from extrapolating the viscosity data to high temperatures.[1] This is what is implied under the general description of low molecular weight fluids.

Statistical theories, mostly based upon the bead and spring model of Rouse,[2] have been developed to describe the dynamical behaviour of long-chain polymer molecular in dilute solution. The Rouse theory assumes that there is no hydrodynamic interaction between the polymer molecule and the flowing solvent, the velocity of the solvent flowing through the molecule being unaffected by the presence of the molecule. Zimm[3] extended this to the non-free draining case, in which account is taken of the hydrodynamic interaction of the polymer. The resulting equations have been widely used for comparison with experimental results with considerable success, albeit often in a region where the assumption of dilute solution is clearly invalid. In making such comparisons, it is important to realise that, in certain solvents, a significant part of the contribution which the polymer makes to the viscosity of the solution is not involved in viscoelastic relaxation, as was first demonstrated by Lamb and Matheson.[4] More recently, Edwards and Freed[5,6] have looked at the problem afresh and in a most original fashion which shows considerable promise for further expansion to the dynamics of dense polymer solutions. They rightly criticise the restrictions of the Rouse model, the limitations of which are also acknowledged by experimental workers who, nevertheless, find the qualitative description of the model useful. The Rouse equations have a surprising application, even to a description of the behaviour of undiluted polymer melts.[7,8]

We are not here concerned specifically with long chain polymers, the behaviour of which could, in principle, be described in terms of a statistical theory, but with the properties of smaller molecules. Nevertheless, it transpires that the viscoelastic relaxation attributed to small segments of longer molecules is, in certain circumstances, akin to the corresponding behaviour of individual molecules of relatively low molecular weight. The basic concepts of the Rouse theory are useful in this comparison. It is assumed that the polymer chain can be divided into N equal segments, each comprising q monomer units, the length of a segment being such that its ends are spatially disposed according to a Gaussian distribution. The modes of motion of the molecule are specified by the mode number, p. Mode $p = 1$ has the longest relaxation time and involves co-ordinated motion of the whole molecule with a central node. Higher mode numbers involve correspondingly reduced numbers of segments until a limiting value of mode number is reached, at which the spacing between nodes is equal to the q monomer units comprising a Gaussian segment. This, therefore, represents the limit of applications of statistical theory: indeed, the upper limit for p normally taken in summing the contributions of the modes to the complex shear modulus[2,9] is $p = N/5$. This leads to a calculated value of about 5×10^5 N/m^2 (5×10^6 dyn/cm^2) for the largest value of shear modulus which could be predicted if the Rouse equations are applied without question to the case of an undiluted polymer.

If the summation is extended to the absolute limit imposed by the sub-molecule ($p = N$) then[9] the maximum calculated value is 1.6×10^6 N/m^2 for a polyisobutylene sample having a weight-average molecular weight of 1.56×10^6. However, limiting values of shear modulus determined experimentally[8] are of the order 10^8 to 10^9 N/m^2, so that considerable contributions to the relaxation spectrum arise from those parts of a long chain molecule which correspond to lengths of the backbone within a few segmental units. Typical values for the number of monomer units, q, comprising a segment are[7,8] from 5 to 10 (10 to 20 chain atoms per segment). It follows, therefore, that a discussion of the viscoelastic behaviour of low molecular weight polymers having a molecular weight less than M_c is also applicable to those contributions to the complete behaviour of a long chain polymer ($M \gg M_c$) which arise from motions within a small number of Gaussian segments along the backbone. Since the relaxation of a polymer melt is conveniently delineated experimentally by varying the effective frequency of oscillation in oscillatory shear, we anticipate that, in measurements of this nature, little difference will be found at sufficiently high frequencies between the measured properties of relatively long and comparatively short molecules of the same type of polymer.

This review is mainly concerned with conclusions and interpretations which are adduced from the results of measurements of the properties of low molecular weight fluids in cyclic shear. The experimental variables are the temperature and pressure of the liquid and the frequency of alternation of the applied shear stress. The inherent advantage is that the external shear is sufficiently small that it merely perturbs the existing equilibrium without sensibly affecting the time constants of molecular flow or local reorientation. Alternatively, one can employ methods of shear creep recovery pioneered successfully by Plazek and his co-workers,[10,11] the results of which are likewise amplitude independent in contrast to other techniques involving shear rate or shear stress dependence of viscosity and associated normal stresses. Our concern will be entirely with linear viscoelastic behaviour.

2. SHEAR WAVE AND CREEP NOMENCLATURE

According to linear viscoelastic theory[12] the expressions for the complex compliance in cyclic shear at angular frequency, ω, and for the creep function are, respectively:

$$J^*(j\omega) = J_\infty + (j\omega\eta)^{-1} + J_r\chi(j\omega) \tag{1}$$

$$J(t) = J_\infty + t/\eta + J_r\,\psi(t). \tag{2}$$

The normalised functions $\chi(j\omega)$ and $\psi(t)$ are related through the spectrum of retardation times, $N(\tau)$; thus

$$\chi(j\omega) = \int_0^\infty \frac{N(\tau)d\tau}{1 + j\omega\tau} \tag{3}$$

and

$$\psi(t) = \int_0^\infty N(\tau)[1 - \epsilon^{t/\tau}]\,d\tau. \tag{4}$$

Limiting values are:

i)   as $\omega \to 0, \chi(j\omega) \to 1$; as $t \to \infty, \psi(t) \to 1$

ii)   as $\omega \to \infty, \chi(j\omega) \to 0$; as $t \to 0, \psi(t) \to 0$.

The equilibrium value of total compliance is therefore $J_e = J_\infty + J_r$ whilst J_∞, often denoted elsewhere by J_g, is the "instantaneous" or limiting high frequency value for the compliance. The shear modulus is simply the inverse of the compliance so that in alternating shear:

$$G^*(j\omega) = G'(\omega) + jG''(\omega) = [J^*(j\omega)]^{-1} = [J'(\omega) - jJ''(\omega)]^{-1}. \tag{5}$$

Experimental techniques operating at frequencies below a few thousand Hz generally provide direct measurement of G' and G'' whilst various piezo electric methods employed at higher frequencies, above about 20 kHz and in the megahertz range, yield measurement of the mechanical shear impedance of the liquid, $Z_L = R_L + jX_L$. Connecting relationships between these parameters are obtained by solving the equations of motion for an elemental volume of fluid, which shows that

$$J^*(j\omega) = \rho/Z_L^2, \tag{6}$$

where ρ is the density. Hence:

$$J'(\omega) = \frac{\rho(R_L^2 - X_L^2)}{(R_L^2 + X_L^2)^2} \quad : \quad J''(\omega) = \frac{2\rho R_L X_L}{(R_L^2 + X_L^2)^2} \tag{7}$$

$$G'(\omega) = (R_L^2 - X_L^2)/\rho \quad : \quad G''(\omega) = 2R_L X_L/\rho. \tag{8}$$

These equations are quoted since, under certain experimental conditions, it is necessary to deal in terms of the impedance components. For example, experimental difficulties often preclude the measurement of the reactive component of the shear impedance and only the resistive component can be determined with acceptable reliability. This is the situation which pertains for all measurements to-date at high pressures. However, at sufficiently high values of "effective frequency" the initial predictions and the results of experiment show that X_L tends asymptotically to zero with increasing frequency. Under these conditions, $J''(\omega)$ and $G''(\omega)$ tend to zero and $J'(\omega) \simeq 1/G'(\omega) \to \rho/R_L^2$, giving the limiting high frequency compliance $J_\infty = 1/G_\infty$. At low values of effective, frequency, the purely viscous or Newtonian region is reached with decreasing frequency where $R_L \simeq X_L \to (\pi f \eta \rho)^{\frac{1}{2}}$. The equilibrium compliance $J_e = J_\infty + J_r$ in Eq. (1) is then entirely swamped by the viscosity term $(1/j\omega\eta)$. It is only possible to obtain values of J_e from the higher frequency methods, which provide direct measurement of R_L and X_L, if extreme precautions are taken in the experimental work and then only for a restricted range of values of J_e.[13] For polymer liquids, determination of $J'(\omega)$, and hence of J_e, is more readily obtained from the low-frequency techniques which yield direct measurement of G' and G''.

3. NON-POLYMERIC LIQUIDS

We distinguish between two regions of viscous flow of a liquid as a function of temperature. At high temperatures, where the relative free volume exceeds about 15%, the viscosity is primarily governed by the energy required for a molecule to jump from one site to an adjacent site which is readily available. The dependence of viscosity upon temperature is then governed by the Arrhenius equation:

$$\ln \eta = A + E/RT, \tag{9}$$

so that a plot of $\ln \eta$ against $1/T$ gives a straight line of slope E/R. In general, these conditions pertain over the temperature range such that the viscosity is lower than about 10^{-2} Pa s (10 cP).

On cooling below this region, the majority of liquids crystallise but there are many liquids which can be readily supercooled. In this lower temperature region the chief factor governing viscous flow is the availability of free volume and, over a substantial range of temperature, the viscosity data fit a modified free volume equation[9, 14–16]

$$\ln \eta = A + B/(T - T_0). \tag{10}$$

Constants A, B and T_0 in this equation are computed by a least squares fit to the measured values of η as a function of T. There is no sharp division between the high temperature behaviour represented by Eq. (9) and the low temperature region to which Eq. (10) applies. For purposes of reference, we specificy a temperature T_A — the Arrhenius temperature — below which the plot of $\ln \eta$ against $1/T$ deviates from the straight line fit of Eq. (9). It happens that all available observations of viscoelastic relaxation using methods of alternating shear correspond to the lower temperature supercooled region, where the viscosity–temperature dependence is given by the modified free-volume Eq. (10). This arises from the fact that such measurements are so far confined to frequencies below about 1 GHz. Thus, the viscoelastic relaxation process is centred roughly about the region $\omega\tau_m = 1$, where τ_m is the characteristic Maxwell relaxation time given by the first two terms of Eq. (1). Ignoring the important retardational term of Eq. (1), the viscoelastic response of the simple Maxwell model is expressed by:

$$J^*(j\omega) = J_\infty + 1/j\omega\eta = J_\infty [1 + 1/j\omega\tau_m],$$ (11)

where $\tau_m = \eta J_\infty$. Now, a typical value for J_∞ is 2×10^{-9} m^2/N, so that for $\eta = 10^{-2}$ Pa s at T_A the corresponding value of τ_m is 2×10^{-11} s. The characteristic frequency, f_c, at which $\omega_c\tau_m = 1$ for these conditions is therefore $f_c \simeq 8$ GHz. Inevitably, we are led to work at lower temperatures where the steady flow viscosity is sufficiently increased to lower the region of relaxation into that which is experimentally accessible. This is merely an order of magnitude calculation: it transpires that the effect of the retardation term $J_r\chi(j\omega)$, which has been neglected in Eq. (11), is to broaden the extent of the viscoelastic relaxation observed in pure supercooled liquids. Experimental results discussed subsequently all refer to this supercooled region and liquids which crystallise at temperatures close to T_A are precluded from present consideration.

Any model which purports to explain the basic process of viscoelastic relaxation is subject to two limiting boundary conditions which must be satisfied. These are that the model must yield the behaviour of a purely viscous liquid at low frequencies ($\omega\tau_m \ll 1$) and that of an elastic solid at sufficiently high frequencies ($\omega\tau_m \gg 1$). The limiting high frequency modulus will be somewhat less than that of a typical solid, due to the presence of structure weakening holes, whilst the limiting value of G''/ω at low frequencies must be equal to the steady flow viscosity obtained from other conventional viscometric techniques. It must also be capable of yielding expressions for the functions $\chi(j\omega)$ and $\psi(t)$, which are in agreement with experimental results. Phillips, Barlow and Lamb[17] have developed theoretically a model first put forward by Glarum[18] based upon the diffusion of defects in one dimension, taking into account the effect of second nearest defects to a site where the stress is relaxed on arrival of a defect. This leads to derivation of an equation first put forward on empirical grounds by Barlow, Erginsav and Lamb[19] for the complex compliance:

$$J^*(j\omega) = J_\infty [1 + (j\omega\tau_m)^{\frac{1}{2}}]^2/j\omega\tau_m$$

$$= J_\infty [1 + 1/j\omega\tau_m] + \frac{2J_\infty}{(j\omega\tau_m)^{\frac{1}{2}}}$$ (12)

Comparing this result with Eq. (1) gives for the B.E.L. equation:

$$J_r\chi(j\omega) = 2J_\infty/(j\omega\tau_m)^{\frac{1}{2}}.$$ (13)

This equation provides a surprisingly good fit to a host of experimental data obtained from experiments in alternating shear, as has been reported extensively.[19-22] It has also been found to apply to results obtained as a function of

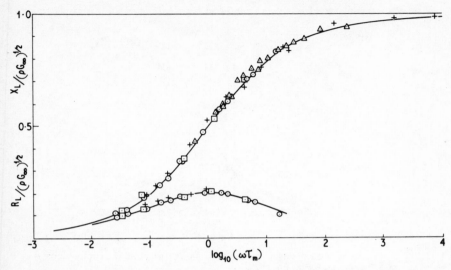

FIGURE 1 Normalised plots of the resistive and reactive components of the shear mechanical impedance for a number of liquids. The curves are calculated from the B.E.L. Eq. (12).

Δ, bis (m-(m-phenoxy phenoxy) phenyl) ether obtained at 30°C as a function of pressure. Measurements at atmospheric pressure as a function of temperature over the frequency range 6 to 450 MHz:
+, bis(m-(m-phenoxy phenoxy)phenyl)ether
o, di-isobutyl phthalate
□, di-octyl adipate.

hydrostatic pressure.[23] A typical example of the agreement found for a large number of liquids is shown in Figure 1.

Since the model is based upon defect diffusion, it is not surprising to find the $\omega^{-\frac{1}{2}}$ dependence of the second term in Eq. (12). This has the effect of broadening the relaxation curves beyond the single Maxwell relaxation which extends over approximately one decade of frequency. As can be seen from Figure 2, the B.E.L. equations for the components of the complex shear modulus, corresponding to the impedance curves of Figure 1, exhibit a relaxation region covering over three decades of frequency (e.g., from $G'/G_\infty = 0.1$ to 0.9).

Despite the good agreement found between experimental results in cyclic shear and predictions obtained from the B.E.L. equation, the latter does not meet all the criteria listed previously for an acceptable model and can therefore only be an approximate representation. The reason for this is that when transformed into the time domain, the corresponding recoverable creep strain increases without limit as $t^{\frac{1}{2}}$, which is unacceptable on physical grounds. Comparison of Eqs. (13) and (3) shows that the amplitude, $N(\tau)$ of the retardation spec-

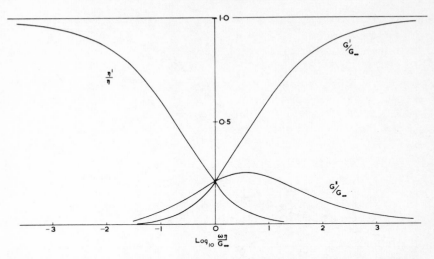

FIGURE 2   Variation with frequency of the dynamic viscosity and the components of the shear modulus for the B.E.L. Eq. (12).

trum for the B.E.L. equation is proportional to $\tau^{\frac{1}{2}}$: the spectrum must therefore be terminated at long times in order to satisfy purely physical limitations. From the standpoint of alternating shear, the effect of this is most likely to be found in the experimentally ill-defined region between viscous Newtonian behaviour and the onset of viscoelastic relaxation: otherwise predictions based upon the B.E.L. equation would not have agreed so well with experimental results.

Similar problems have been encountered in dielectric relaxation and were recognised by Glarum[18] in his defect-diffusion model, which failed to fit experimental results at low frequencies. To overcome this, he introduced a second independent relaxation process, to which he assigned an exponentially-decaying relaxation function with an arbitrary time constant and he associated this in physical terms with the relaxation of a site in the absence of a defect arrival. It has been shown by Davidson[24] that, under certain conditions, Glarum's model leads to loci for the components of the complex permittivity which are sensibly indistinguishable from the skewed-arc plots of Davidson and Cole.[25] The empirical expression for the complex dielectric constant $\epsilon^*(j\omega)$ first put forward by these authors is:

$$\epsilon^*(j\omega) = \epsilon'(\omega) - j\epsilon''(\omega)$$
$$= \epsilon_\infty + (\epsilon_0 - \epsilon_\infty)/(1 + j\omega\tau_D)^\beta, \qquad (14)$$

where the subscripts denote limiting values at very high and very low frequencies, respectively, τ_D is a characteristic time constant of the distribution and β is a constant, $0 < \beta < 1$. τ_D is the longest time constant, defining the cut-off or termination of the spectrum.

By using a further development of the shear wave impedance technique operating at 30 MHz under carefully controlled conditions, Barlow and Erginsav[13] have achieved an order of magnitude improvement in the absolute accuracy of measurement over previous systems employed in the Author's laboratory or elsewhere. Measurements of R_L and of X_L were made as a function of temperature in the supercooled region and results obtained show that the retardational compliance can be fitted within experimental error by the empirical Davidson and Cole expression:

$$J_r(j\omega) = J_r/(1 + j\omega\tau_r)^\beta = J_r^*(j\omega) = J_1(\omega) - jJ_2(\omega). \tag{15}$$

The total compliance is therefore expressed by:

$$\frac{J^*(j\omega)}{J_\infty} = (1 + 1/j\omega\tau_m) + \frac{J_r/J_\infty}{(1 + j\omega\tau_r)^\beta}. \tag{16}$$

For the supercooled liquids measured, β is typically 0.5 with values ranging from about 0.4 to 0.6 in different liquids. A representative plot for benzyl benzoate[26] is illustrated in Figure 3, with corresponding variations of $J_1(\omega)$ and

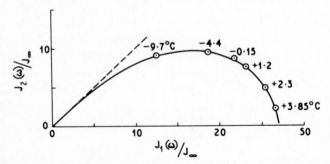

FIGURE 3   Relative variation of the components of the retardational compliance for benzyl benzoate at 29.426 MHz. The curve is drawn in accordance with the Davidson–Cole function of Eq. (15) with $\beta = 0.5$. Taken from Reference 26 by permission of the Chemical Society.

$J_2(\omega)$ with normalised frequency shown in Figure 4. In the region where $\omega\tau_r \gg 1$ the retardation term in Eq. (16) can be taken approximately as $(J_r/J_\infty)/(j\omega\tau_r)^\beta$ and with $\beta = 0.5$ this becomes $(J_r/J_\infty)/(j\omega\tau_r)^{\frac{1}{2}}$. In all such cases, the equal values of $J_1(\omega)$ and $J_2(\omega)$ in this region agree with the value $(2/\omega\tau_m)^{\frac{1}{2}}$ given by the B.E.L. Eq. (12). $\tau_r/\tau_m = (J_r/2J_\infty)^2$ for $\omega\tau_m > 1$ and this value for benzyl benzoate is 182. Thus, in the region $\omega\tau_m > 1$ we find, in general, that $\omega\tau_r$ is sufficiently greater than unity for these approximations to apply. This

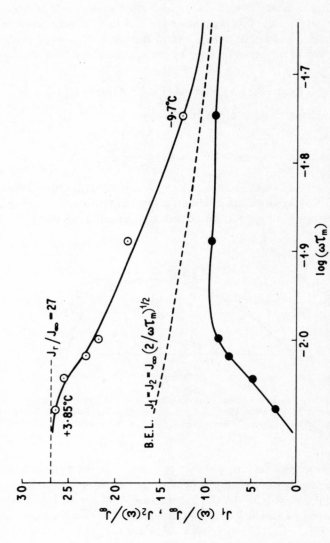

FIGURE 4 Measured values of $J_1(\omega)/J_\infty$ and $J_2(\omega)/J_\infty$ for benzyl benzoate plotted as functions of $\log(\omega\tau_m)$.

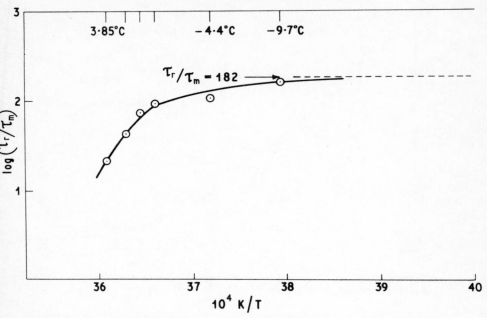

FIGURE 5   Temperature dependence of (τ_r/τ_m) for benzyl benzoate. The limiting asymptotic value at low temperatures is in accordance with the B.E.L. equation, $\tau_r/\tau_m = (J_r/2J_\infty)^2$ = 182.

also explains why deviations from the B.E.L. equation at low frequencies remained un-noticed in the earlier work.

Assuming the form of Eq. (16) we can evaluate τ_r at each temperature corresponding to the result given in Figure 3. The variation of (τ_r/τ_m) with $1/T$ is shown in Figure 5, demonstrating the points made above.

Implicit in the normalisation procedures used in plotting experimental results in the manner of Figures 1, 3 and 4 has been the assumption that the value of J_∞ is known throughout the ranges of temperature and pressure involved in the experimental work. As mentioned in Section 2, J_∞ is obtained from the limiting value of ρ/R_L^2 at effective frequencies such that X_L can be neglected in comparison with R_L, that is above the region of viscoelastic relaxation ($\omega\tau_m \gg 1$), where the liquid behaves elastically and ρ/R_L^2 is ascertained to be independent of frequency at a given temperature and pressure. However, since the same techniques are employed in this measurement as are used to delineate the viscoelastic response, it follows that J_∞ can only be measured at low temperatures approaching the glass transition temperature, T_g. In practice, convenient frequencies of 10, 30 and 450 MHz are used and the liquid is cooled until (ρ/R_L^2) becomes independent of frequency. In over fifty or more liquids which have been studied

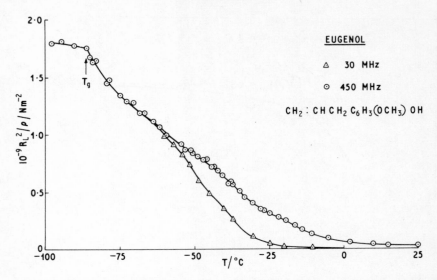

FIGURE 6   Measured values of R_L^2/ρ vs. temperature for eugenol at 30 and 450 MHz. Taken from Reference 27 by permission of the Chemical Society.

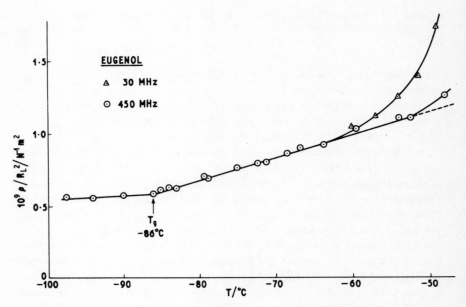

FIGURE 7   The results of Figure 6 plotted inversely as ρ/R_L^2 vs. temperature, showing the linear dependence of J_∞ upon T in accordance with Eq. (17). Taken from Reference 27 by permission of the Chemical Society.

in this way it has been found that J_∞ is a linear function of temperature

$$J_\infty(T) = J_0 + C(T - T_0). \qquad (17)$$

Results for eugenol[27] are shown in Figures 6 and 7, firstly in the form of R_L^2/ρ (Figure 6) showing that below about $-65°C$, the limiting modulus, G_∞, is not linear with T, whilst, for the corresponding plot of ρ/R_L^2 of Figure 7, J_∞ clearly is a linear function of T in accordance with Eq. (17). Indeed, at 450 MHz, this linear region extends over some $30°C$ until the relaxation region is entered at this frequency for temperatures above $-55°C$. The assumption is made that, at higher temperatures, where for the same frequencies of measurement the liquid no longer behaves elastically, the linear dependence of J_∞ on T given by Eq. (17) can be extrapolated into the temperature range of the relaxation measurements. Considerable data have been accumulated but no evidence has been found to suggest that this assumed extrapolation is not justified. Tabulated values of the constants, J_0, C and T_0 are to be found in the paper by Lamb,[28] to which others can now be added[26,27] and are listed in Table I.

TABLE I

Retardation parameters for liquids

	T/K	$\eta/Pa\ s$	$10^9 J_\infty/m^2 N^{-1}$	J_r/J_∞	β	τ_r/τ_m
squalane	242.7	2.600	1.90	21.8	0.6	53.6
tri-2-chloroethyl phosphate	243.2	2.778	0.89	26.0	0.4	609
tri(o-tolyl) phosphate	271.5	3.308	0.96	8.1	0.5	16.4
tri(m-tolyl) phosphate	258.7	3.175	1.18	7.6	0.5	14.4
di(n-butyl) phthalate	229.7	5.100	1.72	4.2	0.5	4.4
di(iso-butyl) phthalate	230.7	74.00	1.23	4.0	0.5	4.0
2-phenylethyl-chloride	213.2	1.62	1.15	3.14	0.39	3.18
benzyl benzoate	243.2	0.456	1.63	27.0	0.5	182
eugenol	243.2	0.82	1.42	1.0	0.5	0.5
tri(α-naphthyl) benzene	337.4	$10^{11.35}$	0.81	2.2	0.3	7.76

When pressure is employed as a variable at constant temperature, it is found unambiguously that G_∞ is a linear function of pressure:[23,29−31]

$$G_\infty(p) = G_\infty(0) + Dp. \qquad (18)$$

Here $G_\infty(0)$ is the measured value at zero gauge pressure of 1 atm. and D is a constant for a given temperature for a particular fluid. Since the value of $G_\infty(0)$

is known, the use of Eq. (18) in the relaxation region is not subject to the extra-polation procedure mentioned above with respect to $J_\infty(T)$.

It is not proposed here to deal extensively with the viscoelastic behaviour of mixtures of liquids, but with reference to subsequent discussions of polymeric liquids, it is appropriate to state that a modification of the B.E.L. Eq. (12) has been found to fit the results observed in mixed systems. This involves the intro-duction of a further constant, K, as a multiplier in the retardation term:[21]

$$\frac{J^*(j\omega)}{J_\infty} = (1 + 1/j\omega\tau_m) + 2K/(j\omega\tau_m)^{\frac{1}{2}}. \tag{19}$$

$K = 1$ gives the original B.E.L. Eq. (12), whereas for $K = 0$ the equation reduces to that of Maxwell.

For binary mixtures it has been found that if the component liquids are either isomers or of approximately equal molecular weight, then $K = 1$ irrespec-tive of composition. K is also unity at a 0.5 mole fraction composition of any two supercooled liquids. However, if one component has a molecular weight sig-nificantly lower than the other, a characteristic curve for K against mole fraction is found[21] (Figure 8). In seeking an explanation for this behaviour, we recall that the relatively wide distribution of the observed effects of viscoelastic relaxation in pure supercooled liquids is atrributed to the time delay in the diffusion of defects through a fluid in which the amount of available free volume is restricted. For binary mixtures, addition of over 5 mole percentage of the lower molecular weight constituent results in a value of K less than unity, the minimum value being $K_{min} = 0.25$ at 0.2 mole fraction. The viscoelastic relaxation curves are less broad than for either pure component ($K = 1$) and are "sharpened" towards the Maxwell curves ($K = 0$). The molecules of the lower molecular weight com-ponents would appear to be able to relax the stress in the local regions around the larger molecules more rapidly than would be the case if only one type of molecule were present. Such considerations apply only to mixtures which strongly consist of a major fraction of the higher molecular weight component (0.8 mole fraction at $K_{min} = 0.25$), but the results suggest that in this region relaxation of a "site" ocurs not only by diffusion of defects to the site but also by the creation of similar defects, due to reorientation and/or outward migration from the neighbourhood of a site of the molecules of lower molecular weight and assumed smaller size. This effect is progressively cancelled as the concentra-tion of the lower molecular weight component increases from 0.2 to 0.5 mole fraction, at which latter value K is again equal to unity when neighbouring mole-cules alternate in the fashion ABAB . . .. The reverse situation holds if we trace the behaviour with increasing concentration of the higher molecular weight com-ponent from 5 mole % upwards. Increasing numbers of higher molecular weight molecules inhibit the diffusion of defects in a composition which is rich in lower molecular weight material, and, moreover, the larger molecules diffuse more slowly than those of lower molecular weight. K, therefore, increases with increasing mole fraction of higher molecular weight species, reaching a maximum $K_{max} = 1.8$ at a concentration of approximately 0.2 mole fraction of the latter. This corresponds to the broadest distribution of the relaxation curves. Further increase in concentration of the higher molecular weight components beyond 0.2 mole fraction towards the 0.5 mole fraction, ABAB . . . distribution, allows defects to diffuse more rapidly than at the concentration corresponding to maximum K and the result is the curve shown in Figure 8 (taken from Reference 21). Comment is not offered on the nature of the sharp peaks in the concentration regions 0 to 0.05 mole fraction of either component, since the reason for these is not clear. The curve of Figure 8, was formed as a result of measurements on four sets of binary mixtures, this study being undertaken because it was suspected that relatively small amounts of impurity had a marked influence on the shape of the relaxation curves.

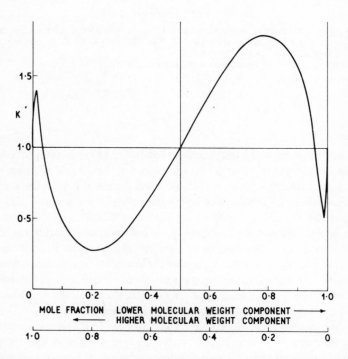

FIGURE 8   Derived curve of the factor K in Eq. (19) as a function of composition for binary mixtures of supercooled liquids. Taken from Reference 21 by permission of the Royal Society.

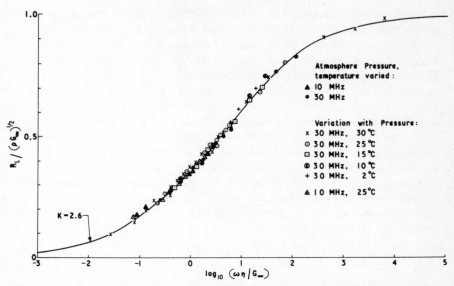

FIGURE 9   Normalised shear resistance of castor oil as a function of reduced frequency.
The curve is calculated from the modified B.E.L. Eq. (19) with $K = 2.6$. Results are
obtained at different pressures and temperatures.
Taken from Reference 21 by permission o the Chemical Society.

An extreme case in which the modified B.E.L. Eq. (19) has been found to
apply within experimental error is that of Castor Oil. This liquid is a multi-
component mixture, for which a calculated curve with $K = 2.6$ fits the measured
values of normalised shear resistance over wide ranges of temperature and pressure[31] (Figure 9).

In view of the fact that the Davidson—Cole empirical distribution function,
employed extensively in the analysis of dielectric relaxation data, has been
found to fit the results of measurements of the complex retardational compliance, $J_r\chi(j\omega)$, it seemed useful to compare the dielectric and viscoelastic
results for the same liquid. Dielectric relaxation measurements[32] have been
carried out on three of the polar liquids for which the components $J_1(\omega)$ and
$J_2(\omega)$ of the retardational compliance are known (Eq. 15). The comparison
is inconclusive for one of these liquids, —tri-2-chloroethyl phosphate — since
the dielectric loss curves for this show evidence of bimodal behaviour. However,
for the other two liquids studied, — di-n-butyl phthalate and tri-o-tolyl phosphate — there is a correlation between the dielectric and viscoelastic retardation

processes, as can be seen from Figure 10. The dielectric measurements were made at lower frequencies and hence at lower temperatures th..n the visco-elastic measurements. Precise quantitative agreement is not obtained between the two sets of results in terms of numerical values for the characteristic times of the separate processes, neither is the distribution of times the same but, from a qualitative point of view, the correspondence of behaviour in the two cases is striking. On this account, it is argued that, since the dipole relaxation is essen-tially a reorientation of the molecule in the force field of surrounding molecules, then the retardational viscoelastic process must likewise be one of local reorienta-tional adjustment. In this connection, we recall the argument of Glarum,[18] when he introduced a second independent relaxation process into his analsyis associated with the relaxation of a site in the absence of a defect arrival: both this mecha-nism and relaxation caused by the diffusion of defects would contribute to the relief of shear stress in a specific local environment. Likewise, the time constants associated with dipole reorientation are broadly subject to the same physical processes, but there are differences in detail since the dielectric relaxation time and the viscoelastic retardation time for the same form of Davidson—Cole distri-bution are not the same. This point is exemplified further by the analysis of dielectric and viscoelastic measurements on benzyl benzoate,[26] the results of which are shown in Figure 11. Thus, in the low-temperature supercooled region below about $-10°C$, the shapes of the retardation and dielectric relaxation time spectra are the same, since $\beta = 0.5$ for both processes (Eqs. 14 and 15). However, in this region, $\tau_r \simeq 16.5\tau_D$, the B.E.L. equation becomes the limiting approxima-tion since $\omega\tau_r \gg 1$ and, as the temperature is lowered, both τ_r and τ_D retain a constant ratio to $\tau_m [\tau_r/\tau_m = 182$ (Figure 5) and $\tau_D/\tau_m = 11]$. The surprising feature is not that τ_r and τ_D differ but that the extent of the difference is more than an order of magnitude, in view of the similar nature of the processes which are presumed to operate in the respective cases. This point is emphasised subse-quently in considering similar comparisons for a low molecular weight poly-propylene gycol.

Also shown in Figure 11 is a single result for the relaxation time τ_c obtained from investigations of the depolarised doublet of scattered light. It has been suggested[33] that the narrow depolarised doublet components of light scattered from liquids arises from the shear retardation processes. Values of the shear compliance found from light scattering are similar to the shear retardational compliance. A distribution of time constants would be expected but the value of τ_c plotted in Figure 11 is derived on the assumption of a single time. Unfor-tunately, the temperature ranges of the two sets of experiments do not overlap and further work is needed to resolve this question.

It transpires that, for the majority of liquids investigated, the retardational time, τ_r, of the Davidson—Cole distribution (Eq. 15) is found to be greater than the Maxwell relaxation time, $\tau_m = \eta J_\infty$, and, in retrospect, it was primarily on

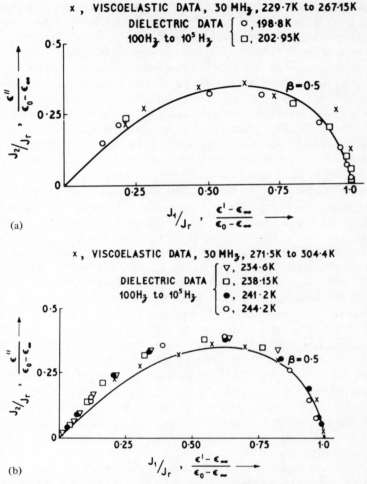

FIGURE 10 Comparison of dielectric relaxation and viscoelastic retardation in (a) di-n-butyl phthalate and (b) tri-o-tolyl phosphate. Curves are drawn through the viscoelastic compliance results in accordance with the Davidson−Cole distribution function of Eq. (15) with $\beta = 0.5$ in each case.
Taken from Reference 32 by permission of the Chemical Society.

this account that the B.E.L. equation was found to fit the earlier experimental results. However, there is no *a priori* reason why this should be so and at least for one liquid this clearly does not hold. Experimental results for eugenol (4-allyl-2-methoxy phenol)[31] are fitted by taking $\tau_r = 0.5\,\tau_m$ and $J_r = J_\infty$ (Figure 12). The effect of the retardation process is principally observed at frequencies around

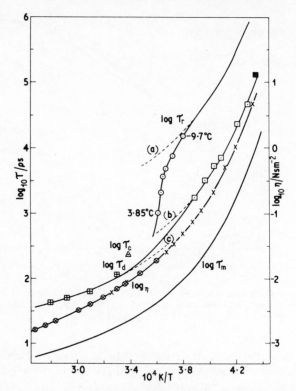

FIGURE 11   The steady-flow viscosity, dielectric relaxation time (τ_D (Eq. (14)), shear retardation time, τ_r (Eq. (15)). and the time τ_c derived from light scattering experiments for benzyl benzoate. Curve (a) $\tau_r = 182\tau_m = 16.5\tau_D$ below $-10°C$; (b) $\tau_D = 11\tau_m$; (c) $\tau_D = 5.5\tau_m$, above $21°C$.
Taken from Reference 26 by permission of the Chemical Society.

$\omega\tau_m = 1$, in which region the calculated B.E.L. equation does not agree with experiment.

4   POLYMER LIQUIDS

4.1   Polypropylene glycols

The viscoelastic properties of two polypropylene glycols have been measured by Barlow and Erginsav[34] at frequencies of 30 and 450 MHz and over a wide temperature range. The two liquids, designated P400 and P4000 have number average molecular weights of 377 and 3034 respectively and for each liquid $\bar{M}_W/\bar{M}_N$ is less than 1.16. Thus, in the formula HO$[C_3H_6O]_n$ C$_3$H$_6$OH, n is

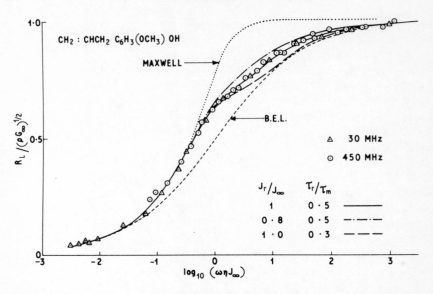

FIGURE 12   Normalised values of the shear resistance of eugenol. The fit of the full line curve to the experimental data is obtained using Eq. (15) with $J_r/J_\infty = 1.0$ and $\tau_r/\tau_m = 0.5$. Taken from Reference 27 by permission of the Chemical Society.

approximately 5 for P400 and 50 for P4000. Viscosity data for P4000 were fitted by the modified free volume Eq. (10) and likewise for P400 below 50°C. The linear dependence of J_∞ on temperature is shown in Figure 13. The relative variation of the normalised components of the retardational compliance is given in Figure 14 for P400 and in Figure 15 for P4000. In each case, the skewed-arc plot corresponding to the Davidson–Cole Eq. (16) is drawn through the experimental points. For P400, $\beta = 0.45$ and $J_r/J_\infty = 17.4$, whilst for P4000, $\beta = 0.76$ and $J_r/J_\infty = 85$. As for supercooled liquids, the temperature dependence of J_r and J_∞ has been assumed to be the same, and the value of τ_r can be determined for each point plotted on the curves of Figures 14 and 15. For P4000 the ratio $\tau_r/\tau_m = 15.4$, and is constant over the range 40 to 80°C: by analogy with previous results this ratio would be presumed to remain constant at lower temperatures. Substituting the above values back into Eq. (16), together with the known variation of J_∞ and η, allows curves for the normalised components of R_L and X_L to be calculated using only the results for 30 MHz corresponding to the points plotted on Figure 15. As shown by Figure 16, this calculated curve for $R_L/(\rho G_\infty)^{\frac{1}{2}}$ is in excellent agreement with results of measurements made at 454 MHz. In effect, in this particular instance, the complete behaviour extending over some six or more decades has been correctly predicted from measurements of R_L and X_L extending over less than one decade of effective frequency. This would not, however, have been possible if the measurements had not been made precisely over this particular region.

The situation with regard to P400 is more complicated, since τ_r/τ_m increases with $1/T$ over the range of measurement corresponding to the results of Figure 14. τ_r/τ_m = 30 at 31.8°C and increases to 155 at 4.4°C, reaching a limiting value of 170 at temperatures below about 0°C (Figure 17). Since at temperatures above 0°C τ_r does not have the same dependence upon temperature as does τ_m, it follows that superposition is not possible here: a separate curve for the normalised variation of $R_L/(\rho G_\infty)^{\frac{1}{2}}$ with log $(\omega\tau_m)$ is required for each temperature. The lower calculated curve of Figure 18 is drawn for J_r/J_∞ = 17.4, β = 0.45 and τ_r/τ_m = 30, these values corresponding to a temperature of 31.8°C and this agrees with the experimental results at 30 and 454 MHz for this temperature. The upper curve is plotted for the same values of J_r/J_∞ and β but for τ_r/τ_m = 170, this being the limiting value below 0°C (Figure 17). This provides a good fit to all the 454 MHz results below 0°C and is in close agreement, as expected, with the 30 MHz result at 4.4°C. Curves calculated for intermediate temperatures between 31.8°C and 0°C with the appropriate value of τ_r/τ_m from Figure 17 agree satisfactorily with the 454 MHz data, but are not drawn on Figure 18 to avoid confusion. As in the previous analysis for P4000, only the results obtained at 30 MHz are used in deriving each curve: the validity of the analysis is confirmed by the agreement found between the calculated and measured behaviour at 454 MHz.

The behaviour of the lower molecular weight material P400 is similar to that of other non-polymeric liquids in all respects and it is considered that this liquid shows no evidence of polymeric characteristics. Similar conclusions have been drawn from results of proton spin—lattice relaxation experiments,[35] which show that at least 10 monomer units are required for the onset of chain segmental motion. By comparison, P4000 has a much greater value of J_r/J_∞ (=85). The normalised component G'/G_∞ is plotted for each liquid in Figure 19 for a temperature of 0°C, thus avoiding the region where τ_r/τ_m is temperature dependent for P400. The emergence of a "plateau" region for P4000 suggests that this material exhibits a polymer character. The difference in shear modulus behaviour between P4000 and P400 is almost completely accounted for by a single relaxation time process with τ_a = 0.93 μs and G_a = 5.2 x 10^6 N/m^2. Alternatively, we can equally well analyse the results in terms of retardation and the calculated curve for $J_1(\omega)$ for both liquids, based upon the experimental parameters given previously, is shown in Figure 20, again for a temperature of 0°C. The difference between these curves for $J_1(\omega)$ versus log ω for P4000 and P400 corresponds very closely to the variation due to a single retardation time process having a time τ_x = 0.11 μs and a compliance J_{r_x} = 132 x 10^{-9} m^2/N. Small differences in the high-frequency behaviour of the two liquids are evident from the results of Figures 19 and 20, possibly reflecting the relative difference in effect of the end groups as between the longer and shorter molecule of the same species. Com-

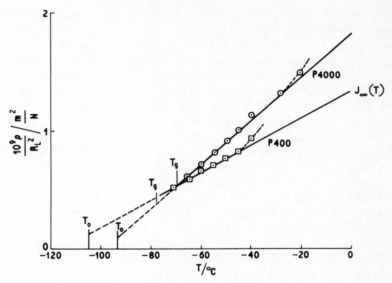

FIGURE 13   Temperature dependence of ρ/R_L^2 for polypropylene glycols P400 and P4000 in the elastic region. Values of T_g shown correspond to $\eta = 10^{12}$ Pa s.
Taken from Reference 34 by permission of the editors of Polymer.

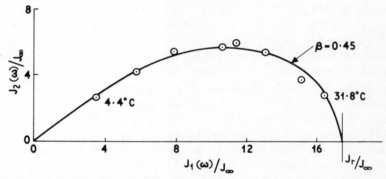

FIGURE 14   Relative variation of the components of the retardational compliance measured at 30 MHz for polypropylene glycol P400.
Taken from Reference 34 by permission of the editors of *Polymer*.

parative numerical data are listed in Table II for both liquids at a temperature of 0°C.

    The salient feature of these results is that an additional process is required to account for the behaviour of the P4000 material and it is inferred that this is due to the polymeric nature of this liquid whereas P400 is sensibly in the category of non-polymeric liquids discussed in Section 3. In terms of compliance this addi-

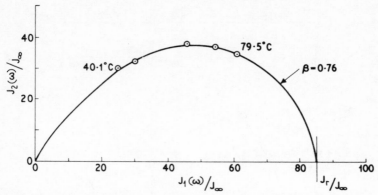

FIGURE 15 Relative variation of the components of the retardational compliance measured at 30 MHz for polypropylene glycol P4000.
Taken from Reference 34 by permission of the editors of *Polymer*.

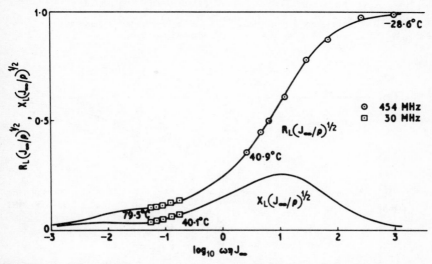

FIGURE 16 Normalised components of the shear impedance for polypropylene glycol P4000. The curves are calculated from Eq. (6) and (16) using $J_r/J_\infty = 85$, $\beta = 0.76$ and $\tau_r/\tau_m = 15.4$, which are obtained solely from results of measurements at 30 MHz.
Taken from Reference 34 by permission of the editors of *Polymer*.

tional process can be represented by the addition of a single retardation term $(J_r)_x/(1 + j\omega\tau_x)$ to the complex retardational compliance $J_r^*(j\omega)$ measured in P400 or, in terms of shear modulus, by the addition of the single relaxation term $G_a/(1 + j\omega\tau_a)$ to the complex modulus of P400, with due account taken of the difference in viscosity between the two liquids. These alternative methods of analysis are equivalent and self-consistent.

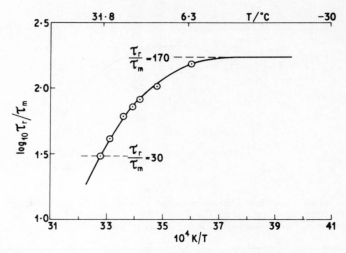

FIGURE 17   Temperature dependence of τ_r/τ_m for polypropylent glycol P400. Taken from Reference 34 by permission of the editors of *Polymer*.

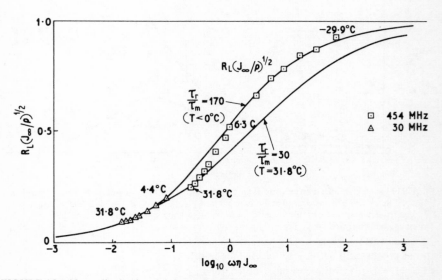

FIGURE 18   Normalised values of shear resistance for polypropylene glycol P400 plotted against $\log(\omega\tau_m)$. The lower curve is calculated for a temperature of 31.8°C using $\tau_r/\tau_m = 30$ whilst the upper curve is for temperatures below 0°C with $\tau_r/\tau_m = 170$. For both curves $J_r/J_\infty = 17.4$ and $\beta = 0.45$ in Eq. (16).
Taken from Reference 34 by permission of the editors of *Polymer*.

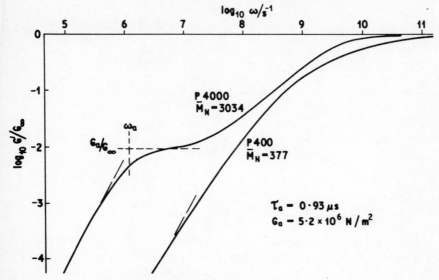

FIGURE 19   Frequency dependence of the normalised shear modulus component G'/G_∞ for polypropylene glycols P400 and P4000 at 0°C.
Taken from Reference 34 by permission of the editors of *Polymer*.

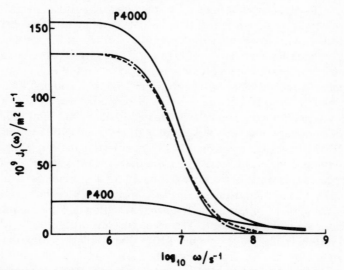

FIGURE 20   Frequency dependence of $J_1(\omega)$ for polypropylene glycols P400 and P4000, at 0°C. The dashed curve is the difference between the values for the two liquids. The broken curve is for a single retardation time with $(J_r)_X = 132 \times 10^{-9}$ m²/N and $\tau_X = 0.11$ μs.
Taken from Reference 34 by permission of the editors of *Polymer*.

TABLE II

Comparison of data for two polypropylene glycols at 0°C

	P400	P4000
$\eta/\text{Pa s}$	0.46	5.0
G_∞/Nm^{-2}	7.5×10^8	5.5×10^8
τ_m/ps	616	9100
$\tau_r/\mu\text{s}$	0.105	0.140
τ_r/τ_m	170	15.4
$J_\infty/\text{m}^2\text{N}^{-1}$	1.34×10^{-9}	1.82×10^{-9}
J_r/J_∞	17.4	85
$J_r/\text{m}^2\text{N}^{-1}$	23.4×10^{-9}	155×10^{-9}
β	0.45	0.76
$\tau_a/\mu\text{s}$	−	0.93
G_a/Nm^{-2}	−	5.2×10^6
$\tau_x/\mu\text{s}$	−	0.11
$(J_r)_x/\text{m}^2\text{N}^{-1}$	−	132×10^{-9}

The additional single relaxation mode in P4000 could be regarded as the first of the series of configurational polymer modes postulated in the Rouse theory but the observed "mode strength", G_a is 7 times the value for the first Rouse mode ($= \rho RT/M$) and the observed relaxation time, τ_a, is $1/7$ times the calculated Rouse value for the first mode, taken as $\tau_1 = \eta M/\rho RT$. Since, however, the molecule contains only some 50 repeat units, several of which are required to form a Gaussian segment, it could be argued that the observed stiffness should exceed that calculated from statistical theory for a long chain molecule and that the effect of the end groups would also enhance the stiffness of the smaller molecule. Unfortunately, it has not been possible to pursue this investigation by studying the behaviour of longer chain polypropylene glycols, since these are not available in the molecular weight range of interest, from 10,000 to 100,000.

Bauer and Stockmayer[35] have measured the dielectric properties of a comparable series of polypropylene glycols, designated P1025, P2025 and P.H.M.W. with number average molecular weights of 900, 1800 and 3700 respectively. Sample P2025E was prepared from P2025 by converting the end hydroxyl groups to methoxyl. A principal dispersion region is observed for which the maximum loss, ϵ'', occurs at essentially the same frequency for a given temperature in all four samples. In addition, there is a small secondary loss occurring at lower frequencies, the position of this in the frequency spectrum and its amplitude being dependent upon molecular weight. This secondary loss process is barely resolvable in the lowest molecular weight sample, P1025, but is still present in the methylated sample, P2025E. We have observed similar secondary loss peak behaviour in P4000 but have found no evidence of its existence in P400. A comparison of observed relaxation times and viscoelastic retardation times is shown in Figure 21 for sample P400. Here, τ_D is the dielectric relaxation time for the "principal" dispersion obtained from the maximum in $\epsilon''/(\epsilon_0 - \epsilon_\infty)$, τ_m is the Maxwell relaxa-

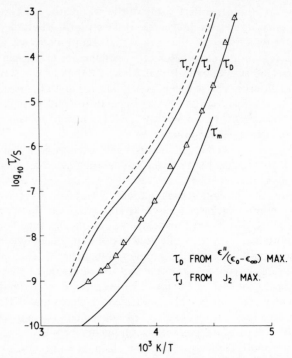

FIGURE 21   Comparison of viscoelastic retardation times and dielectric relaxation times in polypropylene glycol P400.

tion time, τ_r the viscoelastic retardation time of the Davidson—Cole distribution and τ_J is obtained from the maximum of $J_2(\omega)$. This behaviour is similar to that shown in Figure 11 for benzyl benzoate and, again, there is an order of magnitude difference between the viscoelastic retardation time and the dielectric relaxation time. The small amplitude secondary dielectric loss process in the higher molecular weight samples is attributed by Bauer and Stockmayer to local segmental motion in the polymer chain and the conclusion is drawn here that this is also to be associated with the additional process responsible for the much larger viscoelastic response of the P4000 material over that of the lower molecular weight P400. Clearly, viscoelastic measurements are far more sensitive to the effects of polymer segmental motions than are corresponding dielectric loss experiments.

4.2   Polystyrenes

Measurements of shear creep and elastic creep recovery on a series of polystyrene melts of narrow molecular weight distribution have been carried out by Plazek and O'Rourke.[10] A complementary study of the viscoelastic behaviour of these materials in alternating shear is in progress in the author's laboratory and results are now available on a low molecular weight material of molecular weight 3600. Ostensibly, this is from the same batch of nominal molecular weight 4000, produced by Messrs. Pressure Chemicals Inc., as the 3400 molecular weight sample of Plazek and O'Rourke. The sample was held in vacuum in the freeze-dried condition for an extended period (24 h) at a temperature of 60°C, slightly below T_g (= 64.7°C)† in attempts to remove any residual traces of solvent or monomer: it was further melted under vacuum at 140°C.

The components of the complex modulus have been measured at frequencies in the range 2×10^{-2} to 120 Hz, using a torsional rheometer designed by Harrison.[37] Results for $G'(\omega)$ and $G''(\omega)$ are shown in Figure 22 for a temperature of 92.4°C: relaxational behaviour is apparent with $G'(\omega)$ increasing with frequency from 10^3 to 10^7 N/m² over the range of measurement. Experiments have been carried out by Dr. R. W. Gray at four temperatures, the dynamic viscosity $\eta' = G''/\omega$ is plotted as a function of frequency in Figure 23 and corresponding values for the real part, $J'(\omega)$, of the complex compliance, $J^*(j\omega)$, are shown in Figure 24. The zero-frequency asymptote of $J'(\omega)$ in Figure 24 gives the equilibrium compliance, J_e, at each of the four temperatures, thus enabling a direct comparison to be made with the values obtained by Plazek and O'Rourke from creep recovery. The curve of Figure 25 is reproduced from data in Table II of their paper, with results of the present measurements included for the sample of molecular weight 3600. These agree with the creep recovery measurements within experimental error and show without doubt that the equilibrium compliance decreases markedly as the temperature approaches T_g. Measurements of J_∞ for this material give $J_\infty = 1.3 \times 10^{-9}$ m²/N[38] at 100°C, whereas J_e is of the order of 10^{-7} m²/N. Thus, the retardational compliance, $J_r (= J_e - J_\infty)$ is sensibly equal to the measured equilibrium compliance over the temperature range of the present measurements. The ratio J_r/J_∞ also increases markedly with temperature from T_g to $(T_g + 40)$K.

A non-polymeric liquid of comparable viscosity would show no evidence of viscoelastic relaxation in the frequency range around 1 Hz and the observed behaviour of the 3600 polystyrene is attributed to segmental motions in the small polymer chain, possibly of a "quasi-Rouse" nature. It is suggested that the decrease in available free volume as the temperature approaches T_g is responsible for restricting these local segmental motions, thereby decreasing the total equilibrium compliance.

A review of the origin of the method of reduced variables has been given by Ferry,[39] who argues on grounds of general hypothesis that the modulus is proportional to $T\rho/T_0\rho_0$, which is the form obtained from the Rouse equations,

---

† Unpublished result by courtesy of Dr. M. Richardson, National Physical Laboratory.

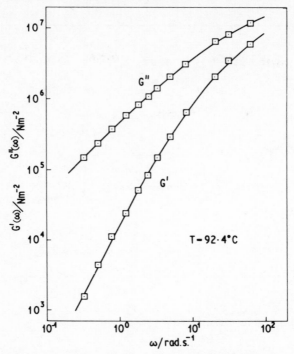

FIGURE 22 Experimental results for the measured components of the complex shear modulus of a polystyrene melt of molecular weight 3600 at a temperature of 92.4°C.

suffix zero denoting the reference temperature: the reduced frequency is $\omega \eta T_0 \rho_0 / \eta_0 T \rho$. This hypothesis has been widely tested but, as pointed out by Ferry, it should be carefully examined each time it is used. The curves of Figures 23 and 24, or the equivalent curves for modulus components, do not reduce in this manner, the reason being that, contrary to previous findings, the equilibrium compliance, J_e, does not decrease with increasing temperature according to $T_0 \rho_0 / T \rho$ but, in fact, *increases* as the temperature increases. We therefore reduce the experimental data by using *measured* values of J_e and η. Moduli are multiplied by $b_T = J_e^T / J_e^{T_0}$ and the frequencies by $a_T = b_T \eta^T / \eta^{T_0}$. In this way, good reduction is found, reduced curves for the components of the modulus being given in Figure 26 and for the components of the compliance in Figure 27. This is the procedure followed previously by Plazek and O'Rourke, who found no evidence of a decrease in compliance with increase of temperature in polystyrenes of higher molecular weight, up to 94,000. Indeed, the higher molecular weights exhibit a temperature-independent equilibrium compliance above about $(T_g + 10)K$ with a dramatic decrease below $(T_g + 10)K$.

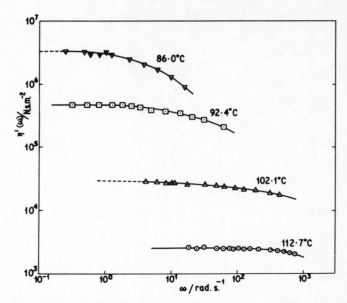

FIGURE 23   Measured values of dynamic viscosity of polystyrene 3600 as a function
of angular frequency at four temperatures.

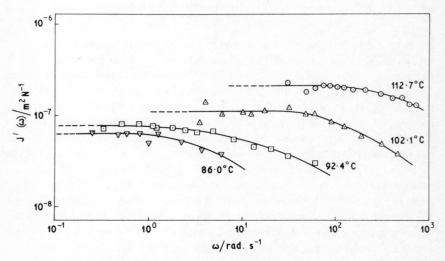

FIGURE 24   Experimentally determined values for the real part of the complex compliance
for polystyrene 3600 as a function of angular frequency at four temperatures.

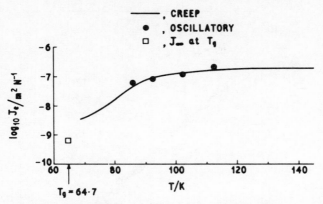

FIGURE 25 Variation of equilibrium compliance with temperature for polystyrene 3600. Points obtained from measurements in oscillatory shear. Curve plotted from creep recovery data of Plazek and O'Rourke. (Ref. 10.)

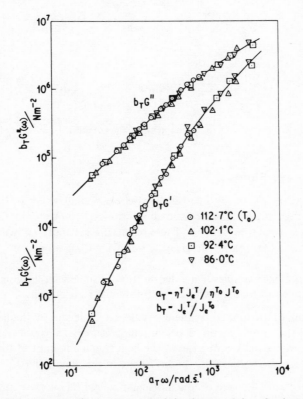

FIGURE 26 Reduced curves for components of the shear modulus of polystyrene 3600.

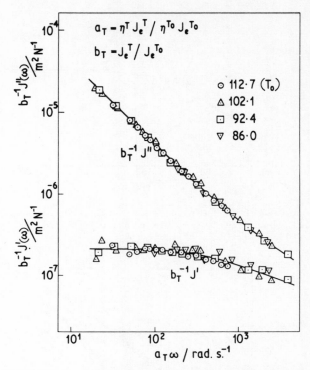

FIGURE 27   Reduced curves for components of the complex compliance of polystyrene
3600.

4.3   Siloxane Fluids

Polydimethyl siloxanes   Studies of the viscoelastic behaviour in alternating
shear of a range of polydimethyl siloxanes were carried out by the author and
his colleagues some time ago.[40,41] These results are generally known and will
not be reviewed here, except to give the following brief summary.

(i) The polymer samples had a distribution of molecular weights which could
be represented approximately by a most probable distribution.

(ii) Owing to the limited frequency range available and particularly to the
fact that crystallisation occurred at low temperatures, it was not possible to
measure the viscoelastic properties at the high frequency limit of the spectrum.
Maximum recorded values of G' and G'' were in the region of $10^7 \, N/m^2$.

(iii) Three samples were of low molecular weight below the critical entanglement region and three samples were of higher molecular weight. The lowest

molecular weight fluid contained an average number of 85 repeat units and the highest 920.

(iv) Calculations based upon an extended form of the Rouse theory to take account of polydispersity provided a good fit to the experimental results for G' and G'' as a function of effective frequency, taking a Gaussian segment as comprising 10 monomer units.

(v) For the higher molecular weights, well above the onset of entanglement, an enhanced friction coefficient was employed to account for relaxation of those lowest order modes for which the average distance between entanglement points exceeded the distance between the nodes of motion measured in repeat units along the polymer backbone. Agreement between calculations and experimental results then extended over some seven decades for the 100,000 cSt liquid.[41]

In view of our present interest in low molecular weight liquids and polymers, measurements have been made recently on a series of linear short-chain poly-dimethyl siloxanes containing, respectively, 5, 7, 9, 10 and 11 silicon atoms per molecule.[42] The temperature dependence of ρ/R_L^2 for these fluids is given in Figure 28, showing the linear dependence of J_∞ upon temperature. The slope of

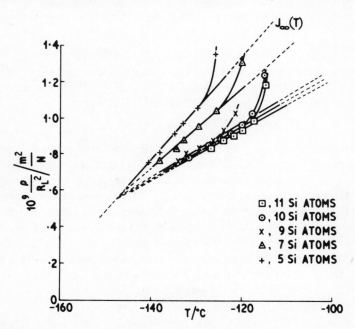

FIGURE 28   Temperature dependence of ρ/R_L^2 for short chain dimethylsiloxanes, showing the linear variation of J_∞ with T.
Taken from Reference 42 by permission of the editors of the *Journal of the Acoustical Society of America*.

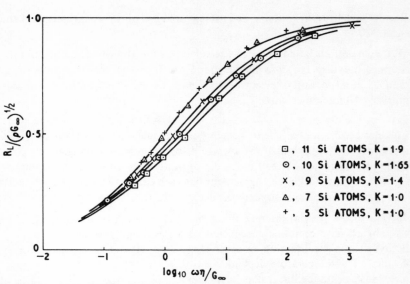

$R_L/(\rho G_\infty)^{1/2}$

$\log_{10} \omega\eta/G_\infty$

□ , 11 Si ATOMS, K = 1·9
○ , 10 Si ATOMS, K = 1·65
× , 9 Si ATOMS, K = 1·4
△ , 7 Si ATOMS, K = 1·0
+ , 5 Si ATOMS, K = 1·0

FIGURE 29   Normalised shear impedance of short chain dimethylsiloxanes.
Taken from Reference 42 by permission of the Editors of the *Journal of the Acoustical Society of America*.

this line, measured by the constant C in Eq. (17), $J_\infty(T) = J_0 + C(T - T_0)$, decreases progressively with increasing chain length. Normalised values of the shear resistance, $R_L/(\rho G_\infty)^{\frac{1}{2}}$, are plotted in Figure 29 and the measured behaviour in each liquid can be described by the modified B.E.L. equation:

$$J^*(j\omega) = \rho/Z_L^2 = J_\infty + 1/j\omega\eta + 2K(J_\infty/j\omega\eta)^{\frac{1}{2}}. \tag{19}$$

The relaxational behaviour of the 5 and 7 silicon-atom siloxanes is adequately described by the original B.E.L. equation with $K = 1$ and hence these may be regarded as typical non-polymeric liquids. The corresponding behaviour of the remaining three siloxanes is described by Eq. (19) with the value of K increasing with length of the molecule from 1.4 for 9 silicon-atoms to 1.9 for 11 silicon-atoms (Figure 29). J_∞ and dJ_∞/dT also tend to become constant with increasing molecular weight (Figure 28). This is a known property of longer chain polymers[8] and it would seem that the broadening of the relaxation region with increasing lengths of the molecule arises from the internal flexibilty despite the comparatively small chain length (molecular weights 680, 754 and 828 for 9, 10 and 11 silicon-atoms, respectively).

Comparing this behaviour with that described previously for the two poly-propylene glycols, it is suggested that the initial onset of the effects of internal segmental motions is a broadening of the relaxation. As the molecule becomes progressively longer, then it is possible to detect the presence of a separate process due to the polymeric nature of the liquid as, for example, in the P4000 poly-propylene glycol having 50 repeat units.

Polymethylphenyl siloxanes   In analysing previous results of viscoelastic relaxa-tion in a series of polyethyl- and poly-*n*-butyl acrylates[8] it was found that the complete behaviour could be interpreted in terms of additive contributions to the complex shear modulus comprising:

   i) a relaxation process at relatively high effective frequencies which follows approximately the same form as the B.E.L. equation for supercooled non-poly-meric liquids.

   ii) a limited summation of Rouse modes giving rise to relaxation processes at lower frequencies.

The polymer modes accounted for the major part of the steady-flow viscosity whilst the high frequency relaxation processes (i) were mainly responsible for the limiting shear modulus, G_∞. The viscosity associated with the high frequency relaxation was, therefore, only a small part of the steady-flow viscosity, the ratio of these being obtained from the shift of the B.E.L. curve along the axis of effective frequency to give agreement with the upper part of the normalised shear resistance curve plotted versus $\log(\omega \eta J_\infty)$.

   Alternatively, in the light of more recent work, the analysis can be cast in the form of contributions to the complex compliance (Eq. 16) which in the range of higher frequencies where $\omega \tau_r \gg$ can be written as the B.E.L. Eq. (12). However, in analysing the high frequency behaviour of a polymer in a region where polymer modes under (ii) above are no longer able to relax, it is convenient to employ the modified form given in Eq. (19), where the constant K incorpor-ates the known fact that the partial viscosity governing behaviour in this regime is much less than the measured steady flow viscosity from which $\tau_m = \eta J_\infty$ is calculated. This procedure has been adopted by Kim[43] in his analysis of results obtained on two polymethylphenyl siloxanes. Each of these fluids has approxi-mately 25% phenyl group substitution. Sample MS550 is of low viscosity (350 cSt grade), has a relatively low molecular weight and a reasonably narrow distribution ($\bar{M}_W/\bar{M}_N = 1.56$). Sample MS350 is of viscosity grade 15,600 cSt with a wide molecular weight distribution, $\bar{M}_W/\bar{M}_N = 26.9$ and is of higher molecular weight, $(\bar{M}_W)_{350}/(\bar{M}_W)_{550} = 114$.

   Measurements were made over wide ranges of temperature and of pressure at frequencies of 30 and 450 MHz and results for the normalised shear resis-tance were fitted to the equation

$$J^*(j\omega) = \rho/Z_L^2 = J_\infty \left[1 + 1/j\omega\tau_m\right] + \frac{2KJ_\infty}{(j\omega\tau_m)^\beta}, \tag{20}$$

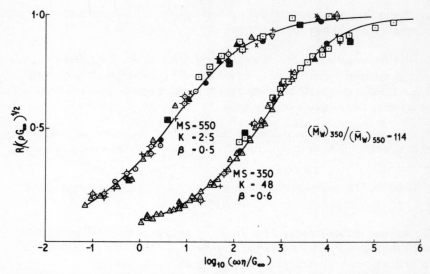

FIGURE 30   Normalised plots of the shear resistance of two polymethylphenyl siloxanes
MS550 (350 cSt grade) and MS350 (15,600 cSt grade). The curves are calculated from
Eq. (20) with:

MS550, K = 2.5, β = 0.5

MS350, K = 48 , β = 0.6.

Taken from Reference 43 by permission of the Chemical Society.

which takes account of the fact that β may differ from the value β = 0.5 of the
modified B.E.L. Eq. (19). The curves of Figure 30 are drawn on this basis with
the following values:

MS550   : K = 2.5, β = 0.5

MS350   : K = 48 , β = 0.6.

Thus, the lower molecular weight material shows a somewhat broader relaxation than would be observed in a pure non-polymeric liquid. This could either
be attributed to internal molecular motions, as has been suggested for the short-
chain polydimethylsiloxanes, or to the presence of a range of molecular weights,
as for a mixture. The observed behaviour is very similar to that of castor oil
but most likely both of the above effects contribute to the observed value of
K = 2.5. However, the measured behaviour of the higher molecular weight
material MS350 is undoubtedly attributable to motions of only short chain segments, the main region of polymer mode relaxations falling at frequencies below
the present range of measurement. These latter will give the major contribution
to the steady flow viscosity and the high value of K (= 48) reflects the fact that the

results are plotted against $\log(\omega \eta J_\infty)$. Indeed, the upper part of the curve for this polymer above $R_L/(\rho G_\infty)^{\frac{1}{2}} = 0.5$ is of very similar shape to the corresponding part of the curve for MS550 if this is shifted along the frequency axis by roughly two decades in frequency.

The conclusion is drawn that the high frequency behaviour of long chain polymer melts is similar in character to that of relatively low molecular weight liquids of the same polymer type since, at these frequencies, the modes of motion of the whole polymer molecule are inactive. When the time scale of measurement is sufficiently short, the polymer modes of Rouse type are "frozen" and only small elements of the polymer chain are able to respond to the rapidly applied shear stress. Experimental results entirely support this conclusion.

Acknowledgements

It is a pleasure to acknowledge the contributions which my colleagues have made to the work which I have reviewed. My thanks are especially due to colleagues of long standing, Dr. A. J. Barlow and Dr. G. Harrison. I would also like to express my appreciation of the contributions made by Dr. T. Alper, Dr. S. Fewster, Dr. R. W. Gray and Dr. M. G. Kim during their periods as Research Fellows in our Group. I am further indebted to them for permission to quote results of their work in advance of formal publication.

The research at the University of Glasgow has been supported by a Grant from the Science Research Council. Certain items of equipment have also been provided by Messrs. Imperial Chemical Industries Ltd.

Note added in proof

In the interval between presentation of this paper in early 1975 and its subsequent publication in 1978 the measurements on polystyrene melts have been completed and a full account published: R. W. Gray, G. Harrison and J. Lamb, *Proc. R. Soc. Lond.*, **A356**, 77, (1977).

Measurements of the volumetric properties of polystyrene have also been published: M. J. Richardson and N. G. Savill, *Polymer*, **18**, 3, (1977).

References

1. G. C. Berry and T. G. Fox, *Adv. Polymer Sci.* **5**, 261 (1968).
2. P. E. Rouse, *J. Chem. Phys.* **21**, 1272 (1953).
3. B. H. Zimm, *J. Chem. Phys.* **24**, 269 (1965).
4. J. Lamb and A. J. Matheson, *Proc. R. Soc. Lond.* **A281**, 207 (1964).
5. S. F. Edwards and K. F. Freed, *J. Chem. Phys.* **61**, 1189 (1974).
6. K. F. Freed and S. F. Edwards, *J. Chem. Phys.* **61**, 3626 (1974).
7. A. J. Barlow, G. Harrison and J. Lamb, *Proc. R. Soc. Lond.* **A282**, 228 (1964).
8. A. J. Barlow, M. Day, G. Harrison, J. Lamb and S. Subramanian, *Proc. R. Soc. Lond.* **A309**, 497 (1969).
9. M. L. Williams, *J. Polymer Sci.* **62** 173 S7 (1962).

10. D. J. Plazek and V. M. O'Rourke, *J. Polymer Sci.* **9**, 209 (1971).
11. E. Riande, H. Markovitz, D. J. Plazek and N. Raghupathi, *I.U.P.A.C. Macromolecules Symposium*, Madrid 1974. To be published.
12. B. Gross, *Mathematical Structure of the Theories of Viscoelasticity* (Herman, Paris, 1953).
13. A. J. Barlow and A. Erginsav, *Proc. R. Soc. Lond.* **A327**, 175 (1972).
14. A. J. Barlow, J. Lamb and A. J. Matheson, *Proc. R. Soc. Lond.* **A292**, 322 (1966).
15. N. T. Laughlin and D. R. Uhlmann, *J. Phys. Chem.* **76**, 2317 (1972).
16. A. Dworkin, *J. Chinie Phys.* **71**, 929 (1974).
17. M. C. Phillips, A. J. Barlow and J. Lamb, *Proc. R. Soc. Lond.* **A329**, 193 (1972).
18. S. Glarum, *J. Chem. Phys.* **33**, 639 (1960).
19. A. J. Barlow, A. Erginsav and J. Lamb, *Proc. R. Soc. Lond.* **A298**, 481 (1967).
20. A. J. Barlow, J. Lamb, A. J. Matheson, P. R. K. L. Padmini and J. Richter, *Proc. R. Soc. Lond.* **A298**, 467 (1967).
21. A. J. Barlow, A. Erginsav and J. Lamb, *Proc. R. Soc. Lond.* **A309**, 473 (1969).
22. J. Lamb, in *Molecular Motions in Liquids*, Edited by J. Lascombe (D. Reidel, Dordrecht Holland, 1974).
23. A. J. Barlow, G. Harrison, J. B. Irving, M. C. Kim, J. Lamb and W. C. Pursley, *Proc. R. Soc. Lond.* **A327**, 403 (1972).
24. D. W. Davidson, *Can. J. Chem.* **39**, 571 (1961).
25. D. W. Davidson and R. H. Cole, *J. Chem. Phys.* **18**, 1417 (1950), and **19**, 1484 (1951).
26. A. J. Barlow and A. Erginsav, *J. Chem. Soc. Faraday Trans. II* **70**, 885 (1974).
27. M. G. Kim, *J. Chem. Soc. Faraday Trans. II.* **71**, 415 (1975).
28. J. Lamb, in *Proc. Fifth Int. Congress on Rheology*, Vol. 4, edited by S. Onagi (Univ. Tokyo Press, 1970) p. 325.
29. W. M. Slie and N. M. Madigosky, *J. Chem. Phys.* **48**, 2810 (1968).
30. J. H. Hutton and M. C. Phillips, *Nature Phys. Sci.* **238**, 141 (1972).
31. A. J. Barlow, G. Harrison, M. G. Kim and J. Lamb, *J. Chem. Soc. Faraday Trans. II* **69**, 1446 (1973).
32. M. F. Shears, G. Williams, A. J. Barlow and J. Lamb, *J. Chem. Soc. Faraday Trans. II* **70**, 1783 (1974).
33. A. J. Barlow, A. Erginsav and J. Lamb, *Nature Phys. Sci.* **237**, 87 (1972).
34. A. J. Barlow and A. Erginsav, *Polymer (London).* **16**, 110 (1975).
35. T. M. Connor, D. J. Blears and G. Allen, *Trans. Faraday Soc.* **61**, 1097 (1965).
36. M. E. Bauer and W. H. Stockmayer, *J. Chem. Phys.* **43**, 4319 (1965).
37. G. Harrison, *Rheol. Acta* **13**, 28 (1974).
38. J. Lamb, *Rheol. Acta* **12**, 438 (1973).
39, J. D. Ferry, *Viscoelastic Properties of Polymers*, Ch. 11 2nd Ed. (John Wiley New York, 1969).
40. A. J. Barlow, G. Harrison and J. Lamb, *Proc. R. Soc. Lond.* **A282**, 228 (1964).
41. J. Lamb and P. Lindon, *J. Acoust. Soc. Amer.* **41**, 1032 (1967).
42. A. J. Barlow and A. Erginsav, *J. Acoust. Soc. Amer.* **56**, 83 (1974).
43. M. G. Kim, *J. Chem. Soc. Faraday Trans. II.* **71**, 428 (1975).

DISCUSSION

T. F. Schatzki (*U.S. Dept. of Agriculture*): The speaker suggests that the K factor in $J^*(j\omega)$ is governed by defect-diffusion. A plot of K vs. composition in a high/low molecular weight mixture indicates that K is increased by the addition of small amounts of high molecular weight material. It is difficult to understand how a small amount of high molecular weight material might affect the defect diffusion. We suggest that some other explanation is required to explain the size of K with increasing high molecular weight component. Would you care to speculate on these observations?

J. Lamb: Professor Schatzki has already commented that he accepts as reasonable the explanation put forward in the paper for the decrease in K with increasing concentration of the lower molecular weight constituent in a binary mixture which is predominantly rich in the higher molecular weight species. He queries the suggested explanation for the increase of K with increasing concentration of the higher molecular weight component in a mixture which is predominantly rich in the species of lower molecular weight (Figure 8).

We therefore accept that the arrival of a defect at a site is required for the local stress to be relaxed and that the time delay in the diffusion of defects through these supercooled liquids is essentially responsible for broadening the spectrum of the viscoelastic relaxation.

It must be pointed out that, if we trace the behaviour with increasing concentration of the higher molecular weight component, the increase in K only occurs when the concentration of higher molecular weight constituent exceeds 5 mole % -K reaching its maximum value at 20 mole %. Professor Schatzki does not suggest any acceptable alternative explanation to the one put forward in the paper, but could it be that in this concentration range the large molecules both inhibit the diffusion of defects because of their size and also act as temporary trapping sites for defects in a situation of dynamic equilibrium.

Finally, it is worthwhile to add that the sharp peaks in K observed for concentrations of either component in the range 0 to 5 mole % can, in the author's opinion, be attributed to relaxation of stress through defect diffusion either at a "site" of a molecule of the majority species or at a "site" of a molecule of the minority constituent.

Relaxation Spectrum of Free-Draining Block and Graft Copolymers

W. H. STOCKMAYER

Dartmouth College, Hannover, New Hampshire 03755, U.S.A.

The problem of computing the viscoelastic relaxation spectrum of block copolymers is shown to be identical to that of the conduction of heat through non-uniform solids. The relaxation times can be computed from those of the parent homopolymers *in the same medium*, but no simple averaging rule can be used for either the terminal relaxation time or the steady-flow Newtonian viscosity. Representative calculations for diblock and triblock polymers have been published.[1] Illustrations for graft copolymers are to be given later.

Reference

1. W. H. Stockmayer and J. W. Kennedy, *Macromolecules* 8, 351 (1975).

DISCUSSION

S. L. Aggarwal (*General Tire and Rubber Co.*): In your treatment, you take account of the relative size of the blocks. One of the important considerations would be the molecular weight of each block, since the effect of the junction points between blocks would depend on the molecular weight of the blocks. Do you not assume in your treatment that the molecular weight of each block is high enough so that its behaviour is similar to that of the corresponding homopolymer?

W. H. Stockmayer: The treatment is based on the assumption that every block is long enough so that the mechanical parameters assigned to it (unperturbed dimensions per repeat unit and friction coefficient per repeat unit) have the same values as for the corresponding homopolymer. However, the force balance condition at junction points is just what makes the normal coordinates and the resulting relaxation times different from simple linear averages of the corresponding properties for the related homopolymers.

R. Simha (*Case Western Reserve Univ.*): The problem for, say, a binary statistical copolymer is challenging, since the division of the molecule into Gaussian subchains looses its validity, unless the reactivity ratios favor the formation of long homo-sequences. Have you given thought to this problem?

W. H. Stockmayer: The theory presented is valid only for true block copolymers, in which each block consists of many repeat units. In a typical statistical copolymer with short run numbers, the viscoelastically important (long-wave) normal-mode coordinates and the distribution of relaxation times would closely resemble those for a homopolymer, but the mechanical parameters could not be fully predicted *a priori*. In favorable cases, predictions could be made for the unperturbed dimensions; but the corresponding friction coefficients might well involve local processes quite different in barrier height or geometry from those occurring in the related homopolymers. Quantitatively, if the run numbers were no greater than about 5 or 10% of the degree of polymerization, the block-polymer theory given here would not be of much use.

Polyethylene Morphologies Generated by Simple Shear

L. A. MANRIQUE, JR.† and ROGER S. PORTER

Materials Research Laboratory, Polymer Science and Engineering, University of Massachusetts, Amherst, Massachusetts 01003, U.S.A.

INTRODUCTION

For a variety of reasons, such as clarity and strength, there is a continuing interest in the preparation of unusual morphologies for the world's largest production polymer, polyethylene.[1-6] It has, nonetheless, not been easy to understand polyethylene behavior even though it possesses nominally the simplest polymer chemical composition. This difficulty is in part due to structural complexities (branching) which are minimized here by performing studies on a high density (linear) polyethylene, Alathon 7050, number and weight average molecular weights of 18,000 and 58,000, respectively.

The conditions used to produce special morphologies can also be, and generally are, complex. These include phenomena in conventional drawing and in the novel drawing process developed in our laboratory which involves a morphological conversion.[4, 5] Since these deformation processes provide polyethylenes of enhanced properties, it is therefore desirable to isolate the molecular mechanisms that may be contributory.

An obvious choice, for distinguishing between deformation processes, is a study of simple shear. For this purpose, a high shear rotational viscometer of custom design is available in this laboratory. It provides a unique opportunity to evaluate the potential for testing the ultimate effects of simple shear on polyethylene melts for the possible development of anisotropic morphologies.[7] Advantages of this instrument for preparing morphologies under controlled conditions include:

1) A homogeneous and high shear field: Indeed, between the concentric steel cylinders of the instrument used, the highest shear rates yet reported, $>10^6$ s^{-1},

†Present address: Johnson and Johnson Research, Route 1, North Brunswick,. New Jersey 08902.

have been developed.[8] The shear rates used here are high but, nonetheless, decades less due to the high viscosity of the polyethylene melt.

2) Excellent temperature control: The temperature of the sample is controlled and measured during continuous shear to 1°C as the sample and cylinders are temperature programmed at variable and defined rates from the melt range for the polyethylene down to temperatures for formation of stable crystals. A unique feature of the instrument is that both the inner and outer cylinders are thermostatted. The sample thickness between the two-inch long cylinders is also small and can be varied by choice of inner cylinder diameter. The sample thickness used in these studies was 0.50 mil (= 0.001 27 cm).

The viscometer method for morphology preparation, with the advantages described above, isolates the effect of simple shear on the resultant polyethylene morphology. No significant pressure is involved in contrast to other processes. Temperature gradients within the polymer must be 1°C or less.[4] Moreover, the shear field is near homogeneous in contrast to flow in capillaries and in wide gap rotational instruments. The shear is also lamellar in contrast to the turbulent shear generation of shish-kebab polyethylene morphologies as in certain stirring experiments.[9]

The pursuit of unique morphologies by simple shear has also gained impetus by the report of dramatic changes in properties, appearance and morphology for a high density polyethylene crystallized in a similar but perhaps less well-defined concentric cylinder experiment[1] and at lower shear rates, nominally about 3 decades lower, than employed here and in thicknesses over 200 x greater.

EXPERIMENTAL

Alathon 7050 (DuPont Company) crystallized at ambient pressure and under shear, 2600 s^{-1}, at 134–135°C, with a melt viscosity of approximately 10,000 poise. Onset of crystallization is shown in Figure 1 by the sharp increase in viscosity with decreasing temperature. With this information, and a computer program previously described,[7] it can be shown that – for the 0.5 mil. melt film thickness between the concentric cylinders – viscous heating causes less than a 1°C increase for rotational speeds below 25 rpm, see Figure 2. Even for the worst case (adiabatic), Figure 2 indicates that the temperature rise due to shear heating is less than 2°C. Thus, as the melt is sheared under these conditions and the film temperature is slowly lowered, crystallization of the film in simple shear should occur.

Filling the cylinder clearance gap with a high-viscosity polyethylene melt was a major difficulty. It was accomplished as follows: With the ring and oil bath of the high shear viscometer held between 140 and 150°C, a special tamping tool

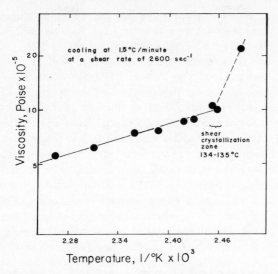

FIGURE 1   High density polyethylene (Alathon 7050) viscosity change on cooling.

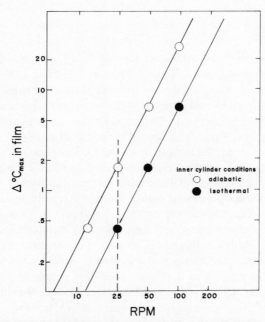

FIGURE 2   Calculated viscous heating during shear of high density polyethylene, Alathon 7050, at 134°C.

was used to coat the inner surface of the ring and fill its sample inlet groove with melted polyethylene pellets. Care was taken to exclude air bubbles. The spindle (#1) inner cylinder was then heated with a hot air gun to $140-150°C$, and then slowly lowered into the ring. The spindle was locked in place and rotated by hand several times while low pressure N_2 ($\sim$ 90 psi) was introduced via the sample inlet system. A polymer melt film of $1-2$ in^2 in area between the cylinders was formed in this fashion. After one half hour for thermal equilibration, shearing was begun at a constant rotational speed and continued until the torque reached a maximum ($\sim$ 10 min), at which time the heating bath and nitrogen pressure were cut off and ambient controlled cooling under shear was commenced. During this operation, the heating bath "T" inlets acted as siphons, removing the heating oil from the cylinder reservoir. Cooling was continued for about 10 mins, at a rate of $\sim$1.5°C/min. Shearing was continued until well below the temperature for onset of crystallization as indicated by an abrupt torque increase, see Figure 1, whereupon rotation was halted and cooling was continued to near room temperature. A small open-end wrench was applied to remove the spindle so that the polyethylene film could be examined. In some cases it was necessary to heat the outer ring somewhat (to $\leqslant 80°C$) in order to facilitate spindle removal.

RESULTS

Properties of the linear polyethylene (Alathon 7050) crystallized at high rates in simple shear are summarized in Table I. The corresponding properties are also given for the same polymer formed by the special capillary rheometer method[4, 5] (shear plus pressure) and for the initial bulk-crystallized (no shear and no pressure) material. Qualitatively, very little enhancement of transparency was achieved by simple shear alone. The mechanisms of transparency in polyethylene have been discussed.[10] In contrast, the samples from the capillary method are highly transparent. The films made by the method described here had a mottled, fibrillar appearance. Inspection under polarized light revealed regions of high orientation dispersed in a less highly-oriented matrix. In general, results indicated a considerable degree of crystalline orientation in the shear direction. The melting point and percent crystallinity, both determined by differential scanning calorimetry (DSC-1B), also placed the properties of the shear-crystallized material intermediate between the transparent (capillary) strand and the bulk-crystallized polyethylene. The thinness of the shear-crystallized sample (0.5 mil) prevented conventional molecular orientation measurements such as those using x-ray diffraction. Birefringence measurements were possible, however, and indicated some orientation, but significantly below that produced by combined shear and pressure in the capillary rheometer process. Because DSC measurements proved

TABLE I
Comparative properties of high density polyethylene (Alathon 7050)
crystallized by three means

Property	Crystallization conditions		
	Simple shear alone[a]	Shear plus pressure[b]	No shear and no pressure[c]
Appearance	Opaque fibrillar	Transparent fibrillar	Opaque
Melting peak,[d] °C	135.7	138.7	131.6
% Crystallinity[e]	80%	85%	70%
Thickness, mil	0.5	50	10
Birefringence	0.005	0.054	—

[a]Crystallized under shear in the concentric cylinder viscometer.[7]
[b]Capillary rheometer.[4,5]
[c]Bulk crystallized (as received).
[d]By DSC at a 5°C/min heating rate.
[e]Determined by area under the DSC curve; some heat loss suspected.

sensitive to the property differences, melting peak vs. heating rate curves were also obtained. Southern has previously demonstrated that the highly-oriented and transparent strands produced in the capillary process show (1) steeper super-heating slopes and (2) little evidence of low heating rate annealing. Figure 3 shows that in heating rate studies the simple shear-crystallized material lies intermediate between the capillary and bulk-crystallized material.

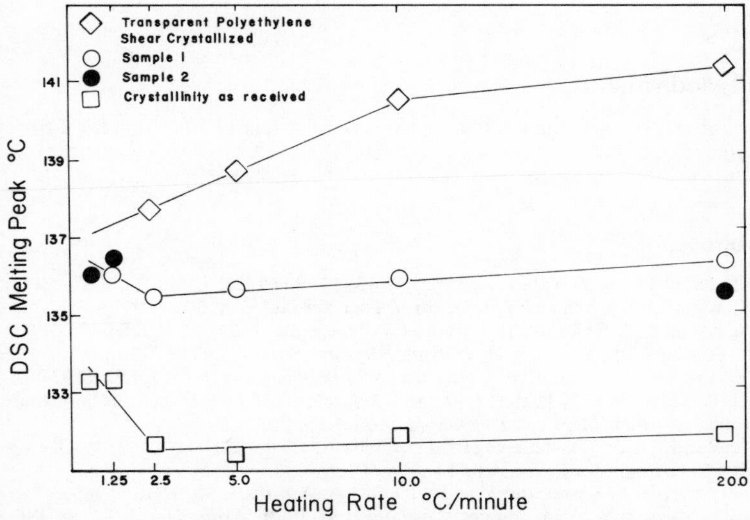

FIGURE 3   Effect of heating rate on the DSC melting peak temperature for high density polyethylene, Alathon 7050.

CONCLUSIONS

Melts of high density polyethylene (Alathon 7050) crystallized under simple shear (2600 s^{-1}) in the absence of pressure exhibit some orientation, but at a level well lower than that achieved by the combination of shear and pressure in the capillary rheometer.[4, 5] This leads to the conclusions that the stress of simple shear is insufficient to produce the highly-transparent and strong polyethylene morphologies since the shear rate is higher (about five times) and more homogeneous in the concentric cylinder viscometer used in these studies. Differences may be attributed to the drawing of partially-crystalline material in the case of capillary extrusion and the elevation of melting temperature due to pressure effects. In any event, simple shear alone produces anisotropic but not highly-oriented material in the case of high density polyethylene crystallized from the melt. It is plausible that the first crystals formed in simple shear are highly anisotropic but that this leads to a rapid and subsequent less-oriented epitaxial growth for the rest of the melt. In any case, it would appear that such processes in simple shear provide some interesting changes in morphology but of a magnitude far less than achievable by other means[4, 5] and less than anticipated from prior results.[1] Kruger and Yeh had reported[1] the preparation, also in a rotational viscometer, of a polyethylene which is both transparent and oriented, yet exhibited an extension 10 x more and a modulus over 10 times less than those for the oriented and transparent morphologies compared here in Table I and described elsewhere.[4, 5]

Acknowledgement

The authors express their appreciation to the National Science Foundation for their support of this study.

References

1 D. Kruger and G. S. Y. Yeh, *J. Appl. Phys.* **43**, 4339 (1972).
2. T. T. Wang, H. S. Chen and T. K. Kwei, *J. Polymer. Sci. B* **8**, 505 (1970).
3. T. K. Kwei, T. T. Wang and H. E. Bair, *J. Polymer. Sci. C* **31**, 87 (1970).
4. J. H. Southern and R. S. Porter, *J. Appl. Polymer. Sci.* **14**, 2035 (1970).
5. N. E. Weeks and R. S. Porter, *J. Polymer. Sci., Polym. Phys. Ed.* **12**, 635 (1974);
    W. G. Perkins and R. S. Porter, *J. Material Sci.*, **12**, 2355 (1977) and citations therein.
6. G. Capaccio and I. M. Ward, *Polymer (London)* **15**, 233 (1974).
7. L. A. Manrique, Jr., Master's Thesis, University of Massachusetts, 1972; L. Manrique
    and R. S. Porter, *Rheologica Acta* **14**, 926 (1975).
8. R. S. Porter, R. F. Klaver and J. F. Johnson, *Rev. Sci. Instr.* **36**, 1846 (1965).
9. A. J. Pennings, J. M. A. A. van der Mark and H. C. Booij, *Kolloid-Z. Z Polym.* **236**, 99
    (1970).
10. R. S. Stein and R. Prud'homme, *J. Polymer Sci. B* **9**, 595 (1971).

DISCUSSION

S. L. Aggarwal (*General Tire and Rubber Co.*): In your discussion, you mentioned the effect of extended-chain conformation of polymer molecules on tensile strength and modulus — at times interchangeably. Which property is most dependent on the molecular alignment of polymer molecules as rods?

R. S. Porter: In our capillary data that has not yet been presented nor published, the tensile modulus of our ultradrawn polyethylene is nearly independent of molecular weight for preparation under the same conditions. Interestingly, however, the tensile strength increases regularly with molecular weight.

R. Johnson (*Borg-Warner Corp.*): In producing orientation of rod-like polymers by extended chain flow, it is always necessary to employ a solvent? If so, are problems encountered in solvent removal?

R. S. Porter: The answers are no and yes, respectively, with comments of Professor Morton Litt being appropriate here.

M. Litt (*Case Western Reserve Univ.*): There are several polymers that have liquid crystal transitions at 100% polymer. One is poly(*p*-hydroxybenzoate), which goes from crystal to liquid crystal at about 350°C. Also, copolymers of poly(ethylene terephthalate), where a portion of the ethylene glycol is replaced by hydroquinone and/or *p*-hydroxybenzoic acid, show shear reduction of viscosity in the melt at about 30 mole % (or larger) replacement of glycol. Extrusion of these polymers routinely gave moduli in the extrusion direction of 2–3 million psi, which shows very high orientation even with little or no crystallinity. The first polymer is produced by Carborundum Corp. The second set of polymers were reported by Tennessee Eastman.

Secondary Loss Peaks in Glassy Amorphous Polymers

J. HEIJBOER

Centraal Laboratorium TNO, Delft, The Netherlands

A short description is given of the determination — as a function of frequency and temperature — of secondary mechanical loss peaks in glassy amorphous polymers.

The molecular motions causing secondary loss peaks are classified. The dependence on frequency of the temperature of the maximum of the loss peak is discussed.

The effects on the secondary loss peaks of structural features such as plasticization, intermolecular steric hindrance and polarity, are illustrated by the example of the β-maximum of poly(methyl methacrylate), emphasis being laid on the difference between a glass transition and a secondary loss peak.

1 WHAT ARE SECONDARY LOSS PEAKS?

Figure 1 schematically shows the dependence of the modulus and damping of an amorphous polymer on temperature. At the *main or glass transition* T_g the modulus drops by a factor of a thousand; in the region of this transition the material changes from glassy to rubbery. This transition is accompanied by a high loss peak, the maximum of tan δ generally lying between 1 and 3. This peak is usually called the α-peak.

Apart from this main transition, which has a tremendous effect on the mechanical properties, there nearly always are minor transitions in the glassy region at which the modulus decreases by a factor of, e.g., not more than two. Nevertheless these minor or *secondary transitions* may have a marked effect on the mechanical behaviour: sometimes a material is brittle below and tough above the secondary transition. In Figure 1 we see that the secondary transition stands out more clearly in the loss curve than in the modulus curve: the loss curve shows a distinct maximum, whereas the modulus curve merely shows a steeper decrease.

When there are more secondary transitions, they are usually labelled from higher to lower temperatures with the successive letters of the Greek alphabet: β, γ, δ, and so on.

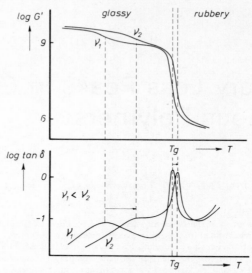

FIGURE 1   Storage modulus G' and damping tan δ as functions of temperature T at two frequencies ν_1 and ν_2 for an amorphous polymer (schematical drawing).

Tan δ is preferably plotted on a logarithmic scale, because the small loss peaks are more prominent and their slopes are more straight than they are in a linear plot. Moreover, this way of plotting is in better agreement with the relative accuracy of the damping measurement itself.

The mechanical properties of polymers are dependent on frequency; as is indicated by the two curves in the parts of Figure 1, ν_2 is a higher frequency

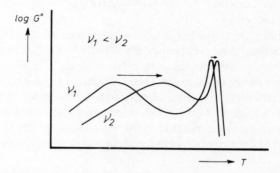

FIGURE 2   Loss modulus G'' as a function of temperature T at two frequencies ν_1 and ν_2 for an amorphous polymer. The curves pertain to the same (hypothetical) data as do those in Figure 1.

than ν_1, the main effect of a frequency increase being a shift of the curve to the right (i.e. to a higher temperature). A secondary transition is shifted more rapidly than a glass transition, and so at higher frequencies the two may merge or coincide. When instead of tan δ one plots the loss modulus $G'' = G'$ tan δ, the separation between the β- and α-maximum is better. This is seen on comparison of Figure 2 and Figure 1.

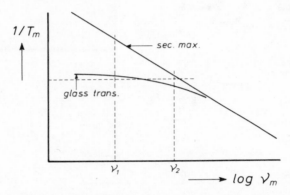

FIGURE 3   Schematic drawing of the course of a glass transition and that of a secondary maximum in a $1/T_m$ vs. log ν diagram.

Figure 3 shows the relative locations of a secondary maximum and a glass transition in an Arrhenius plot. In this plot the slope of a secondary transition is steeper than that of the glass transition. This indicates that the best separation of the loss maxima is obtained at lower frequencies. Moreover, it is seen that by plotting the losses as a function of frequency at a constant temperature, one might be able to separate the α- and β-maxima, even at temperatures near that at which the two maxima coincide.

2   MEASUREMENT OF SECONDARY MAXIMA

Section 1 emphasizes the importance of measurements over a broad frequency range. In dielectric measurements a single instrument, such as a Schering bridge, covers five decades. In mechanical measurements for a similar range of frequencies several instruments are required.

Figure 4 gives a survey of the equipment used at the Centraal Laboratorium TNO. On the horizontal axis is plotted the time (from left to right) or the frequency (from right to left); on the vertical axis the range of moduli covered by the various instruments.

The best known instrument is the torsional pendulum, in which a specimen

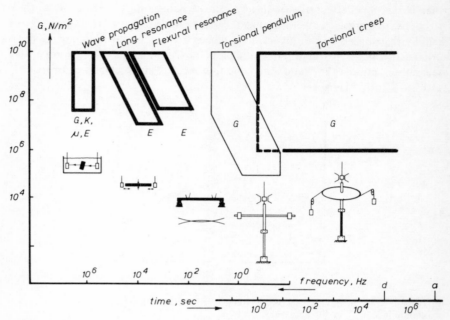

FIGURE 4   Survey of modulus and frequency ranges covered by the instruments used at the Centraal Laboratorium TNO for measurements of modulus and damping.

carries out free vibrations. Application of different moments of inertia allows a frequency range of up to 2 decades to be covered. Normally, with our size of specimens, we obtain data in the region of 1 Hz.

For longer-time experiments we use the torsional creep instrument. This gives the shear compliance J as a function of time t. Creep in torsion has several advantages over creep in tension, one of which is that thermal expansion does not affect the reading, and another that in torsion the material behaves linearly up to higher stresses than it does in tension. The time range covered depends only on the experimenter's patience. The instrument is fully automated, so that measurements can be made overnight, one day being a practical limit. The shear modulus and losses can be calculated from the data by means of the first-order approximations[1]

$$G^*(\nu) \sim \frac{1}{J(t)} \text{ and } \frac{d \ln J}{d \ln t} \simeq \frac{2}{\pi} \tan \delta(\nu) \text{, with } 2\pi\nu = \frac{1}{t} \text{ .}$$

For frequencies between 100 and 10,000 Hz, we have the method of flexural resonance vibration; here the losses are calculated from the width of the resonance peak. For these three measurements we use the same specimen.

Longitudinal resonance vibration extends the frequency range up to 100 kHz; the specimens are cylindrical bars, 10 cm long and 1 cm in diameter, or 1 cm long and 0.1 cm in diameter.

For 0.8 and 5 MHz we used propagating waves. These measurements give the shear as well as the compression modulus, from which we can also calculate Poisson's ratio μ. Waterman[2] found that Poisson's ratio behaves like a compliance; it is a complex quantity whose imaginary part shows a maximum in the temperature region of the maximum of the loss modulus.

Further details about the instruments can be found in Reference 3, which also contains references to the original literature.

Figure 5 shows a representative example of results obtained with this series of instruments. Data are given for poly(cyclohexyl methacrylate) (PCHMA). In the glassy state this polymer has a sharp transition that must be ascribed to the flipping motion of the saturated six-membered ring.

In this figure, G' and tan δ are plotted as functions of temperature for six frequencies. The values of G' for 200, $\simeq 10^4$ and $\simeq 10^5$ Hz have been calculated from Young's moduli, with the Poisson's ratio derived from the MHz-

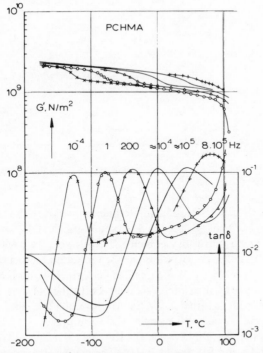

FIGURE 5   Shear modulus G' and damping tan δ as functions of temperature at six frequencies for poly(cyclohexyl methacrylate) (PCHMA).

measurements. The figure is dominated by the γ-maximum, which shifts to higher temperatures with increasing frequency. Below −150°C we also see a part of the δ-maximum. Near −60°C the curve for 10^{-4} Hz shows a faint β-maximum. As expected, each γ-peak is accompanied by a drop in the corresponding storage modulus (G') curve. Figure 6 gives the Arrhenius plot of the γ-maximum. The points fit very well to a straight line over a temperature range of 200°K and a frequency range of 10 decades. (An idea of this time span is gained from the consideration that, if the shortest time were 1 sec, the longest would be about 300 years!) The activation energy is 11.3 kcal/mol. This value agrees very well with the value of 11.4 kcal/mol found from NMR measurements for the activation energy of the chair-chair transition of the cyclohexyl ring.[4,5]

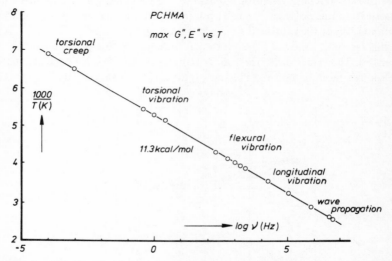

FIGURE 6   Arhenius plot for the γ-loss maximum of poly(cyclohexyl methacrylate). The data are obtained from G'' resp. E'' vs. T-curves.

Figure 7 shows the results of reducing to one temperature the torsional pendulum measurements of G'' for the γ-maximum of PCHMA, by shifting them along the frequency axis until they coincide. The distance over which the data have to be shifted to coincide with those at the reduction temperature is called the shift factor a_T. It is clear that for this loss peak the reduction is justified, as the direct measurements (filled circles) coincide with the reduced data. Although the cyclohexyl loss peak is rather sharp, its width on the logarithmic scale still is about twice that of a peak with a single relaxation time.

Figure 8 shows the relation between the shift factor a_T and $1/T$ for the γ-maximum of PCHMA and for that of PCHA. The graphs are not straight, but slightly curved. This might be explained by the assumption that the apparent

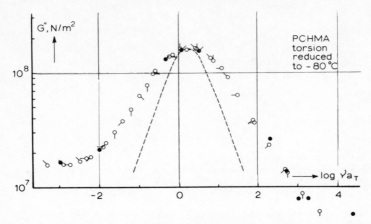

FIGURE 7   Loss modulus G'' for PCHMA at $-80°C$ as a function of reduced frequency νa_T. Filled circles represent direct measurements at the given frequency. Dashed line: maximum for a single relaxation time.

activation energy slightly varies, and that even within a single loss peak there is a distribution of activation energies.

3   THE MOLECULAR ORIGIN OF SECONDARY MAXIMA

In order to deal with the molecular origin of the loss processes, we will consider what types of motion are possible in amorphous polymers below their glass transition temperature. At the glass transition motions of large parts of the

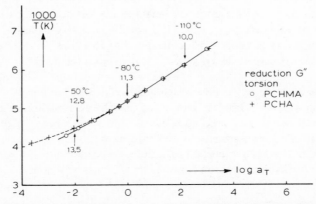

FIGURE 8   Shift factor a_T of G'' plotted vs. $1/T$ for PCHMA and poly(cyclohexyl acrylate) (PCHA). The activation energies, calculated from the local slope, are indicated in kcal/mol.

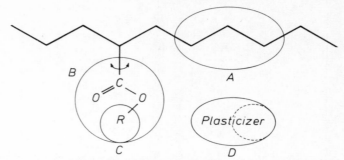

FIGURE 9   The various kinds of groups whose movements give rise to secondary
mechanical loss peaks.

main chain are frozen in. Since these motions require a considerable free
volume, it is reasonable to suppose that below the glass transition the available
free volume is still large enough for the motion of smaller groups.

Figure 9 gives a survey of the four main types of motion (A, B, C and D)
which are possible below T_g and which give rise to secondary loss peaks.

Type A

This is a motion, still within the main polymer chain, but locally much more
restricted than the motion corresponding to the glass transition. A typical
example of a loss peak, connected with such a type of motion is the β-maximum
of rigid PVC,[6, 7] which lies near 210 K at 1 Hz (see Figure 10). It is clear that
this maximum must be attributed to a main-chain motion, since there are no
other possibilities: rotation about the C—Cl bond has no effect, and any motion
of the chain ends would result in the height of the loss peak being dependent on
the degree of polymerization, which it is not.[8] Figure 10 also shows that an
increase in density of about 0.4%, corresponding to a considerable decrease in
free volume, has nevertheless practically no effect on the height of the loss
peak.[9]

Another example of local main-chain motion is found in polycarbonate;[10] a
comparison with dielectric loss data shows that the polar carbonate group is
involved in the secondary maximum. The secondary maxima of polysulphones[11]
and polyesters[12] are also caused by this type of motion.

Type A motion is an important one, since the corresponding secondary loss
peak often marks the transition from brittle to tough behaviour.

Type B

This type is the rotation of a side group about the bond linking it to the main

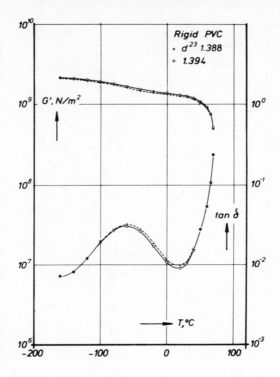

FIGURE 10   Storage modulus G' and damping tan δ at 1 Hz as functions of temperature for two samples of rigid poly(vinyl chloride); one cooled at 50 bar, the other at 1000 bar, from above T_g to room temperature. Measurements at normal pressure, starting at −160°C.

chain (see Figure 9). The side group moves as one whole; its rotation need not be complete, and indeed is more likely to be a transition from one equilibrium position to another. The deformation of adjacent valence angles often forces the main chain to partake slightly in the side group motion. A typical example of this type of motion is found in polymethacrylates.[13] The β-maximum is caused by the motion of the entire —COOR group. Figures 11 and 12 show this β-maximum of poly(methyl methacrylate) (PMMA), together with the much smaller secondary maximum of poly(methyl acrylate) (PMA). Both the glass tempera-ture and the secondary maximum lie at a much lower temperature for PMA than for PMMA. Comparison of Figure 12 and Figure 11 shows that secondary maxima are more pronounced in a loss modulus (G'') plot than they are in a tan δ plot; for PMMA the glass transition is even the minor peak in the G''-plot. In the next section we shall discuss the β-maximum of PMMA in more detail.

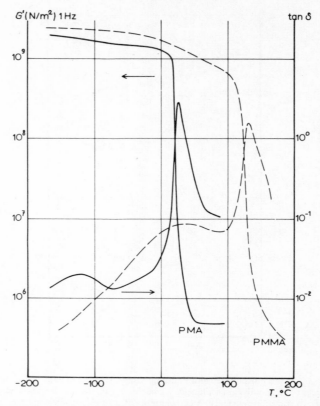

FIGURE 11   Storage modulus G' and damping tan δ at 1 Hz as functions of temperature for poly(methyl methacrylate) (PMMA) and poly(methyl acrylate) (PMA).

Type C

This type of motion is an internal motion within the side group itself, without interaction with the main chain (see Figure 9, type C). Typical examples are the internal motions in R of the —COOR group in polymethacrylates, e.g. that of the n-butyl group in poly(n-butyl methacrylate) and that of the cyclohexyl group in poly(cyclohexyl methacrylate).[14] In Figures 13 and 14 the loss peak of the cyclohexyl group (already shown in Figure 5) is compared with that of the n-butyl group. The two groups have the same length, viz. four carbon atoms, but the n-butyl group has a greater freedom of motion at its end. As a consequence the maximum of the n-butyl group lies about 100°K lower than that of the ring. The higher flexibility of the n-butyl group also results in the T_g of the polymer being about 80 K lower than that of PCHMA.

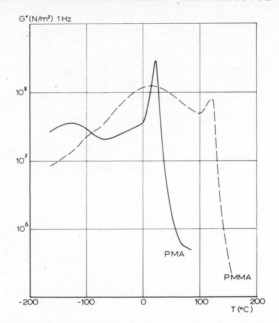

FIGURE 12 Loss modulus G'' at 1 Hz as a function of temperature for poly(methyl methacrylate) and poly(methyl acrylate).

Type D

This is a motion of, or taking place within, a small molecule, dissolved in the polymer. This can be a motion within a plasticizer molecule, e.g. of the n-butyl group in dibutyl phthalate. Figure 15 shows another example: a low temperature maximum that is caused by a motion of the oxo-substituted cyclohexyl ring of an ester, dissolved in PMMA. The maximum of the oxo-substituted cyclohexyl ring lies at a much lower temperature than that of the unmodified cyclohexyl ring, because the barrier to the flipping motion is considerably lower.[14]

Figure 16 shows the losses, measured in creep and in free vibration, of caprolactam dissolved in PMMA. The maximum near $-80°C$, 1 Hz, is due to the caprolactam. The loss peak for this seven-membered ring nearly coincides with that of the six-membered ring of cyclohexyl, the reason probably being that rotation about the bond between CO and NH is blocked.[15] The cycloheptyl ring has a maximum at a much lower temperature, viz. at about $-170°C$, 1 Hz.[16] In Figure 16 the maximum near $0°C$, 1 Hz, is the β-maximum of PMMA.

Small molecules dissolved in polymers, sometimes associate themselves with side groups, particularly if both are polar. When association takes place and there is a combined motion, we get a type C instead of a type D motion. Janacek and

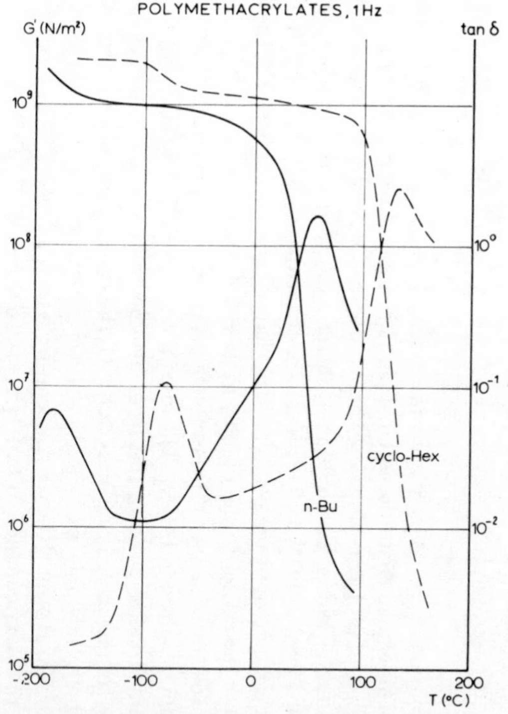

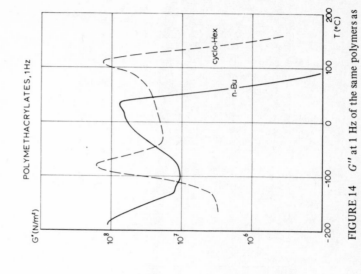

FIGURE 14 G'' at 1 Hz of the same polymers as in Figure 13.

FIGURE 13 G' and tan δ at 1 Hz of poly(n-butyl methacrylate) and poly(cyclohexyl methacrylate).

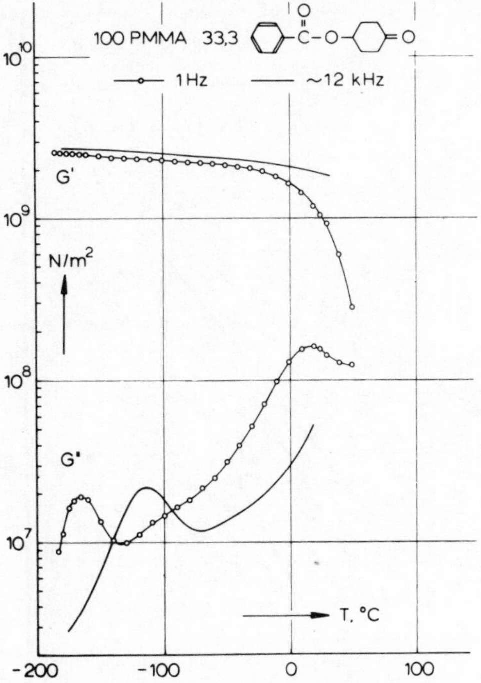

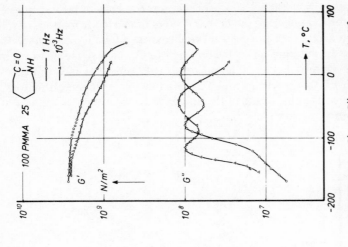

FIGURE 15 G' and G'' at 1 Hz and at about
12 kHz for PMMA containing 25% of
4-benzoyloxycyclohexanone.

FIGURE 16 G' and G'' at 1 Hz and 10^{-3} Hz
of PMMA containing 20% of caprolactam.

coworkers have thoroughly investigated the effect of the interaction of small molecules and side groups on secondary loss peaks in polymethacrylates.[17] In this paper we shall not deal further with these complexes.

We shall now consider the dependence on frequency of secondary loss maxima, originating from the different types of molecular motion. Loss moduli have been determined at a constant frequency as a function of temperature, and from the resulting curves the temperature T_m corresponding to the maximum loss has been read. Figure 17 gives several Arrhenius plots, in which the reciprocal of T_m (in Kelvin) is plotted as a function of the logarithm of frequency ν. The thick parts of the lines indicate the region of the measurements; the thin parts are extrapolations.

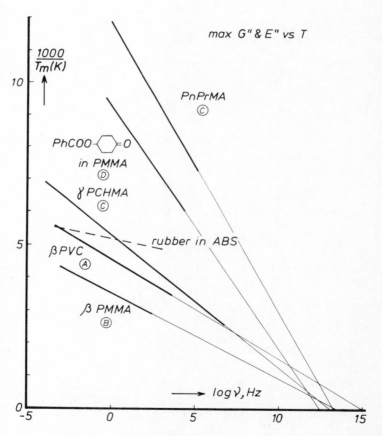

FIGURE 17   Arrhenius plots of secondary maxima. The type of motion is indicated. For purposes of comparison a glass transition (rubber in ABS) is included.

Figure 17 shows the following examples:

Type A: β-maximum of PVC
Type B: β-maximum of PMMA
Type C: γ-maximum of PnPrMA
  γ-maximum of PCHMA
Type D: maximum of 4-benzoyloxycyclohexanone, 25% (w/w) embedded in
  PMMA.

The measured values fit the Arrhenius equation:

$$\ln \nu = \ln \nu_0 - E_a^{\ddagger}/RT_m \qquad (1)$$

$$(\text{or } \log \nu = \log \nu_0 - 0.4343 \; E_a^{\ddagger}/RT_m),$$

where $E_a^{\ddagger}$ is the activation energy and R the gas constant. The straight line corresponding to the loss peak of the type A motion extrapolates to $\log \nu_0 = 15$; the other lines for the types B, C and D motions all extrapolate to 13.5 ± 1.

10^{13} Hz is a reasonable frequency for a molecular vibration. The molecular picture behind an activated process described by an Arrhenius equation, is as follows: suppose the mechanism is one involving two potential wells separated by a barrier much higher than kT (Figure 18).

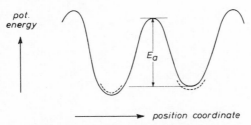

FIGURE 18   Two-well potential model.

Each of the groups is vibrating at its molecular frequency in a well, and sometimes it jumps across the barrier to the other position. If entropy differences between the two positions are neglected, the relative numbers of the groups in each of the two positions in the state of equilibrium are determined by Boltzmann's equation. Application of a stress causes the differences in depth of the two wells to change; the equilibrium is disturbed and a redistribution of the groups over the two positions starts. The rate of redistribution is determined by the temperature and by the height of the barrier, which is therefore the activation energy $E_a^{\ddagger}$. When during the time of redistribution the stress field changes its direction, energy is dissipated. This molecular picture implies that the location of *the secondary loss peak is determined by the local barrier within the molecule*.

Figure 17 also shows a glass transition, viz. that of the rubbery part of acrylonitrile–butadiene–styrene (ABS). Clearly, the behaviour of this maximum is quite different from that of the secondary maxima: the slope of the curve is much less steep, indicating a much higher activation energy. We conclude that *secondary maxima behave quite differently from glass transitions.*

If we assume that in Eq. (1) $\log \nu_0 = 13$ we can estimate the activation energy $E_a^{\ddagger}$ (in kcal/mol) from $T_m(K)$ at a frequency of 1 Hz by the simple equation:

$$E_a^{\ddagger} = 0.060 \; T_m \; (1 \; Hz). \tag{2}$$

For other frequencies we have:

$$E_a^{\ddagger} = (0.060 - 0.0046 \log \nu) \; T_m(\nu). \tag{3}$$

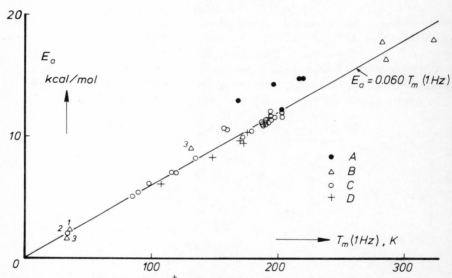

FIGURE 19   Activation energy $E_a^{\ddagger}$ as a function of the temperature of the secondary loss maximum at 1 Hz for different types of molecular motion. The numbered points are literature data. 1: Ref. 18; 2: Ref. 19; 3: Ref. 20.

Figure 19 gives an impression of the validity of Eq. (2). In the figure $E_a^{\ddagger}$ values for secondary maxima of different molecular origin are plotted against their T_m at 1 Hz. The plotted values are rather accurate, because $E_a^{\ddagger}$ is measured over at least 3 decades. Most of the data are derived from our own measurements, mainly on poly(meth)acrylates; only the numbered ones have been taken from the literature. We see that for type B, C, and D motions the relationship is obeyed to within 10%. Only for type A motions are the activation energies higher; the $1/T_m$ vs. $\log \nu_m$ extrapolates to $\log \nu_0$ values significantly higher than 13, e.g. to

15. This is probably due to a considerable entropy contribution to the transition state, which is not surprising for main-chain motions. Nevertheless, even the data for type A motions obey Eq. (2) to within 20%. Equation (2) probably gives a more accurate value for $E_a^{\ddagger}$ than does the usual determination from measurements over a narrow frequency range (say, 1 decade).

$E_a^{\ddagger}$-values for glass transitions are much higher than predicted by Eq. (3), e.g. 200%. The points for the γ-transition in PE and in POM also lie very far from the line in Figure 19 for secondary maxima in glassy amorphous polymers.

From this section it is clear that my viewpoint strongly differs from that of Yohari and Goldstein,[21] who advocate *inter*molecular interactions as providing the potential barriers for secondary loss peaks.

4 DEPENDENCE ON STRUCTURE OF SECONDARY MAXIMA

We will now examine in more detail the effect of molecular structure on a secondary maximum, taking as an example a type B maximum, viz. the β-maximum of poly(methyl methacrylate) (PMMA).[13] The molecular mechanism of this maximum is usually considered to be the (partial) rotation of the oxycarbonyl group (COO) about the C—C bond linking the group to the main chain. The potential barrier is provided by the adjacent methyl groups in the neighbouring mers attached to the main chain; see the structural formula:

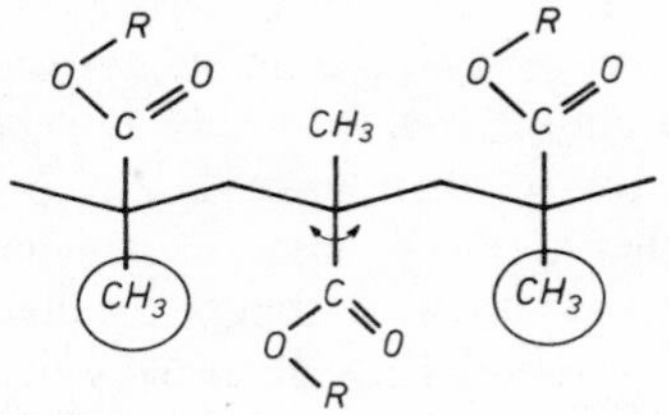

Molecular models (Stewart) clearly reveal the proximity of these methyl groups to the —COO group. It is not yet clear to what extent the main chain takes part in the side group motion.[22] Havriliak[23] even supposes that the β-maximum of PMMA is due to a main-chain motion and the α-maximum to side-chain motions, but for several reasons this idea is rather unlikely.

Strong arguments against it follow from a comparison of the relative sizes of the mechanical and dielectric damping maxima of PMMA and poly(methyl α-chloroacrylate) (PMeClA).[24] (In the latter, the main chain carries Cl atoms instead of CH_3 groups.) (See Table I.)

First, it is not to be expected that mechanically the motion of a relatively small side group would have a more marked effect than an extended main-chain motion; it seems therefore likely that the α-maximum is related to a main-chain

TABLE I
Relative size of tan δ maxima of poly(methyl methacrylate) and
poly (methyl α-chloroacrylate)

		α-maximum	β-maximum
PMMA	mechanical	large	small
	dielectric	small	large
PMeClA	mechanical	large	small
	dielectric	large	large

motion. Secondly, the relative sizes of the dielectric loss maxima of PMMA suggest that the β-maximum must be ascribed to the loosening of the polar part (the side groups) and the α-maximum to the motion of the relatively non-polar main chain. This is confirmed by the relative importance of the dielectric loss maxima of PMeClA: in this polymer not only the side group but also the main chain is polar, and its dielectric α-maximum therefore also is large.

Havriliak's arguments for the molecular mechanism he proposes are based on a careful study of the infrared spectrum in the range of 1050 to 1030 cm^{-1}. He observed that the intensities of two absorption peaks associated with the ester group are independent of temperature below T_g and vary with temperature above T_g, whereas the sum of the intensities remains constant over the entire experimental range. From this observation Havriliak concludes that only above T_g can the ester group rotate. However, an alternative explanation is possible. With respect to rotation there are two favourable positions of the ester group in PMMA, each giving a different infrared spectrum. The spectrum only gives information about the relative populations of the two positions, the jumping from one position to another not being observed. The potential wells of the two positions are determined by the configuration of the main chain in the immediate vicinity of the side group. Above T_g these potential wells will depend rather strongly on temperature, becoming more and more equivalent with increasing temperature, due to the motion of the main chain. Below T_g the motion of the main chain is strongly suppressed and on further cooling the immediate environment of the ester group does not change very much; consequently the relative depths of the wells become nearly independent of temperature. When the difference in depths of the potential wells is not large, this means that the distribution of the ester groups over the two positions is nearly constant below T_g, regardless of the possibility of jumping from one position to the other. By contrast, above T_g the distribution over the two positions changes, becoming more and more equal with increasing temperature. The observed dependence of the infrared spectra on temperature can thus be satisfactorily explained, without it being necessary to assume that the rotation of the ester group is blocked below T_g.[25]

Nevertheless, this discussion illustrates the paucity of our knowledge about the detailed molecular mechanisms of secondary loss peaks: most of the evidence is indirect. Schaeffer's lecture[26] shows that direct observation of mobility by sophisticated NMR techniques can give important contributions to the unambiguous elucidation of the molecular mechanisms responsible for secondary loss peaks.

Returning to the effect of structure on the β-maximum of polymethacrylates, we note (Figure 20) that in the series of tan δ curves of poly(n-alkyl methacrylates) the β-maximum merges with the α-maximum as soon as the side chain reaches a length of three C-atoms. A better separation is obtained in the plot of G'' (Figure 21), which shows that the β-maximum remains near +10°C at 1 Hz. Both figures show that the glass transition shifts to lower temperatures with increasing length of the n-alkyl group. It is clear that these polymers, with their long flexible side chains, do not obey the rule $T_\beta/T_g \simeq 0.75$.[27]

A still better separation of the glass transition from the β-maximum is shown in Figure 22, which gives the loss modulus obtained from creep measurements,

POLYMETHACRYLATES, 1 Hz

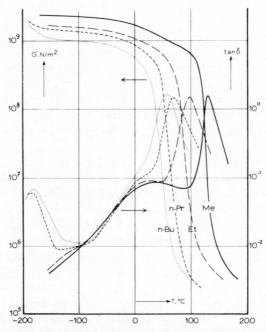

FIGURE 20   Shear modulus G' and losses tan δ at 1 Hz for poly(n-alkyl methacrylates). Me = methyl, Et = ethyl, n-Pr = n-propyl, n-Bu = n-butyl.

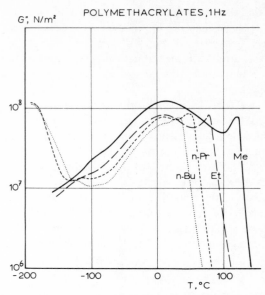

FIGURE 21   Loss modulus G'' at 1 Hz for the same polymers as in Figure 20.

whose equivalent frequency is 10^{-3} Hz; at this low frequency, even the β-maximum of the *n*-butyl ester stands clearly separate from the glass transition. Once more we see that the temperature of the β-maximum is not changed by elongation of the side chain. The shift to lower temperatures of the glass transition by the lengthening of side chains is sometimes called internal plasticization. It is seen that in this respect the β-maximum behaves in a completely different way from the glass transition: internal plasticization does not shift the β-maximum to lower temperatures.

From Figure 3 in Section 1 it was concluded that in a plot of damping as a function of frequency at constant temperature, maxima can be separated at temperatures near those at which they merge. Figure 23 illustrates how powerful this method is: even at $40°C$ a separate β-maximum is obtained for poly(ethyl methacrylate). When we compare Figure 20, and consider the higher frequency of the maximum (about 100 Hz), this separation is quite remarkable.

Figures 22 and 23 show the effect of a bulky *tert*-butyl group on the β-maximum: it is depressed, but remains at the same temperature and frequency. The still bulkier cyclohexyl group suppresses the β-maximum nearly completely. It is often argued that steric hindrance ought to shift maxima to a higher temperature. This is true for the glass transition, but not for this β-maximum. *Intermolecular* interactions do no more than depress the maximum, without shifting it to higher temperatures or to lower frequencies.

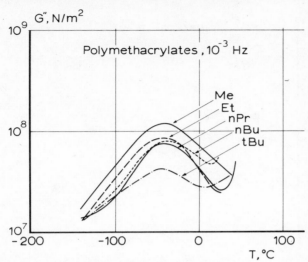

FIGURE 22   Shear modulus G'' at 10^{-3} Hz from creep data for the same polymers as in Figure 20. The curve for poly(*tert*-butyl methacrylate) (tBu) is also given.

External plasticization, which decreases intermolecular interaction, does not affect the location of the β-maximum either. This is illustrated by Figure 24, which is derived from creep data: when PMMA is plasticized by dibutyl phthalate the G''-maximum remains near −40°C, 10^{-3} Hz. The location of the β-maximum is determined by the local *intra*molecular barrier, the *inter*molecular interaction has no effect.

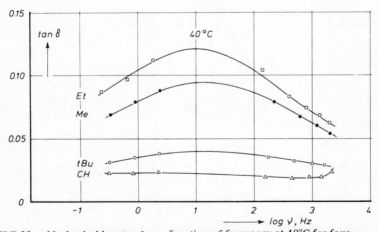

FIGURE 23   Mechanical loss tan δ as a function of frequency at 40°C for four poly(methacrylic esters). Et: ethyl ester; Me: methylester; tBu: *tert*-butyl ester; CH: cyclohexyl ester.

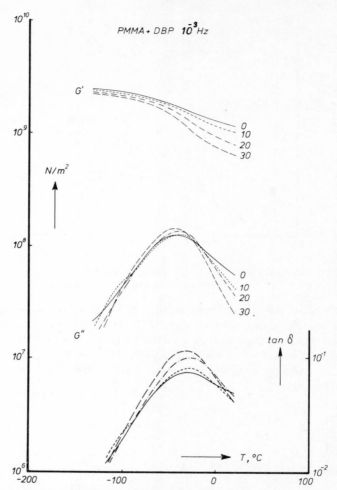

FIGURE 24   Storage modulus G', loss modulus G'' and losses tan δ at 10^{-3} Hz as functions of temperature for PMMA plasticized with dibutyl phthalate (DBP). Parts DBP (w/w) per 100 PMMA are indicated.

Figure 25 shows that this statement also holds for the cyclohexyl maximum (γ-maximum) of a copolymer of CHMA and MMA. This figure brings out the different effects of a plasticizer on the glass and on the secondary transitions: the two secondary peaks are barely moved by the addition of DBP, whereas the α-peak is clearly shifted to a lower temperature. Moreover, there appears a maximum due to the motion of the n-butyl group of the plasticizer. For the cyclohexyl as well as the oxycarbonyl motion it is the local *intra*molecular

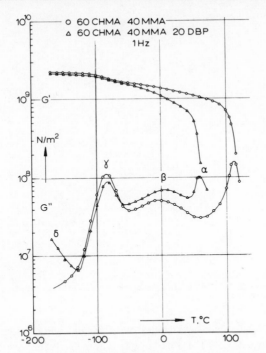

FIGURE 25   Storage modulus G' and loss modulus G'' at 1 Hz as a function of temperature for a copolymer of CHMA and MMA and for the same copolymer plasticized by 20 parts of DBP to 100 parts of copolymer.

barrier, not the intermolecular interaction, which determines the location of the loss peak.

It has been shown before[28] that a decrease of the intramolecular barrier by substituting hydrogen for methyl on the main chain shifts the β-transition of PMMA to a lower temperature, e.g., by copolymerization with styrene, whereas copolymerization with α-Me styrene, in which the methyl is still on the main chain, leaves the maximum at the same temperature.

We can also change the local barrier by attaching Cl instead of methyl to the main chain; this causes the barrier to increase and indeed, Figure 26 shows that the maximum shifts to a higher temperature.

There is one structural feature which affects the location of the β loss peak of PMMA without it being clear that the local barrier is changed, namely the polarity of the ester group. Each part of Figure 27 shows the loss moduli of a series of polymethacrylates containing polar and nonpolar groups of about the same length: the chloromethyl and cyanomethyl ester are compared with the ethyl ester with the n-propyl ester. Although at this low frequency the effect of

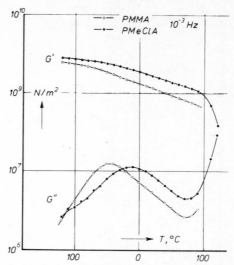

FIGURE 26   Storage modulus G' and loss modulus G'' at 10^{-3} Hz for PMMA and for poly(methyl α-chloroacrylate) (PMeClA). Data from creep measurements.

polarity is only slight, there nevertheless is a shift to higher temperatures and a considerable broadening of the β-maximum. These effects are probably caused by *inter*molecular interactions; so with distant polar ester groups the secondary maximum is not determined exclusively by the local *intra*molecular potential barrier.

Of course, the effect of polarity on the γ-maximum, which is due to a motion within the alkyl group itself, is essentially different; for this motion the polarity contributes directly to the local barrier, and there is a very pronounced shift to higher temperatures with increasing polarity.[29]

Some amorphous polymers show other secondary maxima whose location depends not only on intramolecular interaction but also on intermolecular interactions. A well-defined example is the γ-maximum of 2-methyl-6-alkyl substituted poly(phenylene ethers), studied by Eisenberg *et al.*[30] This maximum, which is supposed to be due to oscillations of the alkyl chain in its planar zig-zag form, shifts from 171 to 224 K (1 Hz) on elongation of the alkyl group from 4 to 9 carbon atoms, whereas the activation energy increases considerably. It is very likely that intermolecular interactions of the alkyl group contribute to an important extent to the potential barrier for this motion.

Kolarik[31] found that the γ-maximum due to the hydroxyethyl group in copolymers of 2-hydroxyethyl methacrylate and MMA, shifts to a lower temperature with increasing MMA content. This can only be understood if, in addition to the intramolecular barrier, there is an interaction with the environ-

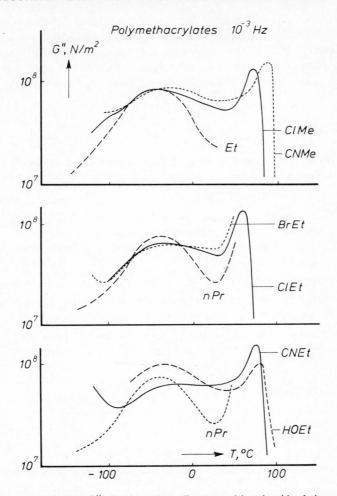

FIGURE 27   Loss modulus G'' of polymethacrylic esters with polar side chains as a function of temperature at 10^{-3} Hz. Ester groups: Et: ethyl, ClMe: chloromethyl, CNMe: cyanomethyl, nPr: n-propyl, ClEt: 2-chloroethyl, BrEt: 2-bromoethyl, CNEt: 2-cyanoethyl, HOEt: 2-hydroxyethyl.

ment. Boyer[32, 33] states that in the secondary maxima of polystyrene a combination of intra- and intermolecular hindrances is involved.

Although it is clear therefore that several secondary loss peaks must in part be ascribed to *inter*molecular interactions, the insensitivity of many of them to annealing leads one to the conclusion that very often it is the *intra*molecular interactions which predominate. This holds especially for the cyclohexyl-maximum[14] and for the β-maximum of PMMA. The minor effect of a strong densification of PVC on the A-type motion (Figure 10) also points in this direction. When considering these effects one should bear in mind that the low-temperature tail of the glass transition is very sensitive to annealing.

CONCLUSIONS

The behaviour of secondary maxima in glassy amorphous polymers is different from that of glass transitions:

a) They obey an Arrhenius relationship over a broad frequency range.

b) For other than local main-chain motions the activation energy $E_a^{\ddagger}$ of the molecular processes underlying the loss maximum, can be calculated from $E_a^{\ddagger} = 0.060 \, T_m$ with an accuracy of 10%. For local main-chain motions $E_a^{\ddagger}$ usually is slightly higher, say 20%.

c) For many secondary maxima the origin of $E_a^{\ddagger}$ is mainly *intra*molecular.

d) Plasticization mostly does not affect the temperature location of a secondary maximum.

e) Sufficiently strong *inter*molecular hindrance mostly decreases the height of the loss maximum, but does not shift it to higher temperatures.

Acknowledgement

My thanks are due to Mr. L. C. E. Struik for illuminating discussions.

References

1. A. J. Staverman and F. Schwarzl, Linear deformation behaviour of high polymers, in *Die Physik der Hochpolymeren*, edited by H. A. Stuart (Springer-Verlag, Berlin, 1956) Bd IV.
2. H. A. Waterman, *Kolliod-Z. Z. Polym.* 192, 1 , 9 (1963).
3. J. Heijboer, Mechanical properties of glassy polymers containing saturated rings, Doctoral Thesis, Leiden, 1972, Centraal Laboratorium TNO Communication 435, Chapter 3.
4. F. A. L. Anet and A. J. R. Bourn, *J. Amer. Chem. Soc.* 89, 760 (1967).
5. G. Binsch, *Top. Stereochem.* 3, 97 (1968).
6. N. G. McCrum, B. E. Read and G. Williams, *Anelastic and Dielectric Effects in Polymeric Solids* (John Wiley, London, 1967) p. 434.
7. a. G. Pezzin, G. Ajroldi and C. Garbuglio, *J. Appl. Polymer Sci.* 11, 2553 (1967).
   b. J. A. Mason, S. A. Iobst and R. Acosta. *J. Macromol. Sci. Phys. B* 9, 301 (1974).
8. a. Y. Ishida, *Kolloid-Z.* 171, 71 (1960).
   b. J. Beautemps, *Rev. Gen. Caout. Plast.* 47 (1), 95 (1970).
9. H. W. Bree, J. Heijboer, L. C. E. Struik and A. G. M. Tak, *J. Polymer Sci. A-2* 12, 1857 (1974).

10. N. G. McCrum, B. E. Read and G. Williams, *Anelastic and Dielectric Effects in Polymeric Solids* (John Wiley, London, 1967) p. 532.
11. M. Baccaredda, E. Butta, V. Frosini and S. de Petris, *J. Polymer Sci. A-2* **5**, 1296 (1967).
12. N. G. McCrum, B. E. Read and G. Williams, *Anelastic and Dielectric Effects in Polymeric Solids* (John Wiley, London, 1967) p. 517.
13. N. G. McCrum, B. E. Read and G. Williams, *Anelastic and Dielectric Effects in Polymeric Solids* (John Wiley, London, 1967) p. 242.
14. See J. Heijboer, Molecular origin of the cyclohexyl loss peak, this volume, pp. 297–310. Gordon and Breach, New York (1978).
15. S.-I. Mizushima, T. Simanouti, S. Nagakura, K. Kuratani, M. Tsuboi, H. Baba and O. Fujioka, *J. Amer. Chem. Soc.* **72**, 3490 (1950).
16. J. Heijboer, *J. Polymer Sci. C* **16**, 3413 (1968).
17. See among others: J. Janacek and J. Kolarik, *J. Polymer Sci. C* **16**, 279 (1967); J. Janacek, *J. Polymer Sci. C* **23**, 373 (1968); J. Kolarik and J. Janacek, *J. Polymer Sci. A-2* **10**, 11 (1972); F. Lednicky and J. Janacek, *J. Macromol. Sci.-Phys. B* **6**, 335 (1971); J. Janacek, *J. Macromol. Sci.-Revs. C* **9**, 1 (1973).
18. J. A. Sauer and R. G. Saba, *J. Macromol. Sci.-Chem. A* **3**, 1217 (1969).
19. K. Shimizu, O. Yano, Y. Wada and Y. Kawamura, *J. Polymer Sci. Phys.* **11**, 1641 (1973).
20. O. Yano and Y. Wada, *J. Polymer Sci. A-2* **9**, 669 (1971).
21. G. P. Johari and M. Goldstein, *J. Chem. Phys.* **53**, 2372 (1970); **55**, 4245 (1971).
22. See J.-F. Jansson, *J. Appl. Polymer Sci.* **17**, 2997 (1973).
23. S. Havriliak and N. Roman, *Polymer [London]* **7**, 387 (1966).
24. K. Deutsch, E. A. W. Hoff and W. Reddish, *J. Polymer Sci.* **13**, 565 (1954).
25. H. Looijenga, Personal communication.
26. J. F. Schaefer, in D. J. Meier, ed., Molecular Basis of Transitions and Relations, Gordon and Breach, London, 1978 p. 103.
27. R. F. Boyer, *J. Polymer Sci. C* **50**, 189 (1975).
28. J. Heijboer, *Proc. Int. Conf. Physics Non-Crystalline Solids* (North Holland Publ. Co., Amsterdam, 1965) p. 231.
29. H. A. Waterman, L. C. E. Struik, J. Heijboer and M. P. van Duijkeren in *Amorphous Materials*, edited by R. W. Douglas and B. Ellis. 3rd Int. Conf. Phys. Non-Crystalline Solids (Sheffield 1970) (Wiley, 1972) p. 29.
30. B. Cayrol, A. Eisenberg, J. F. Harrod and P. Rocaniere, *Macromolecules* **5**, 676 (1972).
31. J. Kolarik, *J. Macromol. Sci.-Phys. B* **5**, 355 (1971).
32. R. F. Boyer, in *Encyclopedia of Polymer Science and Technology*, edited by N. Bikales, Vol. 13 (Wiley–Interscience, New York, 1970) p. 277.
33. R. F. Boyer, *J. Macromol. Sci. Phys. B* **9**, 187 (1974).

DISCUSSION

J. Sauer (*Rutgers Univ.*): You commented that Type A secondary relaxations involve local-mode motions of the main chain and that such transitions occur in PVC, polysulfone, and polycarbonate. Actually, aren't the transitions that occur in these materials of different origin? In polycarbonate and polysulfone, motion of the *p*-phenylene groups are involved in the transition, while no such groups are present in PVC.

J. Heijboer: I only stated that the transitions in PVC, polysulphone and polycarbonate are due to local main-chain motions, that is to say type A motions. Of course, the detailed geometry of these motions can be quite different.

J. F. Jansson (*MIT*): What are some other possibilities for types of side-chain motions, especially those which are strongly coupled with the backbone?

J. Heijboer: Of the type B motions, that of the β-transition in PMMA has been studied in most detail. In my opinion, the β-transitions of poly(α-methylstyrene) and of polystyrene are due to type B motions, which are rather strongly coupled with the backbone.

High-Resolution C-13 NMR Studies of Solid Polymers

J. SCHAEFER

*Corporate Research Department, Monsanto Company, 800 N. Lindbergh Blvd.,
St. Louis, Missouri 63166, U.S.A.*

INTRODUCTION

The single most important feature about a natural abundance ^{13}C NMR experiment is that it is performed on a rare or dilute spin species. This means that homonuclear ^{13}C—^{13}C coupling can be ignored, and that all dipolar interactions involving carbons are necessarily with protons. Consequently, a high-resolution Fourier transform ^{13}C NMR experiment on a solid polymer is both easy to perform and simple to interpret. The experiment is easy to perform since much of the dipolar broadening of the NMR lines can be removed by straightforward heteronuclear decoupling.[1] That is, the protons responsible for the line broadening of the carbons can be continuously stirred by an rf field, of a strength comparable or greater than the proton linewidth of generally 5—10 gauss. The ^{13}C receiver can be tuned in such a way as to ignore this irradiation. The stirring has an effect on the protons not dissimilar from that arising from increased molecular motion (although there are some important differences). The net result of this stirring is the removal of static and near static dipolar broadening of the ^{13}C resonances producing a spectrum with residual dipolar linewidths typically on the order of 100 Hz, substantially less than most ^{13}C chemical shifts for observing frequencies of 15—25 MHz.[2]

The ^{13}C NMR experiment on the solid is simple to interpret for two reasons. First, ^{13}C relaxation parameters are different for structurally different carbons in the polymer, and so can provide uncomplicated direct information about the details of molecular motion for each carbon environment.[3] In particular, spin-lattice relaxation parameters are not averaged by spin diffusion. This averaging process, which removes much of the information from an ^{1}H NMR experiment on a solid polymer, does not occur in the ^{13}C experiment because the necessary mutual spin flips of the dipolar-coupled ^{1}H and ^{13}C spins are not energy conserving, and so are unlikely. The second reason the ^{13}C NMR experiment on a solid polymer is simple to interpret is that the irradiation of the protons can often be tailored in such a way as to select for observation only certain carbons in otherwise highly complicated multiphase systems. In effect, only a part of the system need be considered in any given experiment, with the remainder examined later, in separate experiments.

In this paper, we will briefly describe the determination and interpretation of each of the major ^{13}C relaxation parameters useful in the analysis of polymers. Then, using results of experiments on poly(methyl methacrylate), we will illustrate how this information can be pieced together and so lead to a better understanding of the nature of molecular motions of polymers in the solid state. Finally, we will illustrate the use of a variety of proton decoupling and proton-carbon cross-relaxation techniques in simplifying analyses of the mixed phases of partially crystalline polymers, filled polymers, and some biopolymer systems.

THE SPIN-LATTICE RELAXATION PARAMETERS: T_1 AND THE NUCLEAR OVERHAUSER ENHANCEMENT

The spin-lattice relaxation time, T_1, is the time constant associated with the return of steady-state magnetization along the laboratory static magnetic field, following a perturbation of the magnetization.[4] The mechanism for this return by carbons in a solid polymer is through coupling to the fluctuating ^{1}H dipolar field generated by the microscopic segmental rotational motions of the polymer. The return is most efficient (that is, T_1 is short) when there is motion at about the carbon resonance frequency. Thus, T_1 is generally sensitive to motions in the frequency range 10^6–10^9 Hz. The spin-lattice relaxation time can be measured by a variety of schemes all of which are based on some kind of a disturbance of the carbon magnetization (for example, by an rf pulse), followed by a variable waiting period, after which the extent to which the carbon magnetization has recovered is observed. When experiments of this type are performed on poly-(methyl methacrylate) at room temperature, the T_1's of the various carbons in the repeating unit, — $(CH_2C(CH_3)(COOCH_3))$—, are found to differ by almost three orders of magnitude.[3] (The T_1 measurements are performed while dipolar decoupling the protons so that each carbon of the repeating unit gives rise to its own individual line, with characteristic chemical shift, whose relaxation properties can then be observed.) The T_1 of the α-methyl carbon is about 100 ms, near a theoretical minimum. This means that internal rotation of the methyl group is close to the resonance frequency of about 20 MHz. At the other extreme, the T_1 of the carbonyl carbon of the ester group is about 30 s. A T_1 this long can be attributed to a lack of mobility of the ester group as a whole, as well as to the absence of the strong dipolar interaction of a directly bonded proton. The methylene-carbon T_1 is about 1 s and is one-fourth that of the quaternary-carbon T_1, indicating the former is likely determined predominantly by dipolar interactions with its directly bonded protons, and that these protons are involved in significant, if far from free, torsional rotational motions of frequencies approaching the carbon resonance frequency. Thus, simply from the ^{13}C T_1 experiments at a single temperature, we can, on an *a priori* basis, establish for poly(methyl methacrylate), the presence of high-frequency main-chain segmental motions, as well as decide which side-chain carbons are engaged in rapid internal rotations and which are immobilized.

As a result of continuous irradiation of protons, the Boltzmann distribution of populations of the Zeeman levels of the protons is altered, which, in turn, alters the distribution of populations of those carbons dipolar coupled to the protons. Changing the populations of the carbon Zeeman levels leads to an alteration of the observed intensities of the ^{13}C NMR transitions. The change in integrated intensity of a carbon resonance due to the irradiation of the protons is called the ^{13}C nuclear Overhauser effect (NOE).[4] The NOE is measured by comparison of observed ^{13}C intensities with and without continuous (as opposed to gated) proton irradiation. Because the NOE involves relative populations of Zeeman spin-energy levels, it is linked to what are primarily T_1 processes. In other words, the NOE is determined by dipolar interactions associated with motional correlation times on the order of the resonance frequency. The NOE is a maximum of 3.0 when these dipolar correlation frequencies are much greater than the resonance frequency, is about half maximum near the resonance frequency, and is close to 1.0, when all the T_1-type processes are associated with long correlation times.[5] In practice, it is found that the NOE of polymer systems is rather sensitive to the tails of the distributions of correlation times often necessary to describe the complicated cooperative segmental motions of solid polymers.[6] At the risk of oversimplification, the NOE responds to the width of the distribution of correlation times describing the polymer segmental motion, while T_1 responds to the mean value of the distribution. Thus, in elastomers above T_g, the NOE tends to be lower than one might expect based on T_1 data,[6] while in polymers below T_g the NOE tends to be substantially greater than that expected from T_1 information.[3]

A quantiative description of the NOE in polymer systems, although possible, is actually rather involved. Nevertheless, on a qualitative basis, the NOE is useful in indicating the presence of the equivalent of distributed correlational frequencies. In poly(methyl methacrylate), for example, the NOE's of all the protonated carbons are about the same even though the T_1's indicate substantial differences in rotational mobilities.[3] Thus, we interpret this lowering of what could be expected to be a full NOE for the mobile ester-methyl carbon, and a raising of what could be expected to be a small NOE for the main-chain methylene-carbon, as an indication of the presence of a distribution of correlation times covering a range at least as broad as 10^5–10^{10} Hz.

THE CARBON RELAXATION TIME IN THE ROTATING FRAME, $T_{1\rho}$

The carbon magnetization can be placed in what is called the rotating frame by means of a 90° pulse (which aligns the magnetization vector perpendicular to the rf field, H_1), followed by a fast 90° phase shift of the carbon H_1.[2] The carbon magnetization no longer precesses about the applied dc field H_0, but instead remains locked in phase with H_1 for a time characterized by $T_{1\rho}$, the carbon relaxation time in the rotating frame.[7] (More accurately, as a function of the magnitude of H_1 relative to the local dipolar field, the carbon magnetization, in fact, precesses about H_1.) The net result of a shift into the rotating frame is to permit relaxation experiments with effective fields governed by H_1 (25 gauss) rather than by H_0 (20 kgauss). The ^{13}C $T_{1\rho}$ is measured by placing the carbon magnetization into the rotating frame, holding it there for a variable waiting time by continuous application of H_1 (with no proton H_1), then turning off H_1 and measuring the residual magnetization still remaining parallel to H_1 (with dipolar decoupling).

Examples of the ^{13}C NMR spectra of poly(methyl methacrylate) after the carbon magnetization had been held in the rotating frame for times up to 5.3 ms are shown in Figure 1.[8] The spectra are quite similar differing only in overall intensity and not much in the relative intensities of the various lines (whose partial assignments are also shown in the figure). This indicates that most carbons of the repeating unit have about the same $T_{1\rho}$ of 4–5 ms, with the exception of the carbonyl carbon, which has a significantly longer value. The magnitude of $T_{1\rho}$ is determined mostly by T_1-type spin-lattice processes, only with the important resonance frequency now not being the NMR observing frequency, but rather the rotating frame frequency.[7] Thus $T_{1\rho}$'s are sensitive to those dipolar interactions associated with rotational motions having correlation frequencies of 10^3–10^6 Hz. Since most of the $T_{1\rho}$'s are about the same for poly(methyl methacrylate) at room temperature, it is likely that a major source of relaxation is a broad distribution of torsional intermediate frequency inter-chain motions arising from the necessarily loose packing of a polymer characterized by a large mechanical loss factor.

THE CARBON-PROTON CROSS-RELAXATION TIME IN THE ROTATING FRAME, T_{CH}

If both the carbon and proton magnetizations are simultaneously placed into rotating rf frames with the respective H_1's adjusted so that $\gamma_{carbon}(H_1)_{carbon}$

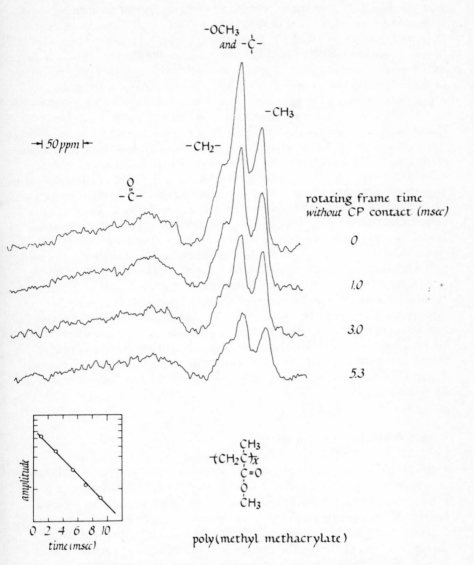

FIGURE 1   Dipolar-decoupled ^{13}C NMR spectra of solid poly(methyl methacrylate) obtained after holding the carbon magnetization in the rotating frame without proton contact for a variable time. The insert shows the $T_{1\rho}$ relaxation behavior for the most intense line of the spectrum. The ^{13}C H_1 was about 25 gauss.

$= \gamma_{proton}(H_1)_{proton}$, then the carbon and proton spin systems are strongly coupled, even though in the static field they are not coupled because of their different resonance frequencies.[9] The coupling occurs because any precession of the protons about their H_1 causes the component of the dipolar field along the direction of H_0 to oscillate at an angular frequency just right to induce transitions of the carbons relative to their rotating field.[10] In this double rotating field experiment, carbon and proton spins are now able to engage in energy conserving mutual spin flips. The time constant associated with the rate at which the two spin systems can exchange polarization is the cross-polarization (CP) time in the rotating frame, T_{CH}. This relaxation time is measured by polarizing the protons, placing protons and carbons in the rotating frames with their respective H_1's differing by a ratio of 4 to 1, waiting a variable time, and then sampling the magnetization of the carbons now aligned parallel to H_1 as a result of contact with the protons.[2] Since this contact is, in effect, spin diffusion, efficient cross polarization occurs for those carbons dipolar coupled to immobilized protons. In other words, T_{CH}'s are sensitive to static dipolar interactions. In poly(methyl-methacrylate), for example, the methylene carbon, with its two directly bonded protons, has the shortest T_{CH}, a value on the order of 50 μs, while the non-protonated carbonyl carbon T_{CH} is an order of magnitude longer.[3]

In addition to providing information about polymer motions having frequencies on the order of 10^0 to 10^3 Hz, the cross polarization experiment is of enormous practical value.[2] This is a result of the fact that the time which must be spent waiting before the carbons can be repolarized and the experiment repeated, is not the carbon T_1, but rather the proton T_1, which is generally much shorter. Thus, cross polarization is well suited to the time averaging necessary in all ^{13}C experiments.

THE CARBON-13 LINESHAPE

Even though dipolar coupling results in carbon linewidths having residual dipolar broadening of only on the order of 100 Hz, chemical shift anisotropy can broaden a line by up to 3 kHz.[2] Chemical shift anisotropy arises from the asymmetric electron density distribution found about many covalently bonded carbon nuclei (especially carbonyl and other unsaturated carbons). The asymmetry about carbon bonds is not averaged by motion in a rigid solid, and so produces a dispersion of chemical shifts resulting from the variety of orientations of a bond relative to the magnetic field. A powder average of all orientations produces a typical lineshape illustrated by the carbonyl-carbon resonance of poly(methyl methacrylate) shown in Figure 1. This characteristic lineshape is tent-like, with sharp sides and an apex, or peak, which is generally off center. Actually, the relative chemical shifts of the two sides of the tent, together with the peak are related to the three unique spatial directions which must be described in order to specify the orientation of any given carbonyl-carbon C—O bond relative to the magnetic field. The isotropic chemical shift is the average of the three values.

While line broadening due to chemical shift anisotropy may occasionally confuse an already complicated spectrum, the anisotropy can be used to diagnose the details of internal motions in a solid polymer. Thus, the presence of the full,[2] approximately 150 ppm chemical shift anisotropy of the carbonyl carbon of the ester group of poly(methyl methacrylate), means that at room temperature any internal motion of the ester group as a whole must be at a frequency much less than the line-width. This puts an upper bound on the ester-group motion of about 100 Hz. If substantial reorientation of the ester group were occurring, then either the width of the chemical shift anisotropy dispersion would be reduced, or the pattern of the lineshape would be changed. (It is true, however, that high-frequency torsional motions can occur with no obvious effect on the lineshape, if the amplitude of these motions is less than, say, 15 degrees.) The latter effect occurs, for example, for methyl carbons where internal free rotation removes one of the unique directions of the carbon-proton bond relative to the magnetic field. The resulting lineshape is tent-like, but with the apex moved over to one of the sides.[2] In general, ^{13}C lineshape information can be used for direct and unambiguous assignment of molecular motions in solid polymers, assuming, of course, the chemical shift anisotropy of a carbon involved in the motion is large enough to be observed. In situations where overlapping anisotropies are only a hindrance to the analysis of a spectrum, they can be removed completely by high-speed mechanical sample spinning.[3] The resulting lineshape, and, to varying degrees, the other carbon relaxation parameters discussed earlier, then become functions of the spinning frequency relative to the correlation frequencies of the various molecular reorientations responsible for relaxation.

THE SELECTION OF CARBONS TO BE OBSERVED IN MULTI-PHASE SYSTEMS BY THE CHOICE OF PROTON IRRADIATION SCHEME

The primary means of selecting certain carbons of a complicated mixed-phase system for observation is the manipulation of the linewidth of the protons which are dipolar coupled to the carbon of interest. For example, with a partially crystalline polymer such as polyethylene, it is possible to use a 5-gauss ^{1}H decoupling field and observe only carbons of the amorphous phase. The carbons in the crystalline phase are coupled to protons having a 15-gauss linewidth, and this interaction is virtually unaffected by the relatively weak decoupling field. Despite the use of only modest decoupling power, the spectrum of the amorphous region is still well enough resolved to enable one to distinguish between the

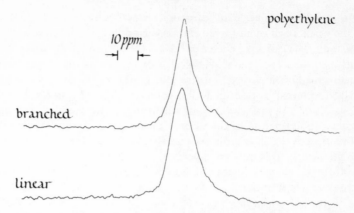

FIGURE 2   Dipolar-decoupled ^{13}C NMR spectra of the amorphous regions of two kinds of polyethylene. The shoulder to the high-field side of the major methylene-carbon resonance of branched polyethylene is due to methyl carbons.

methylene- and the higher-field methyl-carbon lines (Figure 2).[11] Thus, the number of methyl carbons present in structural defects in the amorphous region can be determined for both linear and branched polyethylene.

Analogous experiments can be performed on partially crystalline poly(ethylene oxide), but with an added complication. Even with relatively low-power decoupling, it is not possible to discriminate completely against the carbons in the rigid phase, at least at room temperature. Thus, as shown in Figure 3, both a symmetric amorphous-phase line, and an asymmetric (the asymmetry due to chemical shift anisotropy) rigid-phase line are observed using a 5-gauss

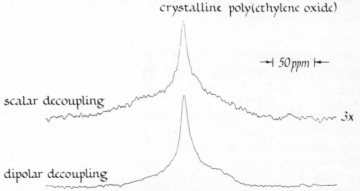

FIGURE 3   Scalar- and dipolar-decoupled ^{13}C NMR spectra of a crystalline poly(ethylene oxide) of molecular weight 20,000. The total integrated intensity of the dipolar-decoupled spectrum is about three times that of the scalar decoupled spectrum. All other conditions of the two experiments were the same.

dipolar decoupling field. Nevertheless, by reducing the decoupling field strength to that commonly used for scalar decoupling in standard Fourier transform experiments (about 0.5 gauss), the asymmetric component disappears, and one observes two symmetric lines superimposed. The narrower of the two may reasonably be associated with the genuinely amorphous phase, while the broad component is probably due to both crystalline and amorphous phases. A more detailed description of this system requires the results of variable temperature experiments.

The variation in linewidths of the protons coupled to carbons can be used in another way to simplify the ^{13}C NMR spectra of mixed phase systems. Protons having short T_2's are generally involved in static dipolar interactions with carbons as well as the protons responsible for the broad proton line in the first place. Hence, these protons are well suited for the cross-polarization transfer of magnetization to nearby carbons. Such carbons will be polarized preferentially relative to those carbons whose only proton neighbors are engaged in what amounts to high-frequency nearly isotropic rotational motion. Thus, in a carbon-black filled rubber, by using cross-polarization techniques one can obtain a ^{13}C NMR spectrum (Figure 4) in which the dominant resonance is a broad asymmetric line covering nearly 6 kHz. This line is due to carbons on the surface (and hence near protons) of the carbon-black particles themselves. (This interpretation can be confirmed from the spectrum of the carbon-black alone.) In addition, the intensities and lineshapes of the resonances due to the rubber are quite different from those observed in a standard Fourier transform experiment on polyisoprene. The latter results are not surprising in view of the fact that the cross-polarization experiment discriminates *against* rotationally highly mobile rubber chains. Presumably we are preferentially observing those chains immobilized by the filler.

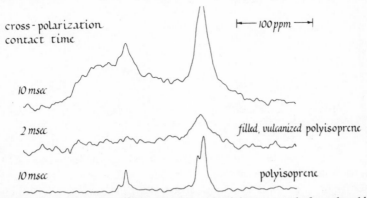

FIGURE 4   Cross-polarization ^{13}C NMR spectra of polyisoprene and of a carbon-black filled polyisoprene (60 pph filler). The broad asymmetric resonance of the filled system is due to carbons on the surface of the carbon-black particles, and has much the same appearance as the resonance observed for solid benzene at low temperature.

Obviously, additional relaxation experiments will be necessary to confirm some
of these qualitative assessments, but the utility of the method to provide new
information is clear.

Mixed phase systems are common in biological materials. Efforts to determine
the oil, starch, and protein composition of crop seeds by dipolar-decoupled
^{13}C NMR spectra are complicated by the fact that the narrow lines due to the oil
in the liquid phase, overwhelm the much broader lines due to the starch and
protein in the solid phase, especially for soybean (Figure 5).[12] Just as in the

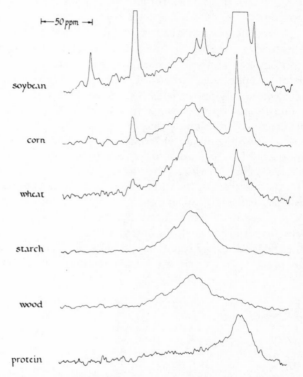

FIGURE 5   Dipolar-decoupled ^{13}C NMR spectra of some biological macromolecular
systems. The intense, sharp lines in the seed spectra are due to the oil components.

situation of the carbon-black filled elastomer, however, the lines associated with
the mobile liquid-like phase can be discriminated against using cross polarization
as the source of magnetization of the carbons. Some examples of cross-polarization ^{13}C NMR spectra of intact crop seeds are shown in Figure 6.[13] The high
protein concentration of soybean is most clearly shown in the broad resonance
at lower field, while the high starch concentration of the corn kernel is evident

cross polarization spectra

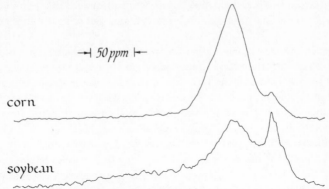

FIGURE 6   Cross-polarization ^{13}C NMR spectra of two kinds of crop seeds.

in the intense resonance near the center of the spectrum. Interferences from intense oil lines are not present in these spectra.

Acknowledgment

Much of the work described in this short review was performed at Monsanto Company, St. Louis, in collaboration with E. O. Stejskal and Rolf Buchdahl.

References

1. F. Bloch, *Phys. Rev.,* 111, 841 (1958).
2. A. Pines, M. G. Gibby, and J. S. Waugh, *J. Chem. Phys.* 59, 569 (1973).
3. J. Schaefer, E. O. Stejskal, and R. Buchdahl, *Macromolecules* 8, 291 (1975).
4. A. Abragam, *The Principles of Nuclear Magnetism* (Oxford University Press, London, 1961).
5. K. F. Kuhlmann, D. M. Grant, and R. K. Harris, *J. Chem. Phys.* 52, 3439 (1970).
6. J. Schaefer, *Macromolecules* 6, 882 (1973).
7. See, for applications to polymers, T. M. Connor, in *NMR, Basic Principles and Progress,* Vol. 4, edited by P. Diehl, E. Gluck, and R. Kosfeld (Springer-Verlag, New York, 1971).
8. J. Schaefer and E. O. Stejskal, paper presented at the 16th Experimental NMR Conference, Asilomar, CA, April, 1975.
9. S. R. Hartmann and E. L. Hahn, *Phys. Rev.* 128, 2042 (1962).
10. D. A. McAruthur, E. L. Hahn, and R. E. Walstedt, *Phys. Rev.* 188, 609 (1969).
11. J. Schaefer, E. O. Stejskal, and R. Buchdahl, unpublished results.
12. J. Schaefer and E. O. Stejskal, *J. Amer. Oil Chem. Soc.* 51, 562 (1974).
13. J. Schaefer and E. O. Stejskal, *J. Amer. Oil Chem. Soc.* 52, 366 (1975).

DISCUSSION

G. Allen (*Univ. of Manchester*): We did a neutron scattering study of rubbers filled with carbon-black in order to learn something about the nature of the rubber—carbon-black interface. However, we were not successful, perhaps because the concentration of "interface molecules" was too small. Have you been able to estimate what fraction of the carbon-black is in contact with the rubber phase?

J. Schaefer: In order to make a quantitative estimate of the fraction of carbon-black in contact with the rubber phase based on ^{13}C cross-polarization experiments, we need to establish the proton and carbon rotating-frame relaxation times as a function of the carbon-black concentration. We have not yet done this. From the pronounced qualitative differences between the intensities and lineshapes of the cross-polarization spectra of polyisoprenes having high, relative to low, levels of carbon-black loading, I suspect we are not dealing with a small effect, but rather with something on the order of 10% of the bulk rubber having what one might describe as a "strong" interaction with filler particles. This does not necessarily mean, however, that 10% of the rubber phase is in physical "contact" with the carbon-black.

C. E. Wilkes (*B. F. Goodrich*): Would you please describe the effects of deuterium substitution on the various relaxation time constants that you discussed.

J. Schaefer: Substituting a deuterium for a proton produces an increase of a factor 5—10 on the directly bonded carbon spin-lattice relaxation time. The effect is greater or lesser depending on the importance of dipolar interactions of the carbon with nearby non-bonded protons. The NOE is generally reduced. The rotating frame parameters, $T_{1\rho}$ and T_{CH} will be lengthened, but perhaps not dramatically so, if the important dipolar interactions are, as may often be the case with glassy polymers, associated with non-bonded protons. If there is no deuterium decoupling, some dipolar line broadening will be observed. The chemical shift anisotropy is unaffected.

S. L. Aggarwal (*General Tire and Rubber Co.*): Am I correct to state that the "tricks" that you can use so advantageously to study the various details of molecular dynamics of polymer chains are available only to pulsed ^{13}C NMR?

J. Schaefer: Carbon-13 relaxation parameters are most easily (and most often) measured by observation of the transient behavior of the carbon spin system; however, there are some continuous wave or cw experiments which will yield the same information and are reasonably practical as well. The proton decoupling part of the experiment can be considered cw since the irradiation is left on for times on the order of 10 ms, or the duration of the carbon observation.

S. L. Aggarwal: With reference to your work on carbon-black filled poly-(isoprene), I presume that you were working with a rubber that had been crosslinked (cured). I would like to know whether you are able to distinguish between permanent crosslinks (covalent bonds) and "pseudo-crosslinks" formed by the rubber chains being immobilized at the carbon-black surface?

J. Schaefer: We have not yet been able to distinguish between the two types of cross links. We still have hope that the carbon $T_{1\rho}$ (which we have not yet studied in detail) may provide a distinguishing characterization.

M. Litt (*Case Western Reserve Univ.*): The rotation of the carboxyl side-chain group is highly restricted to essentially two conformations, as was shown earlier today. How can one apply your NMR techniques to verify such a conformational state?

J. Schaefer: At room temperature, where all of our measurements have been made, the ester side group is effectively immobilized. We have no information about chain conformation. The change in T_1, $T_{1\rho}$ and the chemical shift anisotropy of the carboxyl carbon as the temperature is increased might provide such information, assuming the applicability of relatively simple models relating the observed relaxation parameters and the preferred chain geometries.

W. J. Stockmayer (*Dartmouth College*): What is the method that you use to determine the amplitude of a torsional motion?

J. Schaefer: A strictly qualitative method involves the use of two or more relaxation parameters (for example, T_1, $T_{1\rho}$, and the lineshape or observed chemical shift anisotropy). Simple theories are available which predict values for these parameters, in situations where there is effectively isotropic motion, and when there is only hindered or torsional motion. The latter predictions are in terms of the amplitude of the torsional motion. The observed values then can be compared to the predicted values in order to give an estimate of the amplitude of motion. The whole process should be considered qualitative, since the applicability of the available theoretical models is questionable at this point.

Molecular Motions in Crystalline Polymers

MOTOWO TAKAYANAGI

Department of Applied Chemistry, Faculty of Engineering, Kyushu University, Fukuoka, Japan

Mechanical relaxation of crystalline polymers were briefly reviewed and molecular process of crystalline relaxation was considered as the main subject of this paper. Crystalline polymers were classified into two types: one is flexible to form a lamellar crystal and the other forms a rigid α-helix molecule. As an example of the former type of crystal, the α crystalline relaxation of polyethylene was separated into the α_1 and the α_2 relaxations. Various experimental facts were cited to attribute the α_1 relaxation to the intermosaic block boundaries and the α_2 relaxation to the molecular motion inside the crystal. The latter conclusion was drawn from inspection of the temperature factor and the structural factor of the x-ray scattering, the narrowing of broad component of the NMR absorption, the Gruneisen constant representing the anharmonicity of molecular oscillation and the temperature effect on the dark field image of electron microscope.

The existence of crystalline relaxation in poly-γ-methyl-D-glutamate (PMDG) has been newly confirmed by Kajiyama in the author's laboratory. The density dependence of tan δ peak intensity, the dependence of its temperature location on the interhelical distance, anisotropy of tan δ curve for the oriented α-helix sample, and the thermal stability of α-helix inspected by the decay of x-ray diffraction intensity with temperature clarified the molecular process of crystalline relaxation. The α_1 relaxation, which is mainly observed for the PMDG sample cast from a solution in dichloroethane, is associated with interhelical slip or rotation whereas the α_2 relaxation, which is mainly observed for the PMDG sample cast from a solution in chloroform, with the accordion-like motion of helical chain.

INTRODUCTION

Crystalline polymers exhibit various kinds of relaxation, which can be detected by various methods such as viscoelastic, dielectric or NMR measurements, and also by various structure-analysis methods such as x-ray diffraction measurement or IR spectrophotometry. The effect of pressure on relaxations is one of the promising methods.

When some molecular motion is initiated, it is reflected in the mechanical relaxation curve as a depression of the dynamic storage modulus (dispersion) or as a loss modulus peak (absorption). With increasing temperatures, various kinds

117

of molecular motions are activated by thermal energy, ranging from single bond rotation to motion extending over the whole contour length of polymer. In the case of crystalline polymers, there exist various states of aggregation of molecular chains which affect the molecular motions: some modes of motion are easily activated in the low inter-molecular force field in the amorphous regions, whereas in order that the molecules belonging to the crystalline region become activated to the same motional state, it is necessary to raise the temperature somewhat higher. As far as various modes of motions in the amorphous region of the crystalline polymers are concerned, they are not much different from the corresponding motions in amorphous polymers. Relaxation in crystalline regions is characteristic of crystalline polymers. Therefore, we focus our attention on this problem as the main subject in this paper.

A two-phase model of crystalline texture has been widely accepted for the synthetic polymers which crystallize in the form of lamellae with folding of molecules. In Part 1, we will focus our attention on the crystalline relaxation processes of such polymers. On the other hand, with the increasing interests in life sciences, viscoelastic behaviour of polyamino acid esters has become a center of wide interest for several authors. By selecting appropriate solvents, it is possible to prepare solid state samples in the form of a crystalline state almost wholly composed of molecules with α-helix conformation. It has been recently found by Kajiyama in our laboratories that the crystalline region composed of α-helices show a remarkable crystalline relaxation. From the point of view of molecular motion, the mechanisms of crystalline relaxation of polyamino acid esters will be discussed in Part 2.

It is interesting to classify the difference between both types of crystalline relaxations: one is mainly composed of folded molecular chains and the other is mainly composed of helical rigid molecules embedded in the matrix of side chains. The former has been studied thoroughly in the past by various methods and the latter is a new state, having many problems to be clarified. In some sense, it may be said that the methodology for clarifying relaxation mechanisms established in Part 1 was applied in Part 2 and we have found now a new subject.

1   RELAXATION OF CRYSTALLINE COMPONENT COMPOSED OF FLEXIBLE POLYMERS

1.1   Outline of mechanical relaxations of crystalline polymers

1.1.1   Primary and low temperature relaxations   Various kinds of molecular motion take place in crystalline polymer corresponding to their time constant at the different temperatures. Very short range motions such as methyl group rotation in side chain ends are usually considered to take place at very low temperatures near those of liquid helium or nitrogen. The range of motions gradually increases with increasing temperature to include sizeable lengths of the side chains. In a previous review by the present author on low temperature transitions of polymers,[1] the effect of side chain length on the mechanical relaxation was determined for the series of crystalline isotactic poly-α-olefins, from polypropylene to polyoctadecene-1 with linear side chains. It was concluded that from the lower temperature side the relaxation associated with the twisting motion of long side chain first occurs, then followed by micro-Brownian motion and occurring close to the temperature for micro-Brownian motion of the main chains. Finally crystalline relaxation takes place before melting, which can not necessarily be detected.

In the absence of side chains, polymers with high crystallinity such as polyethylene, polyoxymethylene and polytetrafluoroethylene show a remarkable low temperature relaxation similar to the γ-relaxation for polyethylene. Polymers with intermediate crystallinity such as polyethylene terephthalate and polycarbonates, show a secondary relaxation with intermediate intensity being located at the lower temperature side of the primary relaxation. Vinyl polymers such as polypropylene, polystyrene, polyvinyl alcohol and polyvinyl chloride also show rather shoulder-like secondary absorption. Their relaxation mechanisms have been attributed to the local mode of main chains, whereas another conceivable hypothesis is that the glassy state is in a nodular structure and the motion of the molecules in the internodular region gives rise to the secondary absorption.

Concerning the effect of aggregation state of molecules on relaxation, the present author[2] has indicated in the past that the secondary absorption of polyethylene terephthalate is associated with some molecular motions in defects of lamellar crystals and with the frozen interlamellar non-crystalline regions, whereas the primary absorption is related to the amorphous region, for the reason that the absorption magnitude of the primary relaxation decreases monotonously with increasing crystallinity, extrapolating to zero at 100% crystallinity, but with the secondary relaxation still showing some loss. The side chain motion of isotactic poly-4-methyl-pentene-1 shows a secondary absorption at about −150°C which was separated into the contributions from the defect region inside a lamellar crystal and from the frozen amorphous region.[3] Illers[4] indicated that the γ relaxation of polyethylene is composed of the γ_I, γ_{II} and γ_{III} components in the order of descending temperature, among which the γ_I absorption appears only in the melt-crystallized sample or in single crystals annealed above 110°C and is affected by crystallinity, swelling, cross-linking and chlorination, indicating its being associated with the non-crystalline region, whereas the as-grown single crystals display only the γ_{II} and γ_{III} absorptions. These observations correspond well with the fractions of the narrow and the intermediate components of the NMR spectrum given by Bergman et al.[5]

Pechhold[6] has attributed the γ_{II} absorption to the motion of kinked chains in the crystal and Matsui *et al.*[7] to the motion of screw dislocation line moving between two neighbouring pinning points. The chain ends in the crystal serve as the pinning points. Such type of dislocations in lamellar crystal has not been experimentally detected and their theories remain at present in a hypothetical state.

Hoffman *et al.*[8] derived the γ peak temperature dependence on the number of CH_2's in the n-alkane or fold-length segment of polyethylene based on a model involving rotation of chain segment in neighborhood of row vacancy produced by embedding of chain end inside crystal. Papir and Baer[9] have found the mechanical relaxations for oriented polyethylene at 48 K and 20 K and call them the δ and the ϵ relaxations, respectively. According to them, the relaxation process in crystal phase is caused by the interaction between thermal motions of dislocations jobs and kinds of molecular chains. Further, they showed that these peaks disappear by annealing the sample at 129°C with the ends free.

The primary relaxation associated with glass transition of crystalline polymer is most convincingly explained by the molecular process of conformational changes in a frictionally interacting molecular field in the amorphous region. Molecular theory based on bead–spring model has succeeded in interpreting the transiton zone from the glassy to rubbery states for the amorphous polymers. The amorphous region in a crystalline polymer is different from that in the completely amorphous polymer in that the molecular motions are greatly restricted by the crystalline component.

Intermediate structures such as nematic or smectic structure are observable depending on the conditons of processing or polymer species. Close to the melting temperature, there appears the relaxation characteristic of the crystalline component.

1.1.2   Crystalline relaxations   Very highly crystalline polymers such as linear polyethylene show crystalline relaxation in the temperature region close to their melting temperatures. Schmieder and Wolf[10] were the first who noticed the loss peak characteristic of crystalline phase of polyethylene. The present author and his collaborators[11] ascertained that the same viscoelastic absorption is located in the temperature range where no remarkable crystal melting takes place. By comparing it with the thermal expansion of crystal lattice, this absorption corresponds in some sense to the glass transition phenomenon of the crystal phase. We further clarified that the crystalline absorption originates from the purely crystalline phase by using a sample of single crystal mats which showed thermorheologically simple nature.[12] The latter fact has been reconfirmed by the recent measurements over the extensive frequency range from 0.01 to 110 Hz.[13] Sinnott[14] measured also the crystalline absorption on the single crystal mats of polyethylene. His results that the crystalline absorption magnitude, represented by the peak area, decreases with increasing lamellar thickness for "annealed" single crystals are in sharp contrast to our results[15] which show the crystalline absorption for "as-grown" single crystals increased in magnitude with increasing lamellar thickness. This discrepancy was clarified in the same paper[15] by comparing the effect of lamellar thickness upon the crystalline absorption magnitude for as-grown single crystals with those for the thickened crystal prepared by annealing the as-grown single crystals. It was concluded that the increased lamellar thickness brought by annealing resulted in the decrease of absorption magnitude, perhaps owing to the increase of defects introduced by annealing into the regular crystal.

The effect of lamellar thickness on the temperature location of absorption was found to be the same for both kinds of samples, of as-grown and annealed single crystals, indicating that the length of the fold-length chain segment is a determining factor for the crystalline relaxation time.

The crystalline relaxation can be detected in the melt-crystallized samples for linear polyethylene, polyoxymethylene and polytetrafluoroethylene. It should be noticed that, even for the polymers which do not show remarkable crystalline absorption for the melt-crystallized samples, if measurements are made on the samples of single crystal mats or aggregates of solution grown crystals, the existence of crystalline absorption becomes more apparent owing to the increased crystallinity of the sample of the same polymer. Thus, single crystal mats of polyoxymethylene from cyclohexanol, polyethylene oxide from ethanol, isotactic polypropylene from xylene, isotactic polybutene-1 from decalin, poly-4-methyl-pentene-1 from xylene, nylon 6 from glycerol, and polyvinyl alcohol from diethylene glycol revealed the existence of crystalline absorptions by dynamic measurements.[16] Recently, Sauer et al.[17] have prepared a thorough review on the results in this field, to which nothing need be added.

The inability to detect the crystalline absorption of melt-crystallized polymers with medium crystallinity is attributed to the selective dissipation of mechanical energy in the amorphous region at the measuring conditions. For example, if a system of the amorphous region connected in series to the crystalline region is assumed and the measurements are made perpendicular to the aligning direction of the amorphous and the crystalline regions, then the behavior of the crystalline phase becomes apparent, whereas along the aligning direction no relaxation associated with the crystalline region can be detected.[18] Morphological factors are important in evaluating the relaxation magnitude of the phase in this problem, especially when the related molecular motions extend over distances comparable with the phase dimension.[19,20]

1.2 Multiplicity of crystalline relaxation

Wada et al.[21] indicated on the transition map that the crystalline absorption of polyethylene is separated into the α_2 absorption with higher activation energy and the α_1 absorption with lower activation energy and located at the lower temperature side of the α_2 absorption. Nakayasu *et al.*[22] separated the crystalline relaxation for the melt-crystallized polyethylene into two or three relaxation processes by applying the most reliable orthodox methods in rheology to their data over the extensive range of frequency of 10^{-6}–10^2 s^{-1} and the temperature range of -30 to $+80°$C.

Iwayanagi *et al.*[23] have recalculated the data of Nakayasu *et al.* on melt-crystallized polyethylene to prepare a transition map representing the relationship between $\log f_{max}$ and $1/T$ for the absorption maximum: in the order of ascending temperature, the β, the α_1 and the α_2 absorptions were located in the map as shown in Figure 1. Manabe and the present author[24] have measured the relaxation moduli of single crystal mat of polyethylene over the time scale of 1–10^3 s and the temperature range of 16 to 105°C; the data were converted into the loss moduli in extension, $E\,''$. By using the temperature dependence of the shift factor a_T, the temperature of the absorption maximum was plotted in the transition map as indicated by the open circles in Figure 1.[25] The activation energy of 46 kcal/mol evaluated from the α_2 absorption for the melt-crystallized samples agrees well with the value from the single crystal mat of 46 kcal/mol. Thus, it seems reasonable to consider that the α absorption of melt-crystallized sample includes a contribution from the relaxation process of lamellar crystals called the α_2 absorption, which is equivalent to that of single crystal mats.

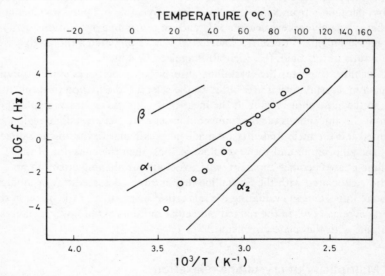

FIGURE 1   Transition map prepared by Iwayanagi[23] based on the data of Nakayasu *et al.*[22] for the melt-crystallized polyethylene sample (—). Open circles (O) are calculated by the present author by using the data on single crystal mats.[24]

Iwayanagi[23] has attributed the relaxation mechanism of the α_1 absorption to interlamellar slip. This view is in contrast to our view that the α_1 absorption is associated with the intermosaic block boundaries which will be mentioned in a later part. Our view is that the interlamellar slip takes place at the temperature region of the β absorption. Thus, in order to solve this controversy, we have analyzed the dynamic data on the melt-crystallized sample by using the data on the single crystal mats.[26] Figure 2 shows the dynamic loss modulus in extension, E'', as a function of frequency at various temperatures concerning the α absorption. Tensile stress is applied along the mat surface or perpendicular to the molecular axis in lamellar crystals. The remarkable feature is found that it is thermorheologically simple, which indicates a single frictional coefficient acting with the various modes of motion of molecular chains in the crystalline field. It reminds us of the molecular theory based on the bead—spring model used for predicting the thermorheologically simple behavior in the transition zone from glassy to rubbery states of amorphous polymers.

On the other hand, Figure 3 shows that the curves of E'' vs log f for the melt-crystallized polyethylene clearly show the multiplicity of the α absorption and the incapability of superposition by a simple shift operation.

By assuming that the relaxation behavior of lamellar crystals in spherulitic structure keeps the thermorheologically simple nature as found in the single crystal mats and has the same activation energy of 46 kcal/mol as that of single

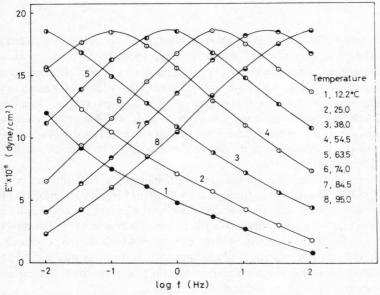

FIGURE 2   Tensile loss modulus, E'', vs frequency for single crystal mats of polyethylene.[13]

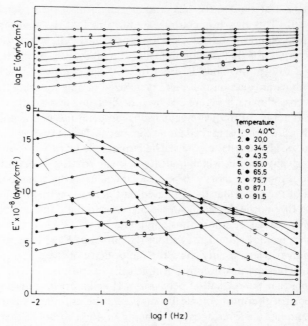

FIGURE 3   Tensile storage modulus, E', and loss modulus, E'' vs frequency for the melt-crystallized polyethylene.[13]

crystal mats, the α absorption of the bulk crystallized sample was separated into the α_1 and the α_2 absorptions. Figure 4 shows an example of the results of analyses by the above method. The upper figure is for the α_1 absorption and the lower figure for the α_2 absorption. The transition map is almost the same as shown in Figure 1. The activation energy of the α_1 absorption newly found is 31–32 kcal/mol, which value well corresponds with the value of 28 kcal/mol for the α_1 absorption evaluated from the data of Nakayasu *et al.*[22] A remarkable feature of the α_1 absorption is the decrease of relaxation magnitude with increasing temperature.

Thus the phenomenological analyses of the multiple α relaxation are considered to be well established through accordance of various data presented by the different authors. The mechanisms of the α_2 absorption has acquired strong support in attributing it to the relaxation of lamellar crystal itself. The direct evidence of the last subject will be given in Section 4 with the data obtained with the independent experimental methods such as the x-ray method, NMR absorption and others. The remaining problem is to explain the structural origin for occurrence of the α_1 absorption.

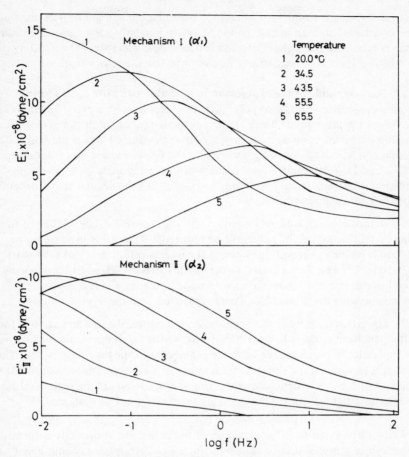

FIGURE 4 E'' vs frequency for the α_1 (upper) and the α_2 (lower) absorptions after separating the total α absorption given in Figure 3.[13]

1.3   The α_1 relaxation

Iwayanagi[23] has attributed the α_1 absorption mechanisms to the interlamellar slip as mentioned before. In contrast to his view, our view is that the interlamellar slip unavoidably induces a conformational change of molecular chains belonging to the interlamellar region, which gives rise to the primary absorption (the β absorption in Figure 1) which is associated with the glass transition. Our view does not necessarily deny the possibility of response of the intermediate region between the amorphous region and the crystalline lattice if such a region actually exists. However another view seems to be more important. The hypothesis that the α_1 absorption is mainly attributable to the mosaic block boundaries, in which the molecular chains are oriented approximately parallel to the molecular chains in the crystal with somewhat distorted conformation and incapable of fitting into the regular crystal lattice, may be supported by the following facts.

1) The magnitude of the α_1 absorption, separated from the α_2 absorption for the melt-crystallized sample of polyethylene, increases with increasing the volume fraction of the intermosaic block region for the samples annealed at $80-100°C$.[27] Fraction of the intermosaic boundary region was evaluated by the data on volume-fraction crystallinity, x-ray long period, and the size of mosaic block evaluated by D_{100} and D_{001} determined by the line profile of wide angle x-ray scattering.[27] Thickness of intermosaic block region was mono- or bi-molecular thick, depending on the annealing condition.

2) Deformation in uniaxial extension and the repeated cyclic loading of the bulk crystallized sample of polyethylene induced a remarkable increase in the magnitude of the α_1 absorption, whereas the α_2 absorption does not change its magnitude.[26,27] This fact means that the applied stress is selectively influencing those intermosaic block boundaries with poor regularity and less rigidity, in comparison with the mosaic block core associated with the α_2 absorption.

3) The strain magnitude at the elastic limit in uniaxial extension at $40°C$ for melt-crystallized samples having different thicknesses of intermosaic boundaries was theoretically predicted by using the geometries of the boundary region. The tilt-sliding model and the simple extension model were employed as deformation processes of the molecular chains belonging to the corresponding regions, although the relative weight of contribution among them could not be evaluated.[26,27]

4) Electron microscopic observation of the uniaxially or biaxially deformed polyethylene samples, using a special technique consisting of very thin metal vapor-deposition perpendicularly to the sample surface prior to stretching, revealed that (1) uniform interlamellar slip takes place at temperatures below room temperature, indicating the occurrence of the β absorption at the corresponding temperatures, (2) decomposition of lamellar platelets into mosaic blocks with lateral size of $15-30$ nm takes place at the temperatures around $50°C$ where the α_1 absorption predominates in the mechanical relaxation curve (especially at low frequency) which indicates that the molecular process of tilt-sliding or simple extension in the intermosaic boundaries takes place, and finally (3) in the temperature range of the α_2 absorption, e.g. at $100°C$, the number of micro-crazes and micro-cracks are greatly decreased and found at intervals of about $150-300$ nm. This fact suggests that the uniform deformation or thinning of lamellar crystals takes place, being accompanied by the tilting of molecules in a crystal along the c-axis. The lamellar crystals are in a viscoelastic state in the

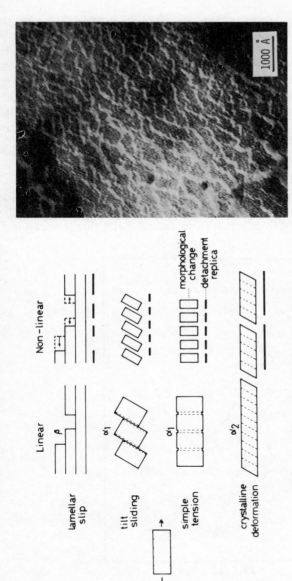

FIGURE 5   Schematic representation of the β-, α_1- and α_2-relaxation mechanisms within linear response (the left-hand column) and the irreversible deformation mechanisms (the middle column) induced by the corresponding relaxation mechanisms. The bold black lines (————) exhibit the schematic representation of the detachment replica for the irreversible deformation. The right-hand side electron micrograph is taken on the sample annealed at 100°C and stretched by 15% in biaxial directions at 55°C, indicating the decomposition of lamellar crystal into mosaic blocks.[28]

temperature region of the α_2 absorption.[28] Figure 5 illustrates the above mentioned points.

5) Upon drawing the melt-crystallized sample of polyethylene in the temperature range of the α_1 absorption, the total deformation can be resolved into the contributions from the strain-sensitive and strain-insensitive regions. By x-ray diffraction methods and evaluating the fractions of undeformed and deformed zones on the lamellar plane with the IR spectrophotometry, it was found that the degree of molecular chain orientation is remarkably high in the preferentially deformed region (having micro-crazes) even at small strain.[28] These analyses are only possible by employing the concept of deformation of intermosaic block boundaries in lamellar crystals.

6) The broad line NMR studies on the single crystal mats of polyethylene conducted by us[25,29] seem to support the model of intermosaic block boundaries mentioned above. Slichter[30] was the first to notice the narrow component emerging in the single crystals of polyethylene annealed at higher temperatures

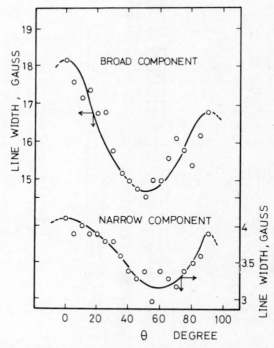

FIGURE 6   Angular dependence of the line width of broad and narrow components of the NMR absorption for oriented single crystal mats of polyethylene annealed at 110°C.[25,29] (Room temperature.)

and attributed it to the defects within the crystal. Odajima et al.[31] supported
Slichter's view based on the results on the n-paraffin crystal. Olf and Peterlin[32]
attributed the narrow component to the molecular motion on the lamellar
surface based on the fact that the mobile fraction of the narrow component decreased by etching the single crystals with fuming nitric acid. Our measurements
were made on the dependence of line width of both broad and narrow components upon the angle between the direction of the main magnetic field and that
of the molecular axis in lamellar crystals. The results are shown in Figure 6. The
angular dependence of the narrow component introduced by annealing at
110°C of as-grown single crystals is the same as that of the broad component.
This fact indicates that the molecular motions in the defect region is directionally
the same as in the crystalline phase and the molecular motion of the conformationally distorted molecules in effect is in a more active state at lower temperatures so as to give rise to the narrow component, although the mode of rotational
motion around the chain axis is similar to those in the regular crystal. Thus, the
chains belonging to the intermosaic block region are conformationally distorted
but oriented on the average along the c-axis in the thin space of a mono- or bi-
molecular layer.

The view that the narrow component is entirely attributable to the lamellar
surface, is conceivable for as-grown single crystals of polyethylene, while it
does not hold for the annealed single crystals. Figure 7 shows that the mobile
fraction corresponding to the narrow component is increased with increasing
long period for annealed single crystals, indicating the increased fraction of
defects of about 4% at infinite lamellar thickness, whereas for the as-grown
single crystals the mobile fraction linearly decreases with the reciprocal of long
period, $1/L$, extrapolating to zero mobile fraction at $1/L = 0$. The last relation
can be explained if the narrow component originates only from the surface region
of lamellar crystals and if the assumption of equal qualities in crystal perfection
for as-grown single crystals prepared at different crystallization temperatures is
made.

7) Takemura et al.[33] measured wide-angle NMR spectra of single crystal mats
of polyethylene under various hydrostatic pressures from 1 to 4500 kg/cm².
Figures 8 and 9 show the narrowing of both broad and narrow components for
the single crystal mats, respectively, with the molecular axis orientation of
$\theta = 90°$ to the direction of the main magnetic field. The arrows indicate the
temperature locations of narrowing at various pressures for the purpose of
comparison. They showed that no narrowing of the broad component was observed in the direction of $\theta = 0°$. This fact was attributed by them to the occurrence of rotational motion of molecular chains in the crystal around their axes.
If the temperature at the completion of narrowing is designated as the narrowing
temperature, the pressure coefficient of it is 10°C/1 kb. The narrowing in the

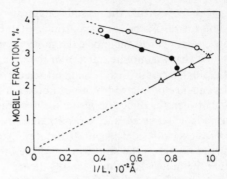

FIGURE 7 Mobile fraction of the narrow component of wide line NMR absorption at room temperature vs reciprocal of long period, $1/L$, for as-grown single crystals (□) and the annealed single crystals prepared at $55°C$ (○) and $74°C$ (●).[25,29]

broad component is attributable to the α_2 relaxation, and will be discussed in more detail in a later part.

It should be noticed here that, almost in the same temperature range as that of the α_2 relaxation, the narrowing of the line width of narrow component is also found at the high temperature side of the narrowing corresponding to the β and γ relaxations. The former narrowing should be attributed to the α_1 relaxation. The pressure coefficient of $10°C/1$ kb and the temperature location of $50-130°C$ is in accord with those found in the broad component corresponding to the α_2 relaxation. According to my view, this accordance may suggest that

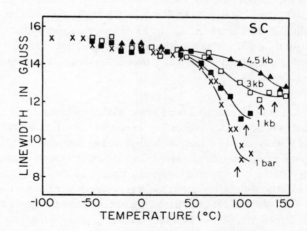

FIGURE 8 The narrowings of line width of the broad component of the NMR absorption as a function of temperature for single crystal mats of polyethylene at $\theta = 90°$ at various pressures (Takemura et al.[33]).

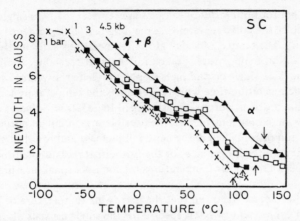

FIGURE 9   The narrowings of line width of the narrow component for the same sample as in Figure 8 at $\theta = 90°$ (Takemura et al.[33]).

the narowing of the narrow component corresponds to the α_1 relaxation, whereas Takemura et al.[33] insist that the latter process is concerned with the lamellar surface because the anisotropy in the narrowing of the narrow component betwen $\theta = 0°$ and $90°$ is negligible. Our opinion about their view is that the disappearance of the anisotropy of the narrow component takes place in the α_2 relaxation temperature region at which the α_1 relaxation has already been completed and the vigorous molecular motions in the intermosaic block boundaries create an almost isotropic narrowing of the narrow component.

Figure 9 shows also the overlapped narrowing of the narrow component corresponding to the β and the γ relaxations which appears at the low temperature side of the α_1 narrowing. This observation predicts that the molecular process associated with the β relaxation should be distinguished from the molecular process associated with the α_1 relaxation in viscoelastic measurements, even for the single crystal mat.

1.4   The α_2 relaxation

There are many reasons to believe that the α_2 relaxation arises from inside the crystalline phase. Various experimental facts on the α_2 relaxation have been accumulated which will be cited in the following, although some facts have been mentioned in the last section.

1) The temperature location of the α_2 absorption maximum increases with increasing long period for both samples of single crystals as-grown at various crystallization temperatures and the same crystals annealed at various tempera-

tures. The data for both series of samples are represented by a single curve, whose shape resembles the curve representing the long period dependence of the melting temperature.[15] The sample annealed at the temperatures sufficiently high above the crystallization temperature is taken for the melt-crystallized sample. Sinnot's data[14] on annealed single crystal mats of polyethylene showed the same tendency as ours.[15] The relaxation time which determines the temperature location of the absorption at the isochronal measurement increases with increasing long period. With increasing long period, the possible maximum amplitude of molecular oscillation of fold—length segment will increase and result in a shift of relaxation time to the longer side for the isothermal measurement or a shift of the loss peak to a higher temperatures for the isochronal measurement.

2) Iwayanagi et al.[34] showed that the narrowing temperature of line width of the broad component of NMR spectra for the single crystals of polyethylene increased with increasing temperature, as shown in Figure 10. The temperature of the absorption maximum is also plotted in Figure 10. The same dependence on long period can be seen in both independent measurements. Hypotheses such as attributing the α_2 absorption or narrowing to the fold surface or the defects inside the crystal can not explain the narrowing of the broad component. Pechhold[35] assumed the motion of "Doppelkinken-paaren" in the crystal, a process will not induce a narrowing of the broad component.

3) Iwayanagi et al.[34] and Takemura et al.[33] showed that the motional narrowing of the broad component of NMR spectrum of as-grown single crystals was observed at $\theta = 90°$ as shown in Figure 8, whereas no narrowing was observed $\theta = 0°$ in the temperature range of the α_2 relaxation. The former authors observed this phenomenon at normal pressure and the latter authors at various hydrostatic pressures. These anisotropic narrowings can be explained only by molecular process in which the rotational motion of molecular chains around their axes takes place inside the crystal.

It should be noticed that the line width after the completion of the α_2 relaxation is increased by hydrostatic pressure, as seen in Figure 8, which indicates that the rotational motions of fold-length chain segments in crystal are greatly depressed by increased molecular interaction from the volume contraction of a crystal with increased hydrostatic pressure. The increased frictional coefficient with increasing pressure makes the relaxation temperature shift to the higher temperature side.

4) Theoretical predictions proposed on the α_2 relaxation are briefly outlined here, focussing on their capability of interpreting the experimental facts.

Sinnott[14] has shown a linear relationship with positive slope when the area of the α_2 absorption is plotted against the reciprocal of long period. For interpreting

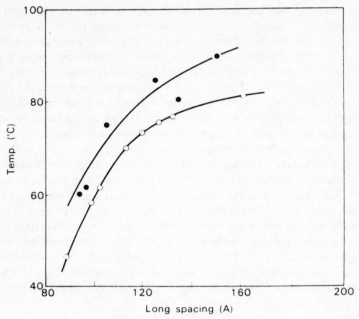

FIGURE 10   Mechanical loss peak temperature of the α_2 relaxation (O) at 110 Hz[15] and the NMR narrowing temperature (●)[34] plotted against long spacing of as-grown single crystals.

this relationship, Hoffman *et al.*[8] assumed that the α_2 absorption is composed of two contributions (1) from a process coming from the fold surface, which gives the positive slope relation against the reciprocal of long period, and (2) from the other process arising from motion inside the crystal and being independent of lamellar thickness. This interpretation does not hold for the sample series of as grown single crystals observed by us.[15] However, Hoffman has succeeded in predicting the chain length dependence of temperature of mechanical relaxation maxima, T_{max}, in n-alkanes based on his rigid rotator model. But, for lamellar crystals of polyethylene, owing to the energetical difficulties for a fold-length chain segment, with larger length than that of n-paraffins, chain twisting mode takes place with lower a potential barrier than as a rigid axial rotator. Thus the T_{max} value tends to level off, deviating from the curve predicted by the rigid rotator model at a number of chain atoms of 60–120. The activation energy also approaches a limiting value. These predictions seem to accord with the experimental results as far as the single crystal mats are concerned. However, as mentioned in a later part, the motional narrowing of the broad component of the extended chain crystals of polyethylene occurs at temperatures around 100°C,[33] which value is far larger than that of Hoffman's prediction.

Now, changing the subject to the relaxation magnitude, the increased relaxation magnitude of the α_2 relaxation with increasing long period for as-grown single crystal mats seems to be better explained by the molecular theory of the incoherent lattice vibration presented by Okano.[39] This theory has been developed more in detail by Wada *et al.*[37] to calculate the α_2 relaxation magnitude. Their concept is that, when the crystal of polyethylene is strained along the *a*- or *b*-axis, the external work is first stored as the change in interchain potential, then the distribution shifts to a new equilibrium with some relaxation time, and finally a new equilibrium is reached between interchain and intrachain freedom. According to the Okano's model, in the temperature range of the α_2 relaxation, the fold-length molecular chain segment in an incoherent molecular field can have various modes of motion such as rotational motion around the molecular axis or translational motion along the molecular axis. Just as with theoretical predictions using the bead—spring model, the mode of the lowest order governs the major part of the motion, and which corresponds with the longest relaxation time. Increasing segment length results in the shift of the α_2 absorption maximum to the higher temperature side for isochronal measurements. As for rotational oscillation, the instantaneous force constant g_0 (in coherent interchain energy field) is relaxed to the time-averaged or smeared-out interchain force constant, g. The ratio of g/g_0 is given by

$$g/g_0 \simeq \exp\{-nkT/5f'\}, \tag{1}$$

where n is the main chain atoms in a fold-length segment and f' is the force constant for twisting the main chain per bond which was calculated to be 1.1×10^{-12} erg/bond for polyethylene chain. The fit to the data for the simple of as-grown single crystals is at present qualitative.

In order to explain the fact that the magnitude of the α_2 relaxation for the annealed single crystals decreases with increasing long period, it is necessary to take into account the increased defects introduced by annealing, which reduce the response of a genuine crystalline lattice. The present author[15] assumed the intermosaic boundaries as the defects to explain the data. Detailed discussion on the role of mosaic block structure on deformation was given in the last section, which concerned the α_1 relaxation.[26—28]

5) Verification of molecular motion inside the polyethylene crystal as a function of temperature using x-ray diffraction methods is effective in constructing a dynamic molecular model.

Iohara and the present author[38] have determined the temperature factor, **B**, based on the precise x-ray structural analyses of polyethylene crystal of melt-crystallized and single-crystal samples over the temperature range from the liquid nitrogen temperature to the melting temperature.

The temperature factor **B** is represented by the symmetric tensor having six independent elements.

$$\mathbf{B} = \begin{pmatrix} B_{aa} & B_{ab} & B_{ac} \\ & B_{bb} & B_{bc} \\ & & B_{cc} \end{pmatrix}. \tag{2}$$

When atomic displacements are harmonic or small enough, **B** is related to the amplitude of atomic displacement **U** by Eq. (3).

$$\mathbf{B} = 8\pi^2\,\mathbf{U} = 8\pi^2 \begin{pmatrix} U_{aa} & U_{ab} & U_{ac} \\ & U_{bb} & U_{bc} \\ & & U_{cc} \end{pmatrix}. \tag{3}$$

Based on the symmetry consideration of crystal structure of polyethylene (Figure 11), carbon atoms 1 and 2 should be taken into account, and $U_{ac} = U_{bc} = 0$ for both atoms and $U_{ab,1} = -U_{ab,2} = U_{ab}$. The atomic displacement parameters, U_{aa}, U_{bb}, U_{cc} and U_{ab} were determined from the structural factor representation for 32 reflections with the measurements of x-ray scattering inten-

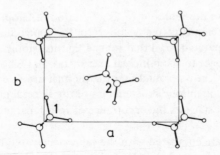

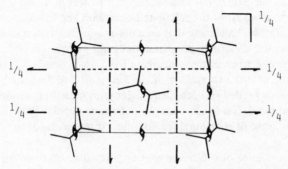

FIGURE 11   Structure and symmetry elements of unit cell of polyethylene crystal.

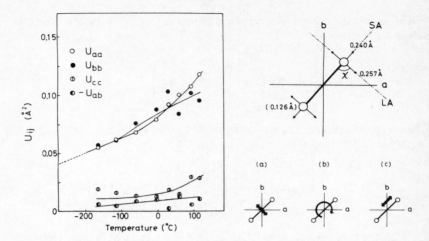

FIGURE 12   Mean square displacements of carbon atoms vs temperature for the melt-crystallized polyethylene (right hand side) and schematic representation of mean square displacement of carbon atom converted to the principal axes of LA and SA with corresponding molecular motions, (a), (b) and (c).[38]

sities. Figure 12 shows the values of U_{ij} as a function of temperature. By extrapolating these tensor elements to absolute zero, the contribution from lattice imperfections was separated from that for the thermal vibration U_{ij}^T. After converting U_{ij}^T with respect to the principal axes of LA and SA as indicated in the right side of Figure 12, it was found that the thermal displacement of carbon atom is largest in the direction perpendicular to the skeletal jigzag plane (LA-direction). This result suggests the occurrence of rotational motion around the molecular axis.

The U_{ij}^T values were compared with the values calculated by the harmonic oscillator model according to Kitagawa and Miyazawa.[39] In the temperature range below 0°C, the agreement was excellent.[41] However, above 0°C, the deviation of the U_{ij}^T values from the calculated ones with the harmonic oscillator model became sizable, indicating that an anharmonic oscillation of molecules takes place around 0°C. At the same time the angle between two diagonals of the ab-plane showed a tendency to decrease from 48° towards 44° at 110°C. It was also found that the IR spectra of the 720 cm^{-1} and 730 cm^{-1} doublet of CH_2 rocking bands tended to become a singlet with increasing temperature. This fact indicates a decrease of anisotropy of the molecular field in a crystal owing to the increase of rotational fluctuations of molecules around their axes.

More direct evidence of rotational motion was obtained from the ratio of the diffraction intensities from the crystal planes of h + k = odd to those from

the planes of h + k = even.[40] Figure 13 shows the interchain energy curves
associated with the central chain rotating among six neighbor chains. Lattice
constants are given at room temperature (full line) and at liquid nitrogen temperature (broken line), the setting angle being fixed at 45°.[37] Angle θ was taken
between the bc-plane and the skeletal jigzag plane of molecule 2 as indicated in
Figure 13. The well I is the position of the molecular chain in the normal crystal

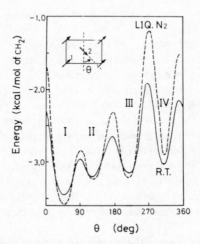

FIGURE 13   Inter chain energy curves for the rotating central chain among six neighbor
chains in polyethylene crystal lattice.[37] Full line is for room temperature and broken line
for liquid nitrogen temperature. Well I is for the regular crystal and well II for the next
stable position.

structure as originally given by Bunn, indicating the most stable state. An almost
similar energy curve has been given by McCullough,[41] based on his own energetic
calculation. Figure 13 shows that there exists the next stable well II positioned
with rotation by as much as 90° from the position of the well I. The potential
barrier between both wells is fairly low, so that the rotation of the fold-length
molecular segment from the I position to the II position is conceivable at
moderate temperatures. Such a rotational transition may give rise to the change
of structural factor in x-ray diffraction, which presents the possibility of inspect-
ing the occurrence of rotational transition by measurements of the intensity of
x-ray diffraction.[43]

It is sufficient for calculation to take into account two kinds of carbon atoms
1 and 2 in the unit cell, as shown in Figure 13. For each chain, two orientations,
I and II, are possible. If the probability for both 1 and 2 chains to take the I
orientation is assumed to be r^2, the possibility for one of two chains to take the
II orientation is $2r(1-r)$ and that for both chains to take the II orientation is

$(1 - r)^2$. Therefore, the average structure factor $\bar{F}$ for crystals having four kinds of unit cell are given by Eqs. (4) and (5).

$$\bar{F}(h + k = \text{even}) = 2f\{\cos 2\pi(hx + ky + lz)$$

$$+ (-1)^{h+k} \cos 2\pi(hx - ky + lz)\} = F_0 \quad (4)$$

$$\bar{F}(h + k = \text{odd}) = 2(2r - 1)f\{\cos 2\pi(hx + ky + lz)$$

$$+ (-1)^{h+k} \cos 2\pi(hx - ky + lz)\} = SF_0 \quad (5)$$

where F_0 is the structural factor of the normal lattice, $S = 2r - 1$, f is the atomic scattering factor of a carbon atom, and x, y and z are the fractional co-ordinates for each atom and for each orientation.

From the derived relationship, it was concluded that (1) the structural factors of the lattice plane having the indices of $h + k = $ even are equal to those for the regular structure, and the rotational transitions have no influence on the diffraction intensities from these planes and (2) the structural factor from the crystal plane of $h + k = $ odd takes the value of F_0 times the fraction, S, which is smaller than unity if the rotational transition takes place. With progressing transition, the S value decreases to take zero at equal probabilities of the I and II orientations, which corresponds to $r = 1/2$. Thus, the diffraction intensities from the $h + k = $ odd planes decrease according to the increased fraction of transformed state.

Figure 14 shows the ratio of the intensities from the $h + k = $ odd planes to those from the $h + k = $ even planes as a function of measuring temperature for the samples of melt-crystallized polyethylene (filled circles) and single crystal aggregates (open circles), indicating that a monotonic decrease starts at about $0°C$. If an assumption is made that all the atoms take the regular I positions in the crystal at the lowest temperature of measurement $-165°C$, the fraction of the atoms taking the II position amounts to 5%. These results are considered to support to the occurrence of a rotational transition inside the crystal beginning at $0°C$, which causes the internal loss in mechanical relaxation. This is the molecular bases for the crystalline relaxation from the view point of x-ray structural analyses. Here we assumed that the fold-length molecular segment is a rigid rotator, but actually various modes of rotational oscillation with higher order may take place at the same time. These various modes of motion give rise to the broadened distribution of relaxation time as shown in Figure 2. The relaxation time distribution might be more broadened by taking into account the long range cooperative motions of a lamellar crystal, similar to the two-dimensional harmonic or anharmonic oscillation of a membrane,[42] if crystal melting does not take place. An increase of the molecular segment length makes possible the motions with larger scale amplitude with a higher activation energy, and which cause the maximum relaxation time shift to the longer time for the isothermal

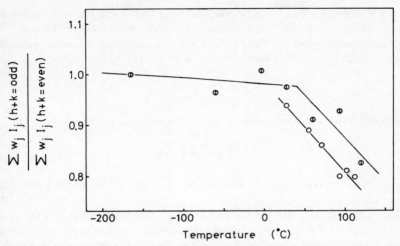

FIGURE 14   The intensity ratio of x-ray diffraction from the h + k = odd plane to that from the h + k = even plane vs. measuring temperature for the melt-crystallized sample (⏀) and for the single crystal mats (○).[40]

dynamic measurements and to the higher temperature side for the isochronal measurements. Figure 14 shows that the temperature at which the rotational transition begins for the melt-crystallized sample with larger long period is lower than that for the single crystal sample with thinner lamellar thickness.

Recently, McCullough[41] has made a more thorough approach to energetic calculation of molecular motions in polymeric solids. He calculated the energy relations in the rigid rotation of chains about their axis by considering the feasibility of coupled chain rotations and longitudinal translations along axes , and prepared the energy contour maps for coupled chain rotations and translations. His conclusion is that the major contributor to the α_2 relaxation (according to the notation in this paper) is the motion along the minimum energy path of partial rotation and translation along the chain axis, being a so-called "flip-flop" mechanism, rather than direct migration of the methylene unit as much as a unit along the c-axis direction. Our experimental observation with x-ray method on rotational transition is not inconsistent with the McCullough's prediction.

6) Another method for detecting the anharmonic motion of molecular chain which induces the lattice expansion and finally viscoelastic loss is to evaluate the values of the Gruneisen constant, γ.[43-47] There have been many papers on the γ values of polyethylene, but almost all of them were concerned with the γ values at room temperature or below. For the purpose of discussing the molecular motion giving rise to the α_2 relaxation, it is necessary to evaluate

the γ values above room temperature. Recently, Kijima and the present author[48] have evaluated the γ value for the samples of single crystal mat (SCM) and extended chain crystals (ECC) of polyethylene by using the values of compressibility, β, and the pressure coefficient of γ as in Eq. (6).

$$\gamma = -\frac{1}{2\beta^2}\left(\frac{\partial \beta}{\partial P}\right)_T - \frac{1}{4} \tag{6}$$

or by using pressure dependence of sound velocity, u, according to Eq. (7).

$$\gamma = \frac{1}{\beta}\left(\frac{\partial \ln u}{\partial P}\right)_T + \frac{1}{2} \tag{7}$$

Eqs (6) and (7) were derived by Broadhurst and Mopsik[52] for two-dimensional Debye solids.

The sample of ECC was prepared by isothermal crystallization at 224°C under 5200 kg/cm² for 3 h. The electron micrograph of the surface replica of the fractured sample revealed a band structure with a thickness of the order of a micron

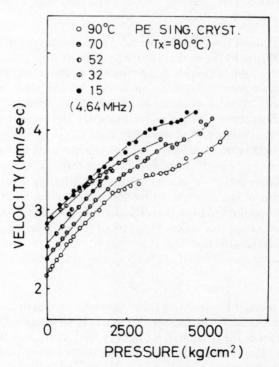

FIGURE 15   Sound velocity vs pressure at various temperatures for the single crystal mats (Kijima et al.[50]).

(the average being 350 nm) and characteristic of the ECC sample. The density was 0.994 g/cm^3 and the melting temperature by DSC was 140.9°C.

The ultrasonic measurement under pressure was carried out inside a capsule inserted in a high pressure vessel, using 4.64 MHz longitudinal wave, by a pulse method with the specimen sheet held between aluminium buffer rods.[50] The isothermal volume—pressure measurements for evaluating the β value was carried out with a high pressure dilatometer including a linear variable transformer which was calibrated with mercury at various pressures and temperatures.

Figure 15 shows the pressure dependence of sound velocity at 4.64 MHz at various temperatures from 15 to 90°C for the mat of single crystals of polyethylene.

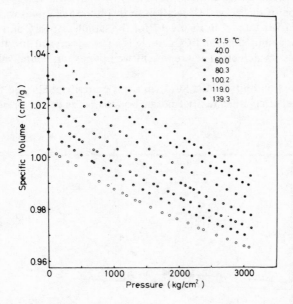

FIGURE 16   Specific volume vs pressure at various temperatures for the extended chain crystals of polyethylene (Kijima *et al.*[48]).

Figure 16 shows the variation of specific volume with pressure at various temperatures relating to the α_2 relaxation. The data were represented by a polynomial expression of P to evaluate the γ value as a function of P.

Figure 17 shows the values of Grüneisen constant at 1 atm for the samples of ECC (open circle) and SCM (filled circle), evaluated from the data given in Figures 15 and 16 according to Eqs. (6) and (7), respectively. Broadhurst and Mopsik[51] have calculated the γ values for the perfect crystals of n-alkanes as a function of temperature and number of carbon atoms per chain, N, according to Eq. (8).

$$\gamma_L = \frac{91 - 28(v/v_{c,0})^3}{28 - 16(v/v_{c,0})^3},\tag{8}$$

where γ_L is the Grüneisen constant related to the lattice potential, $v_{c,0}$ and v are the specific volume of a perfect crystal at 0 K and that of a perfect or imperfect crystal at any temperature, respectively. The assumption was made that the semirigid chains composed of N beads are in a quasiharmonic potential of a two-dimensional hexagonal lattice. Now, for the samples of SCM and ECC of polyethylene, we assumed N = 109 and 2800, respectively, based on the measured values of long period or band thickness. The γ_L values obtained by extrapolation or interpolation of the γ_L vs $1/N$ relation to the desired N values were obtained as shown by curves 1 and 2 in Figure 17 for the samples of ECC and SCM, respectively, assuming that the ECC sample is a perfect crystal and the SCM sample includes 1% defects. They represent well the experimental values as seen in Figure 17.

The increased mobility of the SCM sample is partially attributed to the loop surface of lamellar crystal and intermosaic boundaries as defects inside the crystal.

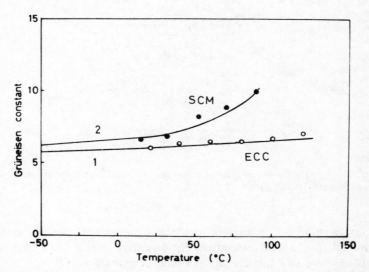

FIGURE 17   Grueneisen constant vs temperature at 1 atm for the extended chain crystals (O) and the single crystal mats (●) of polyethylene.[48] Curve 1 is the theoretical γ_L value for n-alkanes with assumption of perfect crystal of N = 2800 (N: number of methylene units per fold-length chain segment). Curve 2 is the theoretical γ_L value for the imperfect n-alkane crystal, 1% defects in the 0 K volume and N = 109.

All these results on the SCM sample agree well with the conclusions derived
n the last section for the mean square displacements of carbon atoms, U_{ij},
evaluated from the x-ray temperature factor. According to it, the values of
U_{aa}^T and U_{bb}^T increase rapidly above 0°C with raising temperature and deviate
from the theoretical values calculated on the harmonic oscillator model.[42]

7) The γ values of the ECC sample increase more slowly with temperature
compared to the SCM sample. Such a small anharmonicity can conceivably
suppress the α_2 mechanical relaxation, whereas the crystalline relaxation in
mechanical relaxation curve is large, as shown in Figure 18.[52] The rise of the
tan δ curve starts at 40°C, which is higher than the corresponding temperature
for the melt-crystallized sample (broken line) of 0°C. The γ peak at -120°C
which reflects the content of crystal defects is greatly decreased for the ECC
sample compared with that of melt-crystallized sample of polyethylene.

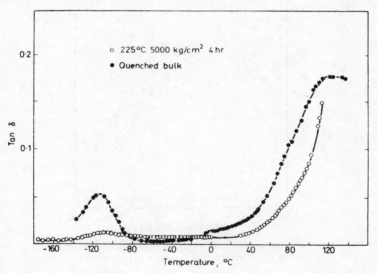

FIGURE 18   Tan δ at 110 Hz vs temperature for the extended chain crystals (○) and
the melt-crystallized sample (●).[1]

The explanation for definite appearance of the tan δ curve in the crystalline
relaxation region for the ECC sample may be attributed to the sensitivity of the
mechanical measurements to the heterogeneity of stress field at the molecular
level; that is, the mechanical energy is selectively dissipated where deformation
occurs owing to the low modulus. Thus, the tan δ curve in the temperature
range from 40 to 100°C for the ECC sample will be attributed to the α_1 relaxation associated with the intermosaic block boundaries in the lamellar crystals

found in the space between the neighbouring ECC bands. In this case, the α_2 relaxation for the ECC sample is expected to be located at the far higher temperature. The last view was confirmed by NMR absorption data.

Figure 19 shows that the narrowing of the line width of broad component of the NMR absorption for the ECC sample occurs at about 100°C or more, which is located far higher than the corresponding temperature for the SCM sample at about 50°C, as shown in Figure 8.[33] Referring to Figure 17, the Grüeneisen value seems to deviate from the theoretical curve at about 100°C.

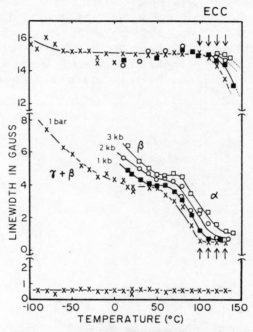

FIGURE 19   Line width of the broad, intermediate and narrow components vs. temperature at various pressures for the extended chain crystals of polyethylene (Takemura *et al.*[33])

By comparing the mechanical data with the NMR absorption data for the ECC sample, it can be concluded that the tan δ curve over the temperature range from 40 to 100°C corresponds to the narrowing of line width of the intermediate component of the NMR absorption in the same temperature range. Takemura *et al.*[33] attributed it to the spherulitic structure generated between the neighboring bands, which they have directly observed at high pressures with a polarization microscope under crossed nicols. This prediction accords with our view.

8) Niinomi and the present author[53–55] have prepared well-developed large

bilayered single crystals by using fractionated linear polyethylene of M_v = 10,000 after Holland.[56] Crystallization was conducted at 70°C from a 0.01% tetrachloroethylene solution. Even for the most regular moiré pattern taken with (110) diffraction beam for as-grown single crystals as shown in Figure 20, the moiré lines are not perfectly straight lines as drawn by using a ruler but slightly wavy. Such a waviness can be reproduced, for example, by superposing the alignments of mosaic blocks of 30 nm edge length riding on a sinusoidal curve with amplitude of 0.05 λ (wave length λ is assumed 100 nm.) Here again, the mosaic structure of lamellar crystal seems to be acceptable.

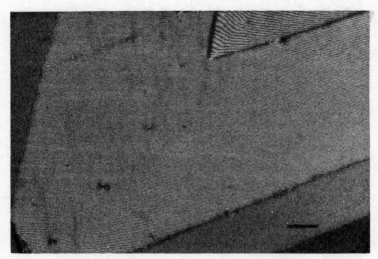

FIGURE 20   Moiré pattern of polyethylene single crystals grown at 70°C, imaged with the (110) diffraction beam.[55] Scale bar is 1 μm.

When we anneal the bilayered crystals having regular interlamellar screw dislocations at a temperature of 85°C, and above the crystallization temperature of the same crystal of 80°C, the dislocation lines are distorted and fragmented as shown in Figure 21.

Figure 22 shows model representation for interpreting the moiré lines and the screw dislocation lines at the interfacial boundary of bilayered crystals. Referring to the origin of dislocation lines, one possible explanation of the distorted or fragmented dislocation lines is that the molecular chains in the intermosaic block boundaries undertake a vigorous micro-Brownian motion and the time rate of change of momentum or force exerted by the molecular segments against the sides of mosaic block crystal push away the block along the direction of the instantaneous net force. Owing to the chains connecting the neighboring mosaic blocks lying along the growth surface, the Brownian motion

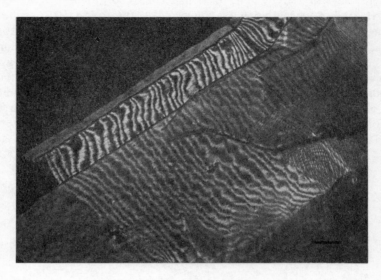

(a)

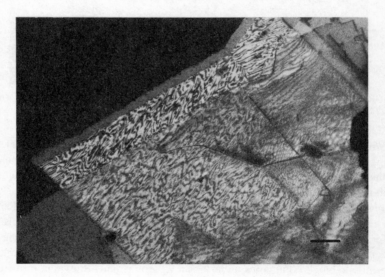

(b)

FIGURE 21   (a) A (110) dark-field image of the interlamellar dislocations of crystals
prepared at 80°C and observed at room temperature. (b) The same area as in (a), observed
at 85°C in a warmed sample stage.[55] Notice that no thickening takes place.

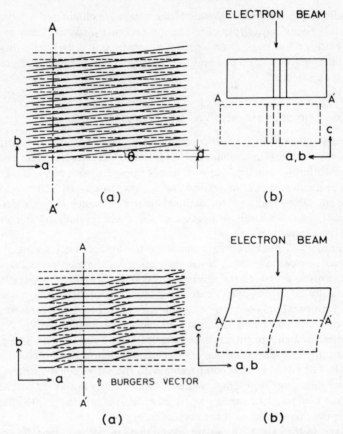

FIGURE 22   Schematic representations of moiré pattern (upper figure) and interlamellar screw dislocation lines at interfacial boundary of bilayered crystals.[55]

of these mosaic blocks will be under some restrictions. By using an ultramicroscope, we can observe the Brownian motion of colloidal particles of gold activated by the net force of thermal motions of water molecules colliding with them. A similar pattern is applicable to explain the distorted pattern of Figure 21 (b).

In conclusion, the crystalline relaxation called the α relaxation can be separated into the α_1 and the α_2 relaxations. The α_1 relaxation takes place on the lower temperature side with a lower activation energy being attributed to the inter-mosaic block region inside a lamellar crystal, but the α_2 relaxation takes place on the higher temperature side with a higher activation energy. The α_2 relaxation is associated with the motions under shear stresses of molecules forming a regular crystal. Once the α_2 relaxation process proceeds to some extent, the characteristics of the inter-mosaic block region are eliminated because distinction between both types of relaxation becomes difficult. Thus, as confirmed by using a hot stage of an electron microscope, the dark field image of electron micrograph is darkened and the contrast is lost as the melting temperature is approached.

1.5   The primary relaxation of single crystal mats

The primary relaxation (the β absorption) is associated with the conformational arrangement of amorphous chains as that for the glass transition temperature. The β absorption for the single crystal mats of polyethylene can scarcely be detected in the mechanical relaxation curve owing to scarcity of loose loops or cilia long enough to make conformational change, although they are detectable for the melt-crystallized sample because of the lower crystallinity in comparison with the former sample.

Nakafuku and Tamekura[33] have measured the motional narrowing of line width of narrow component of NMR absorption for the melt-crystallized sample at various pressures and temperatures, and succeeded in separation of the monotonically narrowing curve, extending from -80 to $20°C$ at atmospheric pressure, into two kinds of narrowing: a low temperature part corresponding to the γ relaxation and a high temperature part corresponding to the β relaxation by measurements in the pressure range of $800-4500$ kg/cm^2. The pressure coefficient of the γ relaxation was $13°C/1$ kb and that of the β relaxation was $31°C/1$ kb. The latter value is comparable with the values for the primary relaxation or the glass transition temperature of some other amorphous polymers, indicating the characteristic sensitivity to pressure in comparison with that for the absorption associated with the crystalline region.

However, in the case of the single crystal mats of polyethylene, Takemura *et al.* could not separate the low temperature narrowing into the β and the γ narrowing. According to our view, the α relaxation should be attributed to the α_1 relaxation as mentioned before. On the other hand, the β relaxation for the as grown single crystal mats cannot be discriminated from the γ relaxation, indicating a lower compressibility of the surface region of the lamellar crystals than that of the amorphous region of the melt-crystallized sample. The surface region of single crystal of polyethylene is in a state of restricted molecular motion compared with that in the melt-crystallized sample.

The most typical example to interpret the relation between the primary mechanical relaxation and the state of molecular chains in the surface region of single crystal is found in the single crystals of *trans*-1, 4-polybutadiene.[57,58] Its tan δ curve shows the peak corresponding to T_g at $-10°C(110$ Hz). Structural analyses give a long period of 10.0 nm, the thickness of crystalline core of 7.2 nm, density crystallinity of 74% and the mobile fraction by NMR absorption of 24%. These data show the existence of loop surface of 1.5 nm thickness on both surfaces of the lamellar crystals, the molecular chains belonging to which are capable of conformational change. If these single crystals are annealed at temperatures above the crystal transformation temperature of 55°C, the primary absorption disappears, the long period increases from 10.0 to 16.0 nm and the narrow component fraction decreases, giving a crystallinity of 91%. These changes are explained by the large contraction of surface loops resulting in tight loops, which are incapable of conformational change associated with the primary absorption. The primary absorption due to the loose loops in the original as-grown single crystal was also largely decreased by γ ray irradiation, which restricts the motions of loose loops by cross-linking.[59] Bromination of double bonds of the loops has the similar effect in making the conformational change energetically difficult. In concluding this section, the necessary condition for appearance of the primary absorption is for some of the main chain atoms to change their spatial positions within a measurable time. Such diffusional motions of atoms are unavoidably accompanied by the frictional force from the surrounding molecular environment.

2 CRYSTALLINE RELAXATION OF POLY AMINO ACID ESTERS

2.6 Outline of the mechanical relaxation of solid polyamino acid esters

There have been reported many investigations on the mechanical relaxations of polyglutamic acid esters.[60,66] In general, two dynamic loss peaks have been observed in the temperature range from −50 to 200°C. The low temperature relaxation around 0 to 20°C has been attributed to the side chain motion and the high temperature one around 100–140°C to the micro-Brownian motion of main chains in distorted conformations.[60,61]

Our detailed inspection of solid state films of poly-γ-methyl-D-glutamate and n-alkyl glutamate cast from the α-helix forming solvents chloroform and dichloroethane (DCE) and from the random coil forming solvent dichloroacetic acid (DCA), revealed the α, β, γ, δ, ϵ and ζ in order of descending temperature.[67] Figure 23 shows the temperature dependence of tan δ at 110 Hz for the solid state films of poly-γ-methyl-D-glutamate (PMDG) cast from the solutions in chloroform (filled circle, curve 1), DCE (open circle, curve 2) and DCA (half-filled circle, curve 3). The data on the solid state film of poly-γ-n-butyl-D-glutamate (PBDG) cast from a solution in chloroform are also shown in Figure 23 (half-filled circle, curve 4).

The mechanisms of these relaxations are assigned as follows. The α relaxation arises from the thermal motions of the α-helices in the crystalline region,

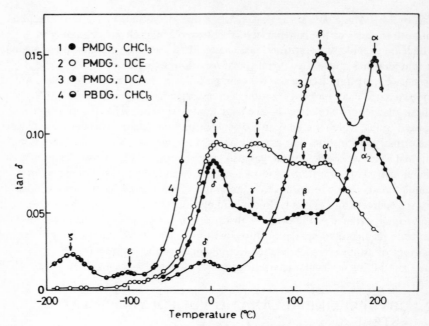

FIGURE 23   Tan δ at 110 Hz vs temperature for the PMDG film cast from the solutions in chloroform (●, curve 1), DCE (○, curve 2) and DCA (◑, curve 3). Curve 4 is for the PBDG film cast from a solution in chloroform. 67).

which is further classified into the α_1 and the α_2 relaxation. The detailed discussion on it will be given in the following sections as the main subject of Part 2.

The β relaxation arises from the micro-Brownian motion of the main chains in the distorted region or in the randomized part of the α-helices. Annealing of the original film of PMDG from a solution in chloroform at 280°C causes a degradation of crystalline component as proved by decrease of intensities of the $(10\bar{1}0)$ x-ray diffraction. With progressing crystal-decomposition, the β peak intensity increases and the α peak intensity decreases. Therefore it can be concluded that the β peak is associated with the amorphous region. By separating the β peak from the α peak by using the annealed samples it becomes possible to measure the frequency dependence of the β peak temperature and to evaluate the activation energy of the β relaxation, which results in the value of 114 kcal/mol. This value is conceivable if the activation energy is to be evaluated with the WLF equation, on the assumption that the distorted region is in an amorphous state.

Another support to this assignment was derived from the remarkable β peak for the PMDG film cast from a random coil- and β-pleated sheet-forming solvent of DCA as shown by curve 3 in Figure 23. The IR band of 650 cm^{-1}, characteristic of the random coil form, was confirmed, The lattice spacing of 1.036 nm characteristic of the α-helix form was not observed. Thus micro-Brownian motion of molecular chains in the disordered region is reasonably considered to be the origin of the β peak.

The γ and the δ relaxations are associated with the motions of entire side chains in the crystalline and the disordered regions, respectively. The ε and the ζ relaxations involve the conformational rearrangements of n-alkyl groups including carbonyl group and of n-alkyl groups alone, respectively. More detailed discussion on these identifications are given in a paper submitted by Kajiyama et al.[67]

2.7   The α_1 crystalline relaxation of polyamino acid esters

Figure 24 shows the temperature dependence of tan δ at 110 Hz for the solid state PMDG film cast from a DCE solution (filled circle), and the same film annealed at 215°C for 30 min. (open circle), together with the corresponding wide-angle x-ray scattering photographs (WAXS). The WAXS of the annealed sample confirms that the molecules in crystal take the α-helical conformation.

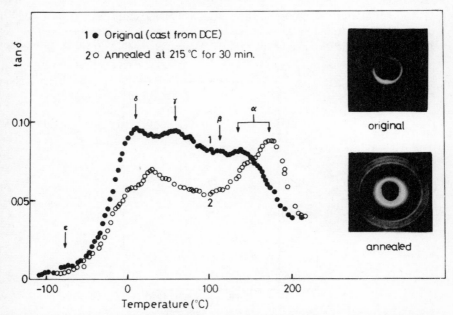

FIGURE 24   Tan δ vs temperature at 110 Hz for the PMDG sample cast from a solution in DCE (●) and the same sample annealed at 215°C for 30 min (○). Corresponding wide angle x-ray photographs are given.[66]

It is noticed that the peak intensity of the α relaxation increases and its peak temperature shifts from 140 to 190°C by annealing. According to the x-ray photographs, the degree of crystallinity clearly increased during annealing at 215°C. The observation that the increase in crystallinity corresponds to increase in the magnitude and the temperature location of the α peak supports the attribution of the α peak to the crystalline relaxation.

This view was further confirmed by the observation that the intensity of the α peak monotonically increased with increasing annealing temperature from 170 to 219°C. Corresponding to the increase of annealing temperature, the sample density, ρ increased from 1.2847 to 1.2892 g/cm^3, the temperature location of the α peak shifted from 141.0 to 170.0°C and the lattice distance of the $(10\bar{1}0)$ plane, $d_{10\bar{1}0}$ varied from 1.095 nm to 1.075 nm, indicating a compaction of packing of α-helices with annealing.

Figure 25 shows the relationship between the height of the α-tan and δ peak and density (open circle), which reveals that a linear relationship between tan δ_{max} and ρ holds for the α peak for the PMDG film cast from DCE solution. Since ρ is directly correlated with the degree of volume-fraction crystallinity, the result in Figure 25 indicates that the α relaxation can be attributed to the thermal motions in the crystalline region, hence appropriately called a crystalline relaxation.

Another interesting feature of the α peak of PMDG sample cast from a solution in DCE was found in the relationship between the α peak temperature, $T_{\tan\delta,max}$, and the distance of the $(10\bar{1}0)$ plane $d_{10\bar{1}0}$. Figure 26 shows this

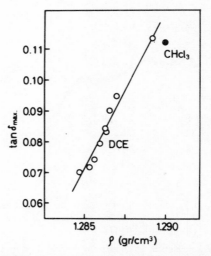

FIGURE 25   The peak height of tan δ at 110 Hz vs density for the PMDG samples cast from a solution in DCE (O) and from a solution in chloroform (●).[66]

relationship for the PMDG sample cast from a solution in DCE annealed at various temperatures (open circles). As the crystal form is hexagonal, the inter-helix distance between two neighboring α-helices is $(2/\sqrt{3})\,d_{10\overline{1}0}$. The temperature of absorption maximum of loss modulus, T, is related to the angular frequency of dynamic measurement ω and the average relaxation time as a function of temperature, $\bar{\tau}(T)$, by Eq. (6).

$$\omega\,\bar{\tau}(T) = 1. \tag{6}$$

Therefore, the fact that the temperature of the α absorption maximum shifts to the higher temperature with decreasing $d_{10\overline{1}0}$ value means that the relaxation time associated with the α relaxation mechanisms becomes longer with decreasing α-helix distance, although in Figure 26 the temperature of tan δ peak was used instead of the loss modulus. However, this does not affect the qualitative story.

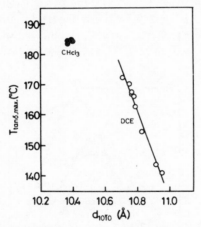

FIGURE 26   The α peak temperature of tan δ at 110 Hz vs the lattice distance of the $(10\overline{1}0)$ plane for the PMDG samples cast from a solution in DCE (O) and a solution in chloroform (●).[66]

In order to explain these experimental facts, we are inclined to suppose a mechanism of α-relaxation where the α-helix chains undertake a mutual slip or rotational motion in the crystalline region. According to this model, the decrease of the $d_{10\overline{1}0}$ value means that the α-helices embedded in the matrix of side chains are packed more closely, and the density of matrix is increased and hence the effective dynamic viscosity is increased. Thus, the migrational motion of α-helix along the helix axis or the rotational motion of α-helix around its axis as a rigid rotator encounters a larger frictional resistance from the surrounding molecular environments. This causes the shift of the tan δ α-peak to higher temperatures. At the same time, the increased internal viscosity in the interhelical space causes the energy loss accompanying the mutual slip or rotation of the α-helix to increase. This is a qualitative interpretation of the α_1 relaxation. Since in a later part another type of a relaxation called the α_2 relaxation will be mentioned, the α relaxation in this section will be designated as the α_1 relaxation. Kajiyama et al.[66] calculated the internal loss based on simple models called the "coaxial vibrating model" and the "torsionally oscillating model", using the data on the packing geometries of α-helix. With some adjusting of parameters, the relationship between the tan δ value and the interhelix distance was predicted. By changing the length of α-helix, we will be able to make a more detailed discussion of this mechanism.

The α_1 relaxation mechanisms proposed in the above section can be further confirmed by the anisotropy found in the tan δ vs temperature curves relating the measuring direction against the stretching direction. The PMDG sample cast from a solution in DCE was stretched to about twice the original length at 100°C and then annealed in vacuo at 219°C for 1 h. Figure 27 shows that the intensity of the α_1-tan δ peak for a specimen cut at 45° to the uniaxially-oriented direction is larger than those cut at 0° and 90°. This result can be rationally explained in

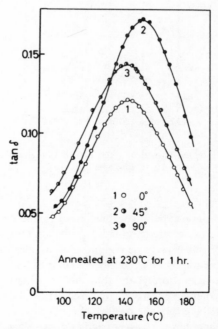

FIGURE 27   Tan δ vs temperature at 110 Hz in the three directions of 0°, 45° and 90° to the drawing direction for the oriented PMDG sample prepared from the film cast from DCE solution.[66]

that the shear stress, τ, exerted on a plane at an angle θ to the direction of uniaxially applied stress σ, is $\sigma \sin \theta \cos \theta = (\sin 2\theta)/2$, according to which the maximum shear stress is expected to occur at 45° to the stretching direction. If the interhelical matrix composed of side chains is to be sheared along the helix axis, the magnitude of the tan δ peak for the sample in which the α-helices are aligned at 45° to the dynamic strain direction, tan δ_{45}, is expected to be larger than tan δ_0 or tan δ_{90}.

In conclusion, the α_1 process is mainly activated by shear along the direction of the α-helix axis, and thermal motions such as coaxial vibration or torsional oscillation of the α-helices may be the molecular process of the α_1 relaxation. It should be noticed that the α_1 relaxation is surely the main contributor to the whole α relaxation, but it does not necessarily mean the exclusion of intrahelical α_2 relaxation processes, which will be mentioned in the next section.

2.8   The α_2 crystalline relaxation of polyamino acid esters

Figure 28 shows the temperature dependence of tan δ for the PMDG film cast from a solution in chloroform (filled circle) and the same film annealed at 215°C for 30 min (open circle). Chloroform is confirmed again to be an α-helix forming solvent (as DCE) from the x-ray diffraction pattern shown in Figure 28. However, there are striking differences depending upon the kind of α-helix forming solvents in that (1) the PMDG film cast from a solution in chloroform shows a sharp α-tan δ peak and well-developed crystallinity before annealing, and (2) the effect of annealing is not remarkable when comparison is made between the original (open circle) and the annealed sample (filled circle).

The density of the PMDG film cast from a solution in chloroform is 1.290 g/cm^3, as indicated by filled circle in Figure 25, which is higher than or comparable to the highest density value of the annealed PMDG film cast from a solution in DCE. The WAXS pattern for the original sample cited in Figure 28 is very sharp, being consistent with its high density value. It was also found that the distance of the $(10\bar{1}0)$ plane, corresponding to the back bone separation of the α-helix, thermally expands linearly with temperature and shows a break point at 145–150°C, above which the data deviate upwards from the linear relation. This temperature corresponds to the temperature range where the α-tan δ curve starts to increase. Such a correspondence suggests that the α peak is the manifestation of molecular motions of the α-helices in the crystalline region.

Another support to this view was obtained by the observation that both the x-ray diffraction intensity from the $(10\bar{1}0)$ plane and the α-tan δ peak intensity decrease with time at 280°C, at which temperature the sample was annealed and the crystalline component was gradually decomposed, leaving no indication of crystalline diffraction after 120 min and disappearance of the α-tan δ peak.

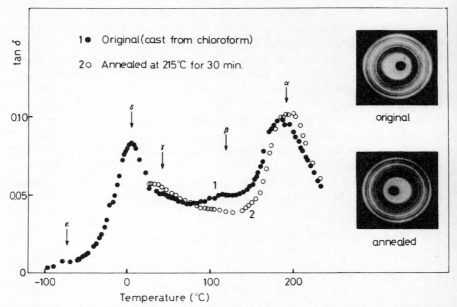

FIGURE 28   Tan δ vs temperature at 110 Hz for the PMDG film cast from a solution in chloroform (●) and the same film annealed at 215°C for 30 min (○), and the corresponding wide angle x-ray photographs.[66]

This means that the α peak is closely related to the fraction of the crystalline component.

In Figure 26, the temperature of the tan δ peak for the PMDG film cast from a solution in chloroform and the annealed films of it are plotted against the lattice distance of $(10\bar{1}0)$ plane, $d_{10\bar{1}0}$. As the interhelix distance is $(2/\sqrt{3})\,d_{10\bar{1}0}$, it can be concluded that the PMDG film cast from a solution in chloroform is composed of more closely packed α-helices in comparison with the one cast from a solution in DCE and its crystalline structure is scarcely affected by annealing at 215°C. It is noticed that the temperature of the α-tan δ peak of these samples do not show the linear relationship as that for the PMDG film cast from a solution in DCE and annealed. These different tendencies found in the $T_{\tan \delta, \max}$ vs $d_{10\bar{1}0}$ plots lead to the postulate that the primary mechanism of the α relaxation may arise from different kinds of molecular process for the PMDG films cast from solutions in DCE and chloroform. The activation energy for the DCE sample was about 105 kcal/mol, whereas that for the sample from chloroform was 186 kcal/mol. Thus, the α peak for the sample from chloroform is considered to be the peak associated with the α_2 relaxation process, being differentiated from the α_1 relaxation assigned to the peak of the sample from DCE.

In order to clarify the feature of the α_2 relaxation, the anisotropy of mechanical loss for the uniaxially drawn samples of the film cast from a solution in chloroform was measured. Figure 29 shows that the intensity of the α_2-tan δ peak decreases in the order of 0°, 45° and 90° where the angle is taken between the tensile stress direction and the α-helix axis direction. This trend exhibits a striking difference from that found for the sample cast from a solution in DCE shown in Figure 27, which exhibits the anisotropy of tan δ_{45} > tan δ_0 > tan δ_{90}. Therefore, the mechanism assigned to the α_1 relaxation is not appropriate to that for the α_2 relaxation. A more reasonable mechanism for the α_2 relaxation may be the thermal molecular motions of the α-helices in the crystalline region which are activated parallel to the α-helix axis. Accordion-like oscillation of α-helix can be considered as one of the corresponding mechanisms. Anharmonic motions of this type will result in some variations in the values of parameters relating the x-ray crystal structure.

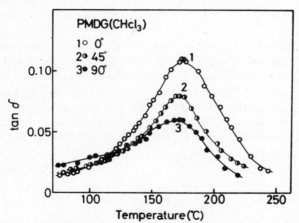

FIGURE 29   Tan δ vs temperature for the oriented film of the PMDG sample cast from a solution in chloroform in directions of α-helix axis taking angles of 0°, 45° and 90° to the stretching direction.[66]

Figure 30 shows a plot of the x-ray diffraction intensity of the $(10\bar{1}5)$ plane vs measuring temperature for the sample cast from a solution in chloroform and annealed at 215°C for 30 min. The (1015) reflection was chosen because it is strong and nearly perpendicular to the helix axis, and its intensity variation will reflect the intramolecular regularity of the α-helix. Figure 30 shows that the relative x-ray intensity from the (1015) plane falls abruptly with increasing measuring temperature, especially above 120°C, and disappears around 170°C. Since the decrease of x-ray intensity in this case cannot be explained by the temperature factor, it seems reasonable to consider that the striking decrease

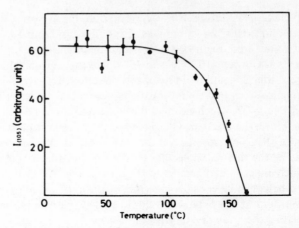

FIGURE 30   X-ray diffraction intensity of the (10$\bar{1}$5) plane vs temperature for the PMDG
sample cast from a solution in chloroform and then annealed at 215°C for 30 min.[66]

of diffraction intensity is caused by vanishing of the regular periodicity of the
helix pitch. In contrast to disappearance of the (10$\bar{1}$5) diffraction, the
equatorial reflection still appears with fairly strong intensity above 170°C. The
x-ray studies indicate that the α-helices can not maintain their regular periodicity
of pitch above 170°C. Thus, it seems that the α_2 relaxation process is associated
with extensive or contractive deformation of the α-helix along the helix axis,
maintaining the periodic interhelix distance. Accordion-like anharmonic oscilla-

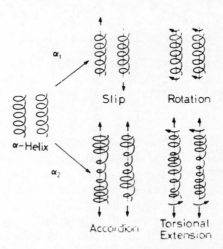

FIGURE 31   Schematic representation of molecular motions for the α_1 and the α_2 relaxation mechanisms.

tion or torsional extension motion are conceivable as its modes of molecular motion.

The main differences between the α_1 and the α_2 relaxations are briefly summarized by mentioning that the α_1 relaxation is concerned with the interhelical relaxation process. For simplicity of explanation, the α_1 and the α_2 relaxations were separately assigned to the α relaxations of the samples cast from the solutions in DCE and chloroform, respectively. However, in the reality, it is conceivable that both relaxation processes will make contributions at the same time to the total α relaxation but with different weights. For example, in the case of the sample cast from chloroform, the interhelix distance is very close and the packing of the side chains is so large that the resisting force for mutual slip overcomes the force delivered to deformation of α-helices and mainly the latter process takes place.

Figure 31 gives the schematic representation of the α_1 relaxation mechanisms and the α_2 relaxation mechanisms. These models may stimulate to develop a mathematically elaborated molecular theory for explanation of the α_1 and the α_2 crystalline relaxation mechanisms in the future.

Acknowledgment

The author expresses his sincere gratitude to Prof. T. Takemura of Kyushu University for kindly presenting the original figures of the NMR absorption and to the collaborators in my laboratories, especially Drs. Y. Kajiyama and T. Kijima, for their presenting fruitful data in their own works.

References

1. M. Takayanagi, *Pure Appl. Chem.* **23**, 151 (1970).
2. M. Takayanagi, M. Yoshino and S. Minami, *J. Polymer Sci.* **61**, S7 (1962).
3. M. Takayanagi and N. Kawasaki, *J. Macromol. Sci.-Phys. B* **1**, 741 (1967).
4. K. -H. Illers, *Kolloid-Z. Z. Polym.* **231**, 622 (1969).
5. K. Bergman and K. Nawotki, *Kolloid-Z. Z. Polym.* **219**, 132 (1967); **231**, 650 (1969).
6. W. Pechhold, *Kolloid-Z. Z. Polym.* **241**, 955 (1970).
7. M. Matsui, R. Masui and Y. Wada, *Polymer J.* **2**, 134 (1971).
8. J. D. Hoffman, G. Williams and E. Passaglia, *J Polymer Sci. C* **14**, 173 (1966).
9. Y. S. Papir and E. Baer, *J. Appl. Phys.* **42**, 4667 (1971).
10. K. Schmieder and K. Wolf, *Kolloid-Z. Z. Polym.* **134**, 157 (1953).
11. M. Takayanagi, T. Aramaki, M. Yoshino and K. Hoashi, *J. Polymer Sci.* **46**, 531 (1960).
12. M. Takayanagi, *Proc. 4th Intern. Congr. Rheology*, 1963, Rhode Island, Part 1 (Interscience, 1965) p. 161.
13. T. Kajiyama, T. Okada, A. Sakoda and M. Takayanagi, *J. Macromol. Sci.-Phys. B* **7**, 583 (1973).
14. K. M. Sinnott, *J. Appl. Phys.* **37**, 3385 (1966); *J. Polymer Sci. C* **14**, 141 (1966).
15. M. Takayanagi and T. Matsuo, *J. Macromol. Sci. -Phys. B* **1**, 407 (1967).
16. M. Takayanagi, *Kobunshi* (High Polymers, in Japanese) **14**, 314 (1965); *J. Soc. Materials Sci., Japan* **14**, 314 (1965);
17. J. A. Sauer, G. C. Richardson and D. R. Morrow, *J. Macromol. Sci.-Revs. Macromol. Chem. C* **9**, 149 (1973).

18. M. Takayanagi, K. Imada and T. Kajiyama, *J. Polymer Sci. C* 15, 263 (1966).
19. M. Takayanagi and T. Kajiyama, *J. Macromol. Sci. Phys. B* 9, 391 (1974).
20. M. Takayanagi, *J. Macromol. Sci.-Phys. B* 9, 391 (1974).
21. Y. Wada and K. Tsurge, *Japanese J. Appl. Phys.* 1, 64 (1962).
22. H. Nakayasu, M. Markovitz and D. J. Plazek, *Trans. Soc. Rheology* 5, 261 (1961).
23. S. Iwayanagi, in *Solid State Physics*, Vol. 14, edited by Seitz-Turnbull (Academic Press, New York, 1963), p. 458.
24. S. Manabe, A. Sakoda, A. Katada and M. Takayanagi, *Mem. Fac. Engng. Kyushu Univ.* 28, 279 (1969); *J. Macromol. Sci.-Phys. B* 4, 161 (1970).
25. M. Takayanagi, *J. Ind. Chem.* 73, 1277 (1970).
26. T. Kajiyama, T. Okada, A. Sakoda and M. Takayanagi, *J. Macromol. Sci.-Phys. B* 7, 583 (1973).
27. T. Kajiyama, T. Okada and M. Takayanagi, *J. Macromol. Sci.-Phys. B* 9, 35 (1974).
28. T. Kajiyama and M. Takayanagi, *J. Macromol. Sci.-Phys. B* 10, 131 (1974).
29. N. Kusumoto, T. Yamamoto and M. Takayanagi, *J. Polymer Sci. A-2* 9, 1173 (1971).
30. W. P. Slichter, *J. Appl. Phys.* 31, 1965 (1960); 32, 2239 (1961).
31. A. Odajima, J. A. Sauer and A. E. Woodward, *J. Chem Phys.* 66, 718 (1962).
32. H. G. Olf and A. Peterlin, *Kolloid-Z. Z. Polym.* 215, 97 (1967).
33. C. Nakafuku and T. Takemura, *Mem. Fac. Engng. Kyushu Univ.* 34, 11 (1974).
34. S. Iwayanagi and I. Miura, *Japanese J. Appl. Phys.* 4, 94 (1965).
35. W. Pechhold, *Kolloid-Z. Z. Polym.* 228, 1 (1968).
36. K. Okano, *Rep. Inst. Phys, Chem. Res. (Tokyo)* (in Japanese) 40, 295 (1964).
37. Y. Wada and R. Hayakawa, *Progr. Polymer Sci., Japan* 3, 215 (1972).
38. K. Iohara, K. Imada and M. Takayanagi, *Polymer J.* 3, 357 (1972).
39. T. Kitagawa and M. Miyazawa, *Preprint of the 16th Polymer Symposium*, Oct. 1967, p. 230.
40. K. Iohara, K. Imada and M. Takayanagi, *Polymer J.* 4, 232 (1973).
41. R. L. McCullough, *J. Macromol. Sci. Phys. B* 9, 97 (1974).
42. N. Yoshioka, Y. Obata and H. Kawai, *J. Macromol Sci.-Phys. B* 1, 567 (1967).
43. Y. Wada. A. Itani, T. Nishi and S. Nagai, *J. Polymer Sci. A* 7, 201 (1969).
44. R. E. Barker and R. Y. S. Chen, *J. Chem. Phys.* 53, 2615 (1971).
45. T. Ito and H. Marui, *Polymer J.* 2, 768 (1972).
46. C. K. Wu, G. Jura and M. Shen, *J. Appl. Phys.* 43, 4348 (1972).
47. C. K. Wu and M. Nicol, *Bull. Amer. Phys. Soc.* 17, 274 (1972).
48. T. Kijima, T. Koga, T. Yoshizumi, K. Imada and M. Takayanagi, submitted to *Proc. 4th Intern. Conf. High Pressure*, Kyoto (1974).
49. M. G. Broadhurst and F. I. Mopsik, *J. Chem. Phys.* 52, 3634 (1970).
50. T. Kijima, K. Koga, K. Imada and M. Takayanagi, *Polymer J.*, in press.
51. M. G. Broadhurst and F. I. Mopsik, *J. Chem. Phys.* 54, 4239 (1971).
52. M. Takayanagi, *Pure Appl. Chem.* 23, 151 (1970).
53. M. Niinomi, K. Abe and M. Takayanagi, *J. Macromol. Sci. B* 2, 649 (1968).
54. K. Abe, M. Niinomi and M. Takayanagi, *J. Macromol. Sci. B* 4, 87 (1970).
55. M. Niinomi and M. Takayanagi, *Progress Polymer Sci. Japan* (Kodansha-Scientific, 1971) p. 199.
56. V. F. Holland, *J. Appl. Phys.* 35, 3235 (1964).
57. T. Tatsumi, T. Fukushima, K. Imada and M. Takayanagi, *J. Macromol. Sci.-Phys. B* 1, 459 (1967).
58. M. Takayanagi, *Pure Appl. Chem.* 15, 555 (1967).
59. T. Mori, K. Iohara and M. Takayanagi, *Rep. Prog. Polymer Phys. Japan* 11, 307 (1968).
60. Y. Hashino, M. Yoshino and K. Nagamatsu, *Rep. Prog. Polymer Phys. Japan* 9, 297 (1966).
61. A. Tsutsumi, K. Hikichi, T. Takahashi, Y. Yamashita, N. Matsushima, M. Kanake and M. Kaneko, *J. Macromol. Sci.-Phys. B* 8, 413 (1973).
62. A. L. Nguyen, B. T. Vu and G. L. Wilkes, *J. Macromol. Sci.-Phys. B* 9, 367 (1974).
63. M. Ichikawa, R. Sakamoto, Y. Abe and K. Makishima, *Kobunshi-Kagaku* (in Japanese) 30, 346 (1973).

64. M. Kuroishi, T. Kajiyama and M. Takayanagi, *Chemistry Letters* 659 (1973).
65. M. Kuroishi, T. Kajiyama and M. Takayanagi, *Rep. Prog. Polymer Phys. Japan* 16, 641 (1973).
66. T. Kajiyama, M. Kuroishi and M. Takayanagi, *J. Macromol. Sci.-Phys.*, in press.
67. T. Kajiyama, M. Kuroishi and M. Takayanagi, submitted to *J. Macromol. Sci.-Phys.*

DISCUSSION

R. H. Boyd (*Univ. of Utah*): You showed some results for high pressure crystallized poly(ethylene) (extended chain) in which the crystal thickness is presumably extremely large. The alpha relaxation is quite prominent and the γ very weak. What are the implications of this in terms of participation of interlamellar regions?

M. Takayanagi: Figure 18 shows the comparison of mechanical relaxation curves represented by tan δ as a function of temperature between the extended chain crystals (ECC; open circles) and the usual melt-crystallized sample (filled circles). As you indicate, the α relaxation is prominent and the γ relaxation very weak for the ECC sample. However, the magnitudes of both the α and the γ peaks of EEC are far smaller than those for the folded chain crystals (FCC). This means that the contribution from lamellar crystal region included in the crystalline texture of ECC sample is still detectable. Referring to the motional narrowing of the broad component of ECC sample (cf. Figure 19), the mechanical relaxation of extended chain crystal itself is expected to be located above 100°C and its intensity should be prominent if the measurement is actually possible. The observed tan δ curve of the α relaxation in Figure 18 is originated from the lamellar crystals generated in the space between neighboring bands composed of extended chains. The very weak γ relaxation is associated with largely decreased fraction of frozen interlamellar region and very few crystal defects in the ECC sample. According to my view, an ideally perfect ECC crystal itself is absent in the γ relaxation and the α_2 relaxation takes place above 100°C, being prominent at the temperatures very close to the melting temperature. A strong support to this view can be found in the comparison of tan δ curves of the solid-state polymerized polytetraoxane samples with that of single crystal mats of polyoxymethylene as shown in Figure 34 of a paper by the same author appearing in *Pure and Applied Chemistry*, 15, 584 (1967). The former sample does not include any folded chain lamellar crystals, but is only composed of extended chain crystals. Quantitative correlation of tan δ values of ECC sample to the small fraction of lamellar crystals included in the ECC sample of poly(ethylene) constitutes a difficult problem to solve owing to the heterogeneous stress field generated in the crystalline texture when the specimen is strained. According to my view a contribution of interlamellar region with conformational change of amorphous chains should appear in the β relaxation process which appears at about $-20°C$ (110 Hz). The α relaxation in Figure 1 is, according to my view, mainly the α_1 process associated with intermosaic block boundaries included in lamellar crystals.

M. Shinohara (*Dow Corning Corp.*): You pointed out the importance of morphology on the local motion. Have you ever studied or observed that solid structure (i.e., supermolecular structure) affects the temperature where local motion starts to take place, particularly in copolymeric materials?

M. Takayanagi: Side chain motion of isotactic poly(4-methyl pentene-1) (P4MPI) seems to be affected by the molecular environment, which is in close relation to the supermolecular structure. The loss modulus peak associated with side chain motion of atactic P4MP1 is found at $-150°C$ at 110 Hz, whereas the same absorption of isotactic P4MP1 is splitted into two: one is located at $-150°C$, but the other takes place at about $-180°C$ (M. Takayanagi and N. Kawasaki, *J. Macromol. Sci.-Phys.* **B1** (4), 741 (1967). Our view is that the latter absorption is split by partially changed molecular environments from amorphous glassy state to the crystalline state in the isotactic sample. In the case of copolymeric materials, they usually are in an amorphous state and, therefore, it is difficult to define the change of molecular environments. Depending on the degree of separation of phases of both components, there is expected a gradual change of molecular environments. Even in such a case, the energetical contributions from conformational (i.e., intramolecular) factor are still important, the weight of which to intermolecular factor may depend on the polymer species. As far as concerned with the primary relaxation, a detailed discussion was made in a paper by the author, S. Manabe and R. Murakami in *Int. J. Polymeric Mater.*, **1**, 47 (1971).

M. Shinohara: Annealing of poly (amino acid ester) samples caused close packing of helices with a decreasing inter-helical distance according to your data. Annealing also depresses lattice defects. Are you saying that both of these effects take place during annealing?

M. Takayanagi: Yes, x-ray diffraction pattern of the PMDG film cast from a solution in dichloroethane is remarkably sharpened by annealing at $215°C$, as shown in Figure 24. This means that the crystalline regularity is largely improved and the crystal defects have been removed. Our discussion on the molecular process of the α relaxation is based on this fact, although we can not construct at present a definite model of a crystal defect. Annealing of the same sample at $280°C$ gradually destroys the crystalline region and increases the distorted region, which gives rise to the increase of the tan δ peak of the β relaxation located at about $100°C$ and the decrease of the α peak at about $180°C$.

R. F. Boyer (*Dow Chemical Company*): Should helical vinyl polymers, such as isotactic poly(propylene) or isotactic poly(styrene), show loss mechanicals similar to those arising from the helical structure of γ-n-alkyl glutamate polymers?

M. Takayanagi: In principle, yes. But the helical conformation of isotactic poly-α-olefins is not stabilized by hydrogen-bonds as in the case of the helical structure of poly (amino acid esters). With increasing temperature, the energetically unstable helix of poly-α-olefins more easily tends to melt rather than to display the distinctive behavior such as the displacement of helix or its accordion-like motion. Thinning of a lamellar crystal composed of folded-chain polymers during deformation and detected by the small angle x-ray scattering is caused by the mutual slip of molecular segments of folded-chain length along the molecular axis. Such behavior well corresponds with the α_1 relaxation process associated with the mutual slip of α-helices in poly (amino acid esters). Thickening of lamellar crystals may include the contribution from the intramolecular modes of motion such as long range accordion-like motion. All crystalline polymers including helical vinyl polymers show crystalline relaxation when solution-grown samples with high crystallinity are employed for measurements. More detailed analyses of anharmonic molecular oscillation in crystals of these polymers may clarify the common points with the α_1 and the α_2 relaxation processes of poly(amino acid esters).

S. L. Aggarwal (*General Tire and Rubber Co.*): You discussed two types of phenomena associated with transitions in crystalline poly(ethylene): (1) inter-lamellar deformation; (2) motions associated with defects in the crystal. You showed the transitions associated with interlamellar deformation. What transitions are associated with defects in the crystals and movements within the crystals of poly(ethylene)?

M. Takayanagi: Relaxation associated with defects in the crystal is called the α_1 relaxation in my lecture and that within the crystal of poly(ethylene) is called the α_2 relaxation. As for the crystal defects related to the α_1 relaxation, we have considered the mosaic block boundaries with mono- or bimolecular layer thickness. The thermal motion of molecular segments with distorted conformation in the mosaic block boundaries takes place more easily and the transition is found at lower temperature or higher frequency sides of the crystalline relaxation in response to the external excitation. Experimental supports to this view have been cited in my article and are not mentioned here again. The α_2 relaxation is concerned with the rotational motion of molecular segments of fold-chain length around their axes and the translational motion along with the molecular axes. Such molecular motions are more difficult to take place than those associated with the α_1 relaxation because the molecular environment in regular crystal lattice more rigidly binds the molecules than that in the inter-mosaic boundaries. Thus, the α_2 relaxation is found at higher temperature or lower frequency side than the α_1 relaxation.

It should be noticed that there exist many types of crystal defects. Edge dis-locations are directly confirmed and the others are only suspected to exist in crystal lattice. Migration of point dislocation along the molecular axis or movement of dislocation line might be the origin of another relaxation process. Interlamellar deformation is related to the β relaxation process, which is called the primary relaxation and corresponds with T_g.

E. Baer (*Case Western Reserve Univ.*): Have you examined any low molecular weight polymers in order to identify the associated motions in short chains?

M. Takayanagi: Your idea of using low molecular weight polymer to identify the associated motions is stimulative. Actually, relaxation measurements were already made on a series of n-paraffine to identify the crystalline relaxations of poly(ethylene). Systematic variations of crystalline relaxation temperature and its activation energy as a function of chain length were observed and theoretical interpretation of the results were tried by several authors, for example, Dr. J. D. Hoffman. Similar measurements are desired on a series of poly(amino acid ester) to confirm and clarify the molecular mechanisms of the α_1 and the α_2 relaxa-tions of poly(amino acid esters) proposed in my lecture. If the tensile measure-ments are adopted as ours, operational difficulties are encountered owing to the brittleness of the sample. We have tried the dynamic measurements on various pure amino acid crystals by embedding them in low loss matrix of polystyrene to identify the side chain motions and succeeded in comparison of the dynamic data with the motional narrowing of the wide line NMR absorp-tion.

J. Sauer (*Rutgers Univ.*): In his NMR studies as a function of pressure, does Takemura see evidence for both the α_1 and α_2 processes? If so, has he determined the pressure shift of these processes?

M. Takayanagi: According to Takemura, the motional narrowing of the broad component found for the sample of extended chain crystal (ECC) above 100°C and for the sample of single crystal mats (SC) above 50°C corresponds to the α_2 relaxation, whereas the motional narrowings of the narrow component for both samples were assumed to be associated with the surface region of lamellar crystals which exist even in the crystalline texture of the ECC sample. According to the author's view, the latter process should be ascribed to the β relaxation for both samples, that is, the interlamellar slip, and the motional narrowing of the narrow component in the same temperature region as that of the α_2 absorption is considered to be associated with the α_1 relaxation concerning the inter-mosaic block boundaries. Several supports to this view were mentioned in the text and need not be cited here. The pressure coefficients of shift of both broad and narrow components of the α_2 and the α_1 absorption were equally about $10°C/kb$. This fact suggests that these narrowing processes are originated by the crystalline nature, because the pressure coefficient of the primary absorption related to T_g for the amorphous polymer is $30°C/kb$, which is far larger than $10°C/kb$.

Polyethylene: Detailed Interpretation of Mechanical Relaxation in a Crystalline Polymer

N. G. McCRUM

Department of Engineering Science, Oxford University

1 INTRODUCTION

It is now ten years since Bryan Read, Graham Williams and I drafted our book on mechanical relaxation. It seemed right to take a broad view of the many hypotheses and theories current at that time. I hope that we were not ambiguous but there were real difficulties in such a new field in taking sides in a logical way. So we described, for instance, both Hoffman[1] and Fröhlich[2] mechanisms for distribution of relaxation times; both the crank shaft[3,4] and the local mode[5,6] hypotheses for γ relaxations in linear polymers; both proposals for the glass transition of linear polyethylene; and both proposals for the structure of linear crystalline polymers, the older crystalline—amorphous model and the newer but short lived crystal defect model with no amorphous fraction. And of all the fourteen chapters, that on polyethylene was the quintessence of our "on the one hand this and on the other hand that" style. Ten years later and particularly at this symposium in honor of Ray Boyer would it be reckless to attempt to put all that right?

Reckless or not it is certainly what I have been invited here to do and it is certainly appropriate. Ray Boyer has never avoided the task of synthesis, of bringing together facts and formulating the best working hypothesis and in the discharge of this duty he has maintained a scrupulous generosity towards views not coinciding with his own. We are all very much in his debt both in this respect and also for the warmth and good sense he brings to our gatherings.

There are good reasons why a synthesis for the linear crystalline polymers should be attempted at this moment in time. In the last ten years the number of new techniques and theoretical developments has been formidable.

167

1) Important advances in the theory of composite materials.

2) Calculation of the elastic stiffness matrix for crystalline polymers.

3) New experiments on oriented specimens with three dimensional single crystal texture.

4) Equipment advances so that pole figures are obtained routinely for oriented specimens using automatic diffractometers and an off-the-shelf sophisticated mechanical spectrometer from Japan.

5) Developments in neutron inelastic spectroscopy.

6) Experiments on specimens formed from single crystals crystallized from solution.

In addition the older allied techniques, NMR, dielectric relaxation and dilatometry have been applied with increasing theoretical sophistication and experimental precision.

I do not intend to concentrate on any new development no matter how sophisticated unless in my view it has contributed to our understanding of mechanical relaxation and phase transitions. I cannot also, partly for this reason, and partly because the work lies outside the scope of my paper deal with crystal defects both line and point. This is not to say that they may not indeed in time be shown to contribute important and even dominant characteristics to the mechanical properties of crystalline polymers. But to date this is not so.

Figure 1 shows mechanical relaxation data on branched and linear polyethylene.[7] BPE exhibits three well resolved relaxations α, β and γ. LPE exhibits two alpha relaxations (in this terminology α and α') a γ relaxation and some evidence for a β relaxation. Now there are close parallels between this pattern of behaviour and that exhibited by polyoxymethylene and polytetrafluoroethylene.[8] The principal characteristic they hold in common is that of the three relaxation ranges in each polymer there is not one which clearly and unambiguously can be termed the glass transition. Another characteristic in common is that for each polymer the concept of amorphous fraction is an indispensable prerequisite for intelligent discussion yet the pure amorphous forms cannot be prepared. The properties of the amorphous polymer have therefore to be obtained by extrapolation. And it must be kept in mind that the extrapolation yields partial information in the thermodynamic sense: it yields information about an amorphous phase finely dispersed in a crystalline matrix. It seemed advantageous therefore, and indeed I have been specifically requested, to include POM and PTFE in this review. I shall confine my discussion to LPE, BPE, POM and PTFE drawing upon work on BPE, POM and PTFE to illuminate the model polymer, which I take to be LPE.

Section 3 is a review of the experimental facts. It is a systematic collection of facts both new and old which appear significant to me. It is preceded in Section 2 by a discussion of experimental matters frequently overlooked or imperfectly understood. The evidence is discussed in Section 4.

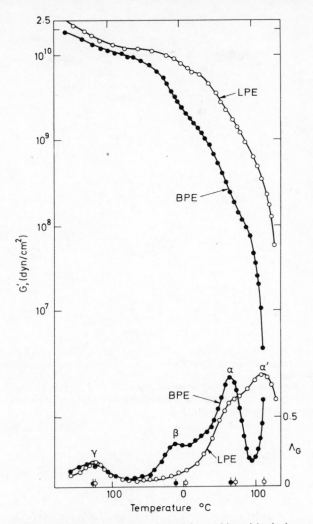

FIGURE 1   Temperature dependence of shear modulus and logarithmic decrement of linear (LPE) and branched (BPE) polyethylene. (After Flocke.[7])

2   EXPERIMENTAL PRINCIPLES

Of the many aspects of experimental technique three only can be touched on here because they have a bearing on our conclusions; (a) experimental artefacts due to thermoelastic stress relaxation produced by a fast rate of cooling, (b)

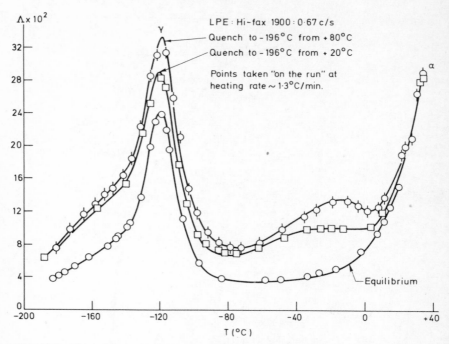

FIGURE 2   Temperature dependence of logarithmic decrement Λ at 0.67 c/s for linear
polyethylene. Equilibrium curve: points taken at successively lower temperatures, cooling
slowly and maintaining the specimen at temperature for *ca* 15 min before measurement.
The other two curves were obtained at successively higher temperatures at a constant heating
rate after the specimen had been cooled quickly to liquid nitrogen temperature from the
indicated temperatures (After Cooper.[9])

failure of time–temperature equivalence, (c) selection of the correct viscoelastic
model.

a) Figure 2 shows Coopers[9] study of the effect of rate of heating on the
temperature dependence of Λ for LPE. The equilibrium data is taken with the
specimen cooled slowly from point to point and left at temperature for *ca* 15
min before measurement. Now if the specimen is quenched rapidly to liquid
nitrogen temperature and warmed up at a constant rate the other two curves are
obtained. It will be seen that these experimental procedures modulate the weight
of the γ peak and particularly the low temperature shoulder and insert what looks
like a β peak. The effect is probably due to the freezing in of thermoelastic stresses
on cooling.[10] Many of the discrepancies in the literature can be attributed to
experimental artefacts of this type. Some workers cool slowly from point to point,
others heat having cooled slowly (and sometimes having cooled quickly). The best

method is certainly to cool slowly from point to point because the data is reproducible and time—temperature equivalence is most likely to hold. The whole point of the temperature scan experiment at constant frequency is of course to slow down or speed up the viscoelastic relaxation processes. Temperature should ideally cause no other effect and under this circumstance time and temperature will be interchangeable (apart from small changes in the limiting compliances).

b) When making a temperature scan it is important that the specimen be unaffected by the temperature in an irreversible way. It must not partially melt or anneal or change in orientation. Branched polyethylene pre-melts at temperatures far removed from the melting point: in my view data taken above room temperature is suspect in BPE unless reversibility can be proved. Specimens should invariably be annealed if experiments are to be performed above room temperature. Oriented specimens are notoriously fickle, even when annealed.[11] If the specimen changes irreversibly during the temperature scan then time—temperature equivalence is certain to fail. The data may still be useful from some points of view but precise and fundamental interpretation is forfeit.

c) The interpretation of a temperature scan must be based on a viscoelastic model. For example, the γ peak of PTFE is shown in Figure 3 for specimens of

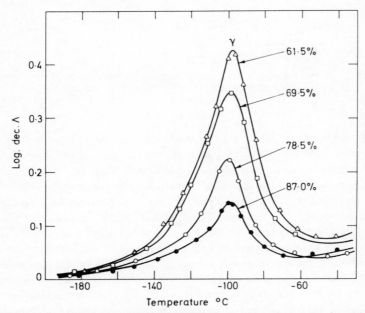

FIGURE 3   Temperature dependence of logarithmic decrement Λ at 0.67 c/s for polytetrafluoroethylene at the γ relaxation with crystal volume fractions as indicated. (After Gray and McCrum.[12])

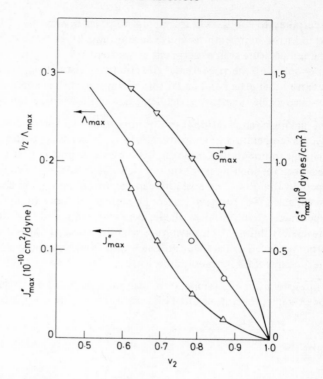

FIGURE 4   Dependence on crystal volume fraction v_2 of the maxima of three loss parameters for the γ relaxation of polytetrafluoroethylene. (After Gray and McCrum.[12])

different crystallinity.[12] Now for this relaxation all models yield the same qualitative result, Figure 4. Plots of G''_{max} (parallel model), J'' (series model) which
are the extreme models, when plotted against crystallinity indicate that at 100%
crystallinity the relaxation disappears. And it is to be expected that one of the
many intermediate models should yield the same result: as seen in Figure 4, Λ_{max}
does indeed extrapolate to zero at 100% crystallinity. So clearly the γ relaxation
in PTFE is of amorphous origin. But all three models cannot be correct. They
only yield the same qualitative result for PTFE because the unrelaxed moduli of
the crystal and amorphous phases are equal.[12] For LPE a similar plot yields confusion, Figure 5.[12] In my view for an isotropic polymer Λ is the only defensible
parameter to plot. The evidence for this stems from experiments Gray and I performed some years ago.[12] For an oriented specimen this is not so and in this case
the analysis is in one sense more complex. Before attempting to interpret viscoelastic measurements it is necessary to hypothesize a detailed mechanism.[13–15]

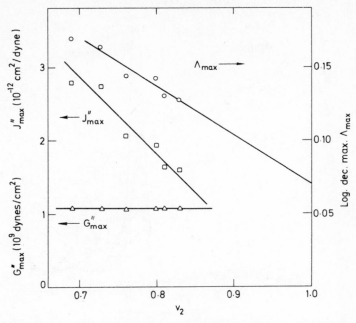

FIGURE 5   Dependence on crystal volume fraction v_2 of the maxima of three loss parameters for the γ relaxation of linear polyethylene. (After Gray and McCrum.[12])

But once this is done and assuming the crystal pole figures are obtainable then the predictions of the mechanism can be realistically compared with experiment.

3   SIGNIFICANT EXPERIMENTAL FACTS

The γ and β relaxations are completely different in origin to the α relaxation. The most powerful demonstration of this came from the chlorination experiments of Schmieder and Wolf,[16] Figure 6. As chlorination proceeds crystallinity is destroyed, the α peak reduces in temperature and finally disappears when the specimen is completely amorphous. Willbourn's[17] experiments show the same result, Figure 7. The α process is eliminated when the crystallinity is destroyed by electron irradiation. The γ and β relaxations dominate. The β processes increases in magnitude as the number of tertiary carbon atoms increase. The same effect is shown if the tertiary carbon atoms are inserted by copolymerization instead of cross linking, Figure 8 (see also Kline, Sauer and Woodward).[18] It will be seen that at 14 CH_3 groups per 100 carbon atoms the β relaxation is dominant and the α relaxation almost eliminated. It may be reasonably inferred from this

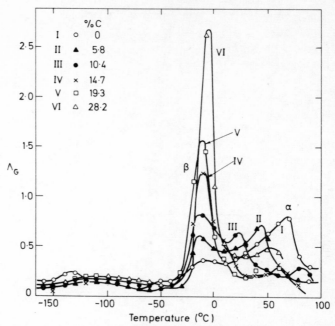

FIGURE 6   Temperature dependence of logarithmic decrement for chlorinated polyethylene. (After Schmieder and Wolf.)

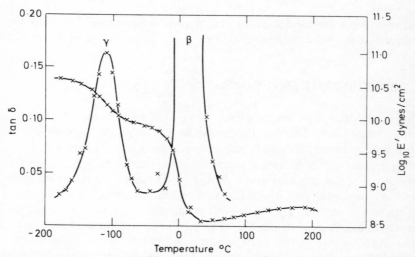

FIGURE 7   Temperature dependence of tan δ and E' for irradiated polyethylene (Alkathene 2). Dose, 10 Mrep at 35°C followed by 1000 Mrep at 140°C. Specimen not detectably crystalline; density 0.89 g/cm³. (After Willbourn.[18])

therefore that the partially crystalline polymer exhibits two amorphous relaxations and a third, the α relaxation, which does not occur if crystallinity is destroyed. The α relaxation may therefore be presumed to occur (a) within the crystal or (b) by means of a relaxation originating in crystal-amorphous interaction. An example of type (a) is the crystal incoherent lattice vibration theory:[19-22] the inter-lamellar shear process is an example of type (b).[23-27,14,15]

An analogous result holds for PTFE and POM.

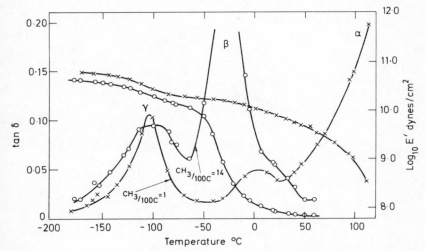

FIGURE 8   Temperature dependence of tan δ and E' for methyl branched polymethylene at the methyl contents indicated. (After Willbourn.[18])

1) All these polymers exhibit a γ relaxation the intensity of which depends on the amorphous content, Figure 9. This relaxation is clearly due to the relaxation of linear sequences of the main chain in the amorphous fraction.

2) The γ relaxation of LPE and POM polymers is highly asymmetric: the low temperature shoulder is extremely broad.

There is good evidence that the low temperature shoulder is in fact a separate but much smaller process (γ_{II} process); this follows from both mechanical and dielectric experiments on bulk or crystallized specimens and on single crystal mats.[28-34]. There is evidence that the γ_{II} process is a relaxation occurring in the crystal fraction although this view is disputed by Stehling and Mandelkern[35] who attributed the γ_I process to the amorphous fraction and the γ_{II} process to an interfacial zone between the crystal and amorphous fractions (see also the interpretation of Hideshima and co-workers referred to below, Table I). The work of Bergmann[36] in this respect has been invaluable.

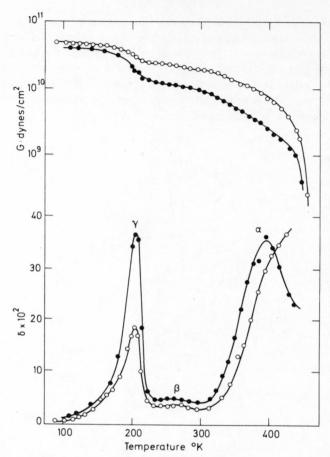

FIGURE 9   Temperature dependence of logarithmic decrement Λ and shear modulus G for polyoxymethylene. (After McCrum.[45])

The temperature of the α-process in single crystal mats increases with crystal thickness. This observation is due to both Sinnott[28] and to Takayanagi and Matsuo.[29] Sinnott's data is shown in Figure 10. As the annealing temperature increases the α peak moves to higher temperature. Now the experiments of these authors disagree over the dependence of α intensity on annealing (see Sauer *et al.*[37]). The explanation of this paradox is probably as follows. The single crystal mats are in reality a polycrystalline anisotropic aggregate. Pole figures from a single crystal mat of LPE showing the effects of annealing at 95° and 120°C are shown in Figure 11.[38] It will be seen that there is a pronounced change in the distribution of crystal poles produced by annealing, an effect which is not unexpected. The effect of this change in texture will be to change the relaxation intensity. For this reason the experiments of Sinnott, and of Takayanagi and Matsuo, cannot yield information on relaxation intensity. Nevertheless, the information concerning relaxation temperature will be acceptable, since this parameter will be unaffected by change in anisotropy.

The experiments of Olf and Peterlin[39] have shown that the α process as observed by NMR to be in accord with the α process as observed dielectrically. Both

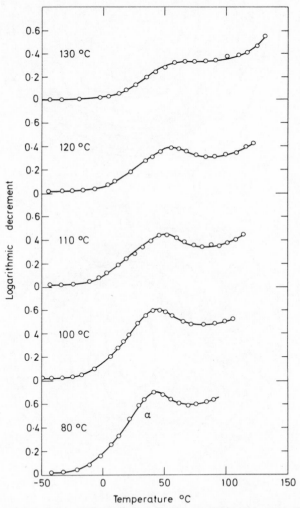

FIGURE 10   Temperature dependence of logarithmic decrement in α region for single crystal mat specimen of linear polyethylene. (After Sinnott.[28])

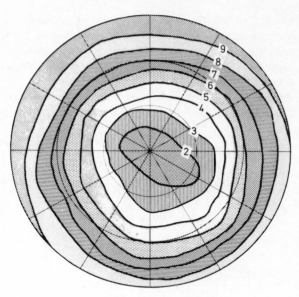

ANNEALED AT 95°C

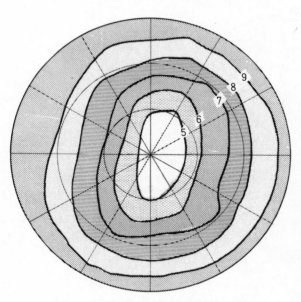

ANNEALED AT 120°C

FIGURE 11   (110) pole figure for a compacted mat of linear polyethylene single crystals showing effect of annealing on the anisotropy. (After Blackadder, Gray and McCrum.[38])

experiments are best interpreted by the hypothesis that the mechanism ("flip-flop") consists of a rotational jump of the molecule (i.e. that portion of the molecule forming a fold period) around its axis by 180° and a simultaneous translation along the axis by one CH_2-group. This process is central to the theories of Hoffman et al.[40] and in the dielectric theory of Tuijnman.[41]

Insensitivity of the γ relaxation of POM to copolymerization with ethylene oxide has been observed by Papir and Baer,[42] Figure 12 (see also Bohn[62]). Note in this figure the increase in magnitude of the β process caused by copolymerization. POM exhibits a pronounced shoulder to the γ peak (γ_{II} peak) although its magnitude will presumably depend on the thermal history of the scan. The γ_{II} peak has been observed dielectrically in POM and LPE by Hideshima and co-workers.[30,43] There can be no doubt that in all three polymers the γ process is a composite of a dominant γ_I mechanism together with at least one subsidiary process. This experimental fact has been verified for LPE in a series of skillfully devised experiments by Illers.[32−34].

The composite nature of the α process is more difficult to establish. We know that the temperature of the α process depends on crystal perfection (annealing conditions and copolymer content being the usual variables). If a specimen is inhomogeneous, for instance if the crystals nucleated at the platten differ from the matrix, and if this difference is not eliminated by annealing then the specimen

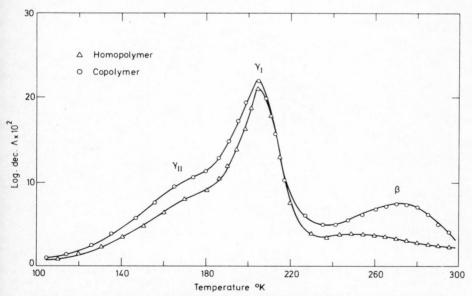

FIGURE 12   Temperature dependence of logarithmic decrement for polyoxymethylene and a random copolymer containing 1.4% ethylene oxide: △ homopolymer; ○ copolymer. (After Papir and Baer.[42])

will exhibit a multiple α relaxation. This argument is due to Stachurski and Ward[27] and is based on experimental evidence from mechanical studies of drawn polyethylene.

If a specimen of LPE is drawn then the α peak appears as shown in Figure 13.[11] As the specimen is annealed at successively higher temperature the α peak becomes more clearly defined. After the final anneal at 126°C the α peak is observed at 50°C (0.67 c/s). It is clearly a single relaxation (the small rise in Λ above 110°C is taken to be an artefact due to the approach of melting). However in an isotropic melt crystallized specimen the α peak is never so well resolved. The argument of Stachurski and Ward is that the α relaxation is a single relaxation, best shown by drawn and annealed LPE, and that the multiple structure observed in melt crystallized specimens is due to inhomogeneity.

The α relaxation in LPE after drawing and annealing is an inter-lamellar shear mechanism. The model of Ward and co-workers[15] is shown in Figure 14. By drawing and annealing under varying conditions it is possible to prepare specimens with

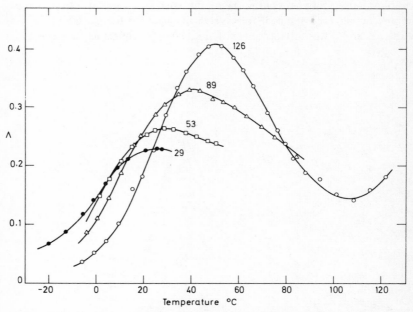

FIGURE 13   Temperature dependence of logarithmic decrement for a biaxially drawn specimen of LPE (torsion axis parallel to draw direction) after annealing at the temperatures indicated. The specimen was first annealed (after drawing) at 29°C and then a temperature scan obtained from −20 up to 29°C. It was then annealed at 53 and a temperature scan obtained from − 20 up to 53° C (and so forth). Frequency 0.67 c/s. (After Buckley and McCrum.[11])

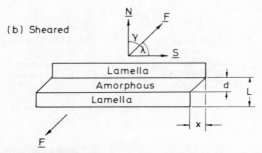

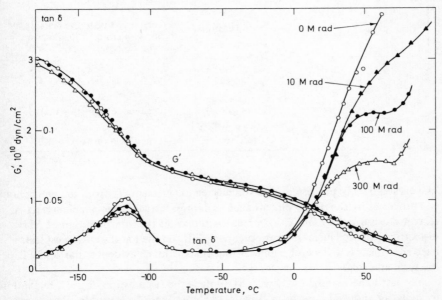

FIGURE 14   Model of a simple shear process due to inter-lamellar shearing: (a) unsheared (b) after shearing. N is the normal to the slip plane (S plane). The external tensile load is applied in direction F and slip takes place in direction S. (After Davies et al.[15])

FIGURE 15   Suppression of the α relaxation in LPE by radiation induced cross linking. (After Illers.[34])

lamellar normals and crystal axes so aligned that the location of the shear plane may be determined. The alternatives are the interlamellar surface, a crystal plane of type (hko) and a crystal plane of type (hoo). The inter-lamellar shear process certainly fits the data better than the other alternative mechanisms.[14−15]

This simplified view is to be contrasted with the detailed experiments of Hideshima and co-workers.[30,43,44] For LPE the activation energies and outline relaxation mechanisms are given in Table I. There are three resolved γ processes.

TABLE I

A tabulation of results and conclusions of Hideshima and co-workers[30,43,44]

	Loss-band	Component relaxations	Activation energy	Location of relaxing chains (mechanism of relaxation)
Descending temperature	α	α_2	45~50 kcal/mol	C (relaxation in the interior of crystals)
		α_3	25~35 kcal/mol	$A_1 + A_2 + C$ (boundary slip relaxation)
		α_1	20~30 kcal/mol	A_2 (relaxation in the surface regions of lamella)
	β	β	~50 kcal/mol	A_1 (corresponding to the relaxation due to micro-Brownian motion in amorphous polymers)
	γ	γ_1	~23 kcal/mol (dielectric)	A_1 (local mode relaxation in A_1)
		γ_2	~13 kcal/mol (dielectric)	A_2 (local mode relaxation in A_2)
		γ_3	~9 kcal/mol (dielectric)	A_3 (a modification of A_1 or A_2?)

The crystalline (C) processes are distinguished experimentally from the amorphous (A) processes by the use of diluent such as carbon tetrachloride which affects A relaxations but not C. This work which encompasses the interpretation of NMR, dielectric and mechanical relaxation experiments on melt and solution crystal-lized specimens is a forceful reminder that there is much evidence that for LPE the term amorphous is too vague.

The last experimental fact I wish to quote is the suppression of the α-relaxation by radiation induced cross links: Figure 15 shows Illers[34] measurements. I turn now to discuss the theories for mechanical relaxation in LPE.

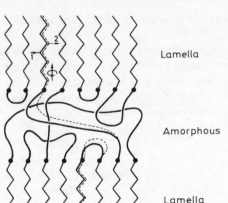

Lamella

Amorphous

Lamella

FIGURE 16   Extension of the model of Hoffman *et al.*[40] for the mechanical α relaxation in linear polyethylene. The α relaxation as observed by NMR and dielectric relaxation is due to rotation of molecular stems in the crystal. This rotation of 180° (for example, from position 1 to 2) causes a displacement of $[00\frac{1}{2}]$. The pinning point at the crystal surface is thus relaxed leading to an additional relaxation of the amorphous fraction (which is already partially relaxed due to the γ_I mechanism). The dotted line illustrates the change in length of a molecular stem in the amorphous fraction. (After Buckley and McCrum.[15])

4   DISCUSSION

β and γ relaxations

It can be stated that the β and γ_I relaxations of LPE are of amorphous origin. The perennial question of which of these is the Tg we leave until later. The β-relaxation is in LPE of almost negligible size and is advanced as a relaxation in this polymer only because (a) copolymerization and chlorination studies point unerringly to the region at *ca* −30°C as being a relaxation region in LPE (b) an analogous relaxation in POM may be enhanced by thermal history[45] and in the region of −30° in LPE thermal history enhances a mechanical relaxation[9] which is interpreted to be the β-relaxation.

Of the proposed mechanisms for the γ_I process it seems to me the most preferable is the platzwechsel theory with molecular process as described for instance by Boyd and Breitling.[46] Whatever the superiority of this detailed Boyd-Breitling mechanism over the earlier models of Schatzki and Boyer the concept of a platzwechsel process is more *a priori* reasonable than the local mode process. The local mode theory, particularly in its lucid reformulation by Hayakawa and Wada,[22] is not without elegance and indeed experimental support. Nevertheless my view is that the best theory will prove to be of the platzwechsel type for the γ_I processes in LPE, POM and PTFE.

In highly branched polyethylene the β relaxation fits all the criteria associated normally with a glass transition: (a) the relaxation is the dominant amorphous relaxation (b) copolymerization with increasing proportions of comonomer causes a gradual shift of the relaxation temperature. In this latter property the β-relaxation differs entirely from the γ relaxation.

α relaxation

Evidence from studies of anisotropic specimens   The evidence from orientation studies is that the mechanical α relaxation is an inter-lamellar shear process[14,15] and not due to shear on crystal planes of type (hko) or (hoo).[40,47,48] The experiments verifying this hypothesis did not study the possibility that the incoherent lattice vibration theory would also fit the data (there being no rationale by which the data could be compared with the theory).

It is necessary to ask how the inter-lamellar shear mechanism for the mechanical α relaxation fits with the observation of the α process by NMR and dielectric relaxation. There is strong evidence that the molecular relaxation observed by these two techniques occurs within the body of crystals,[8] by means of hindered chain rotation of $180°$[8,41,39] accompanied by longitudinal displacement of $[00\frac{1}{2}]$.[8,39] Furthermore, the mechanical α relaxation in polyethylene is known to depend for its existence on the presence of polyethylene crystals. How may the present conclusions, assigning mechanical relaxation to deformation of non-crystalline layers, be rationalized in the light of these facts?

Hoffman et al.[40] proposed that relaxation within the crystals by chain rotation of $180°$ and longitudinal displacement of $[00\frac{1}{2}]$ would be coupled to the fold-containing crystal surface layers when folds were sufficiently tight (containing the minimum of 5 CH_2 units). Their particular model of the fold surface has been seriously challenged by recent experiments, which suggest more mobile folds,[49] but the essence of their argument probably remains valid.

Mechanical relaxation could occur as follows in an isochronal experiment.[15] At liquid nitrogen temperature both phases are immobile. The γ_I relaxation is caused by a rate activated relaxation of the "flip-flop" mechanism in the amorphous fraction (also observed dielectrically and by NMR). At temperatures above the γ region but below the α relaxation the crystals are immobile but the amorphous fraction fully relaxed subject to the pinning of the molecular stems at the crystal surface. At a somewhat higher temperature the hindered rotation of molecular stems in the crystal (observed by NMR and dielectric relaxation) is initiated. Dipoles rotate for instance, from position 1 to position 2 is shown in Figure 16. It is the longitudinal displacement of $[00\frac{1}{2}]$ which initiates the mechanical α relaxation since the pinning of the molecular stems at the surface is now relaxed.

By this means, relaxation within the crystal (the dielectric and NMR α relaxation) increases the compliance of the non-crystalline surface layers and give rise to mechanical relaxation.

The essential feature of the above suggestion is that the two α relaxations, observed by dielectric and NMR techniques on the one hand, and by mechanical tests on the other, are coupled but have different molecular origins. Independent evidence that these two relaxations are not identical is obtained by comparing their temperature/frequency positions. Mechanical and dielectric results in the α region have been compared by Reddish and Barrie[50] and Sandiford and Willbourn.[51] The two techniques yield approximately equal activation energies.[51] Nevertheless, the mechanical α relaxation lies consistently at higher temperatures (lower frequencies) than the dielectric α relaxation.[50,51] Furthermore, McCall and Douglas[52] found the temperature/frequency position of the NMR α relaxation to agree well with its dielectric counterpart, but to lie at considerably lower temperatures (higher frequencies) than the mechanical α relaxation.

Evidence from studies of isotropic specimens   The early proposals for the lamellar boundary slip process were by workers studying isotropic, spherulitic specimens.[23−26] As mentioned above these specimens normally show evidence of more than one relaxation. It seemed from the work I performed with Morris[25] that the effect of irradiation induced cross links was on the α' relaxation and from this and other evidence we attributed this process to inter-lamellar slip. Although there is a confusion in the literature of symbols and the exact identification of relaxations between workers at different frequencies it is clear that in the melt crystallized, spherulitic specimens the inter-lamellar shear process is supported by such evidence. Of all the several proposals for the observed α processes inter-lamellar slip is the only one which is commonly accepted.

The additional proposals include intracrystalline mosaic block and uniform shear of lamellar crystals, incoherent lattice vibration of the crystal, intracrystalline plasticity and the proposals of Hideshima referred to in Table I.[28−30,43,53] Of the new techniques for resolving this complicated problem the most sophisticated and potentially valuable is the dynamic x-ray technique of Kawai and co-workers.[54] The combination of dielectric and rheo-optical studies by Stein and co-workers has also been of value.[53]

5   THE GLASS TRANSITION IN POLYETHYLENE

The literature is now formidable on this topic. The most recent and comprehensive collection is due to Davis and Eby[55] following several earlier reviews. The conflicting proposals are three in number: that the T_g occurs in temperature regions at (1) $-130°C$, (2) $-80°C$, (3) $-20°C$.[†] Proposals (1) and (3) connect the T_g with the γ_I and β relaxations respectively. Proposal (2) has no mechanical analogue. The evidence in its favour is reviewed in the following paragraph. For a detailed exposition see Boyer's review.[56]

The $-80°C$ proposal is due originally to Tobolsky.[57] Taking the T_g of poly-propylene to be $-20°C$ and then from measurements of the T_g of a mole ratio 2/1 ethylene–propylene copolymer Tobolsky extrapolated the T_g of polyethylene (using the Gordon–Taylor equation) to $-81°C$. This proposal was supported in Boyer's major review[4] with further supporting arguments (1) the brittle point of branched polyethylene is ca $-68°C$ (2) this value of T_g is approximately 0.5 times the melting point T_m (in agreement with the Boyer–Beamann rule that $0.5 < T_g/T_m < 0.67$).

Boyer has generalized this hypothesis to take into account the γ and β relaxations.[56] In brief his views may be stated as follows.

1) The T_g of completely amorphous polyethylene occurs at $-80°C$: this temperature is a reference temperature since of course polyethylene cannot be obtained in the completely amorphous state.

2) In the partially crystalline solid, amorphous polyethylene exhibits two glass transitions: $T_g(L)$ at $-80°C$ and $T_g(U)$ at $-20°C$. Both temperatures are stated to rise with crystallinity and are attributed to two types of amorphous material.

3) The $-130°C$ transition is a glass to glass transition and is not a glass transition (T_g).

Stehling and Mandelkern[35] present a lucid argument based on thermal expansion, mechanical relaxation and NMR which supports the earlier contention of Willbourn that the T_g of the amorphous —CH_2— molecular stem is at $-130°C$. It cannot be denied (see Figure 2 for instance) that when LPE is studied without shock cooling the dominant amorphous relaxation is the γ_I relaxation. And for POM and PTFE the γ relaxation is also dominant.

Eby and Davis[55] conclude from rate studies of volume changes following quenching that the T_g of LPE, is the $-20°C$ transition. It can be argued that if shock cooling is avoided LPE exhibits no well resolved mechanical β relaxation. Nevertheless, BPE even at very low branch concentration exhibits a β relaxation. Indeed *any* comonomer acts to increase the magnitude of the β relaxation. This effect also occurs at the β relaxation in POM and at the analogous (α) relaxation in PTFE.[58] Furthermore, thermal history of the temperature scan plays a role in modifying the size of the β relaxation in POM;[45] in LPE the identical parameter

---

[†]In order to ease the presentation of the argument and on taking cognizance of the large errors involved we ignore differences quoted between various workers (of order ± 10°C) and round off the temperatures of the candidate transitions in this way.

inserts a clearly resolved relaxation in the β region.[9] The early work of Cole and Holmes[59] also placed the T_g of LPE at $-20°C$ as do the observations of Rabold using a novel ESR probe technique.[60]

Of the evidence from hypotheses of molecular motion it seems clear that the Boyd–Breitling model for the γ_I relaxation is not the type of motion normally associated with a glass transition. It clearly has more of the characteristics of a sub-T_g relaxation, such as the chair-to-chair relaxation of the cyclohexyl group studied by Heijboer[61] in polycyclohexylmethacrylate. Indeed Boyd and Breitling use precisely this model in order to describe the details of their mechanism. Co-polymerization studies of the cyclohexylmethacrylate system show that the chair-to-chair relaxation remains at its characteristic temperature independent of the proportion of comonomer, the relaxation merely changing in magnitude.[61] There is good evidence that this pattern of behaviour occurs at the γ relaxations in LPE, POM and PTFE. This is not the type of behaviour normally associated with a classical glass transition.

My own view, stated elsewhere,[8] is that the problem is one of definition.

1) The $-130°C$ transition in LPE is the major transition: if T_g is defined operationally as the major transition then there is no doubt that T_g occurs at $-130°C$.

2) The $-20°C$ transition of LPE has physical characteristics (other than its magnitude) which are analogous to those of accepted glass transitions in other polymers: the most persuasive of these is the evidence from copolymerization studies.

Neither of these types of definitions is markedly superior to the other. Why should it be required that in a finely structured lamellar solid such as LPE the T_g should have all the physical characteristics of T_g in completely amorphous polymers. Because of this my answer to the question what is the T_g of LPE is "First tell me what you mean by T_g". This is not a linguistic side step but an acquiescence before the evidence; the amorphous forms of the linear crystalline polymers are not in all respects identical to the amorphous forms of other polymers.

Acknowledgements

The author would like to thank Professors Hideshima, Sauer and Wada for valuable communications.

References

1. J. D. Hoffman, *J. Chem. Phys.* **22**, 156 (1954): **23**, 1331 (1955).
2. H. Fröhlich, *Theory of Dielectrics* (Oxford University Press, Oxford, 1949).

3. T. F. Schatzki, *ACS Polymer Preprints* 6, 646 (1965).
4. R. F. Boyer, *Rubber Rev.* 34, 1303 (1963).
5. K. Yamafuji and Y. Ishida, *Kolloid-Z. Z. Polym.* 183, 15 (1962).
6. N. K. Saito, K. Okano, S. Iwayanagi and T. Hideshima, *Solid State Phys.* 14, 343 (1900).
7. H. A. Flocke, *Kolloid-Z. Z. Polym.* 180, 188 (1962).
8. N. G. McCrum, B. E. Read, and G. Williams, *Anelastic and Dielectric Effects in Polymeric Solids* (Wiley, London, 1967).
9. J. W. Cooper, M.Sc. Thesis, Oxford (1972): see also J. W. Cooper and N. G. McCrum, *J. Mat. Sci.* 7, 1221 (1972).
10. J. M. Hutchinson and N. G. McCrum, *Nature Physical Science* 236, 115 (1972).
11. C. P. Buckley and N. G. McCrum, *J. Polymer Sci.* $A-2$ 9, 369 (1971).
12. R. W. Gray and N. G. McCrum, *J. Polymer Sci.* $A-2$ 7, 1329 (1969), *J. Polymer Sci. B* 4, 639 (1966).
13. M. W. Darlington and D. W. Saunders, *J. Phys. D: Appl. Phys.* 3, 535 (1970).
14. G. R. Davies, A. J. Owen, I. M. Ward and V. B. Gupta, *J. Macromol. Sci. Phys. B* 6, 215 (1972).
15. C. P. Buckley and N. G. McCrum, *J. Mat. Sci.* 8, 928 (1973).
16. K. Schmieder and K. Wolf, *Kolloid-Z. Z. Polym.* 134, 149 (1953).
17. A. H. Willbourn, *Trans. Faraday Soc.* 54, 717 (1958).
18. D. E. Kline, J. A. Sauer and A. E. Woodward, *J. Polymer Sci.* 22, 455 (1956).
19. K. Tsuge, H. Enjoji, H. Terada, Y. Ozawa, and Y. Wada, *Japan J. Appl. Phys.* 1, 270 (1962).
20. K. Tsuge, *Japan J. Appl. Phys.* 3, 588 (1964).
21. K. Okano, *J. Polymer Sci. C* 15, 95 (1966).
22. R. Hayakawa and Y. Wada, *Rep. Progr. Polymer Phys. Japan,* 11, 215 (1968); *Progr. Polymer Sci. Japan* 3, 215 (1972).
23. S. Iwayanagi, *Rep. Progr. Polymer Phys, Japan* 5, 135 (1962).
24. S. Iwayanagi and H. Nakane, *Rep. Progr. Polymer Phys. Japan* 7, 179 (1964).
25. N. G. McCrum and E. L. Morris, *Proc. R. Soc. London* A292, 506 (1966).
26. C. Nakafuku, K. Minato and T. Takemura, *Japan J. Appl. Phys.* 7, 115 (1968).
27. Z. H. Stachurski and I. M. Ward, *J. Macromol. Sci. Phys. B* 3, 445 (1969).
28. K. M. Sinnott, *J. Appl. Phys.* 37, 3385 (1966).
29. M. Takayanagi and T. Matsuo, *J. Macromol. Sci.-Phys. B* 1, 407 (1967).
30. M. Kakizaki and T. Hideshima, *J. Macromol. Sci.-Phys. B* 8, 367 (1973).
31. D. Hideshima, private communication.
32. K. H. Illers, *Colloid Polymer Sci.* 252, 1 (1974).
33. K. H. Illers, *Rheol. Acta* 3, 202 (1964).
34. K. H. Illers, *Kolloid-Z. Z. Polym.* 231, 622 (1969).
35. F. C. Stehling and L. Mandelkern, *Macromolecules* 3, 242 (1970).
36. K. Bergmann, *Kolloid-Z. Z. Polym.* 251, 962 (1973).
37. J. A. Sauer, G. C. Richardson, and D. R. Morrow, *J. Macromol. Sci. C* 9, 149 (1973).
38. D. Blackadder, R. W. Gray and N. G. McCrum, to be published.
39. H. G. Olf and A. Peterlin, *J. Polymer Sci.* $A-2$ 8, 753 and 771 (1970).
40. J. D. Hoffman, G. Williams and E. Passaglia, *J. Polymer Sci. Part C* 14, 173 (1966).
41. C. A. F. Tuijnman, *Polymer [London]* 4, 259 and 315 (1963).
42. Y. S. Papir and E. Baer, *Mater. Sci. Eng.* 8, 310 (1971).
43. M. Kakizaki, Y. Morita, K. Tsuge and T. Hideshima, *Rep. Progr. Polymer Phys. Japan* 10, 397 (1967).
44. M. Morozumi, K. Tsuge and T. Hideshima, in *Proc. Fifth Internat. Congr. Rheology,* Vol. 3 edited by S. Onogi (University of Tokyo Press) p. 325.
45. N. G. McCrum, *J. Polymer Sci.* 54, 561 (1961).
46. R. H. Boyd and S. M. Breitling, *Macromolecules,* 7, 855 (1974).
47. M. Takayanagi, T. Aramaki, M. Yoshino and K. Hoashi, *J. Polymer Sci.* 46, 531 (1960).
48. W. Pechhold and S. Blasenbrey, *Kolloid-Z. Z. Polym.* 241, 955 (1970).
49. A. Peterlin, *J. Macromol. Sci. Phys. B* 3, 19 (1969).
50. W. Reddish and J. T. Barrie, *IUPAC Symposium,* Wiesbaden, 1959, Paper 1A3.

51. D. J. H. Sandiford and A. H. Willbourn, in *Polythene,* 2nd Ed., edited by A. Renfrew and P. Morgan (Ilife, London, 1960).
52. D. W. McCall and D. C. Douglass, *Appl. Phys. Letters* 7, 12 (1965).
53. A. Tanaka, E. P. Chang, B. Delf, I. Kimura, and R. S. Stein, *J. Polymer Sci. Polymer Phys. Ed.* 11 1891 (1973).
54. H. Kawai, T. Ito, S. Suehiro, Y. Kato, *Rep. Progr. Polym. Phys. Japan* 14, 367 (1971).
55. G. T. Davies and R. K. Eby, *J. Appl. Phys.* 44, 4274 (1973).
56. R. F. Boyer, *Macromolecules* 6, (1973) 288.
57. A. V. Tobolsky, *Properties and Structures of Polymers* (Wiley, New York, 1960).
58. N. G. McCrum, *Makromol. Sci.* 54, 561 (1959).
59. E. A. Cole and D. R. Homes, *J. Polymer Sci.* 46, 245 (1960).
60. G. P. Rabold, *J. Polymer Sci. Part [A−1]* 7, 1203 (1969).
61. J. Heijboer, *Kolloid-Z. Z. Polym.* 148, 36 (1956).
62. L. Bohn, *Kolloid-Z. Z. Polym.* 201, 20 (1965).

DISCUSSION

E. Baer (*Case Western Reserve Univ.*): What about the so-called δ and ε transitions? Would you care to speculate about the nature of these transitions?

N. B. McCrum: No. I have excluded these from this discussion.

H. W. Starkweather (*Du Pont*): What is your "ten years later" view of the α relaxation in poly(tetrafluoroethylene)?

N. B. McCrum: I think it is an amorphous transition. The copolymer evidence is very strong and clearly indicates this.

R. S. Porter (*Univ. of Mass.*): This question concerns your good model for poly(ethylene) crystallized from the melt. There would certainly seem to be high concentration of trapped entanglements at folds in lamellar crystals, as suggested by Prof. Marrucci. An estimate of this concentration can come from the entanglement density measured on the melt and from the lamellar thickness measured after crystallization.

N. B. McCrum: A good point; it would be interesting if this could be done.

Y. S. Papir (*Richmond College*): Several investigators have reported in the literature that γ internal friction peak in poly(ethylene) and poly(oxymethylene) can be resolved into at least two distinct processes. In fact the slide of Papir and Baer on poly(oxymethylene) shows a distinct low temperature shoulder. Hejfow, *et al.,* have distinctly resolved to peak maxima with various copolymers of POM. How do these data reconcile themselves with your proposed mechanism?

N. B. McCrum: I have not discussed γ_{II} mechanisms. My detailed comments refer to the γ_I, amorphous process.

H. W. Starkweather (*Du Pont*): There is a practical manifestation of the differences to which Prof. Ward has referred in the relaxations in high and low density poly(ethylene). This is in creep at room temperature. Low density polyethylene has a much smaller short-term modulus, and the dynamic mechanical behavior shows much more internal motion. Nevertheless, creep in high density polyethylene is much more severe and may even lead to catastrophic failure.

M. Takayanagi (*Kyushu Univ.*): I want to make a comment on the method of determining T_g in crystalline polymers, which is to employ the pressure effect. The amorphous region associated with T_g is sensitive to hydrostatic pressure, whereas the relaxation process associated with the crystalline region is rather insensitive to pressure. For example, Professor Takemura of Kyushu University has separated the motional narrowing of the narrow component of low density poly(ethylene) relating the composite narrowing of the γ and β relaxations at atmospheric pressure into the pressure-sensitive β relaxation and the less pressure-sensitive γ relaxation. In the case of single crystal mats of PE, it was not possible to separate both relaxations. The latter fact means that the β relaxation of PE single crystal mats is associated with the state of molecules in the surface region of the lamellar crystals, which molecules are in a rather restricted state. The T_g of PE corresponds with the β relaxation, not the γ relaxation.

N. G. McCrum: I am in complete agreement with this statement.

I. M. Ward (*Univ. of Leeds*): The anisotropy of the α-relaxation in oriented samples of linear polyethylene is consistent with interlamellar shear. The α-relaxation for oriented branched poly(ethylene), on the other hand, shows the character of a c-shear process, where shear is parallel to the c-axis of the crystalline regions on planes containing this axis, and it is the β-relaxation which can be associated with interlamellar shear.

It has therefore been concluded by us that in low density PE the motions in the non-crystalline region, without involving motions in the crystalline region, are sufficient for interlamellar shear to be activated when the polymer is subjected to external stress. There is then a separate relaxation process associated with c-shear. Dielectric measurements on oriented oxidised branched PE are consistent with this view, in that the β-relaxation is isotropic dielectrically and the α-relaxation shows the anisotropy expected for the c-shear process. In the case of linear PE, with its much higher crystallinity, it may well be that motion within the crystal, as proposed by Hoffman, Williams, and others, is necessary for the required mobility in the non-crystalline region so that interlamellar shear can occur.

These considerations suggest that the α-relaxations in linear and branched PE are not identical.

See also:
Z. H. Stachurski and I. M. Ward, *J. Polymer Sci., A–2* **6**, 1817 (1968).
G. R. Davies and I. M. Ward, *J. Polymer Sci., A–2* **10** 1153 (1972).

Effects of Crystallinity on Amorphous Transitions

TURNER ALFREY, JR. and RAYMOND F. BOYER†

The Dow Chemical Company, Midland, Michigan 48640, U.S.A.

INTRODUCTION

Amorphous polymers typically exhibit a major mechanical loss peak in the vicinity of the glass transition temperature (T_G), and in many cases additional minor loss peaks at lower temperatures ($T < T_G$). The "strength" of such a transition can be variously gauged by the area under the loss peak, by the maximum value of G'' or tan δ, or by the change in G' on going through the transition.[1,2]

Partially crystalline polymers may also exhibit loss peaks associated with the amorphous regions (T_G and $T < T_G$). However, various authors[1,2] have pointed out that polymer chains in amorphous regions of semi-crystalline polymers are not necessarily identical, structurally and dynamically, with similar chains in completely amorphous polymers. Some "amorphous" chains may be attached to crystallites as dangling cilia, as tight or loose loops, or as tie-molecules.[3-6] Such attachment would restrict the mobility of a chain, and can be expected to modify the mechanical properties of the "amorphous" material; in particular, to modify the "amorphous transitions".

These considerations lead to the question: "In a partially crystalline polymer, how do the values of T_G and ($T < T_G$), and how do the 'strengths' of the glass transition and the ($T < T_G$) transition depend upon the crystalline fraction, χ_c?" A preliminary treatment of these questions has appeared.[7]

Clearly, there can be no unique answer to this fourfold question. A given value of χ_c can represent various crystalline morphologies, and hence various degrees of constraint on the amorphous chains.

This paper is concerned primarily with a study of the relative strengths of the T_G and $T < T_G$ relaxations. The specific $T < T_G$ process under discussion is an in-chain amorphous phase relaxation designated variously as the "local mode"

---

† Now with the Midland Macromolecular Institute.

process or the β loss peak when T_G is labeled as the α process.[1] In hydrocarbon backbone polymers and some oxide polymers, this relaxation occurs at about 0.75 T_G for a frequency of 100 Hz.[8] Goldstein[9] considers this process to be a necessary concomitant of the glass transition. Numerous examples of its wide-spread occurrence are found in Ref. 1 as well as in specific references to follow.

A comparison of the relative mechanical relaxation strengths of the T_G and $T < T_G$ process as a function of χ_c was first reported by Takayanagi et al.[10] They crystallized PET at different temperatures to achieve a range in χ_c values (presumably with different fold lengths). The tensile loss modulus, E'' at 100 Hz, was plotted against χ_c for T_G and $T < T_G$. They obtained results of the type shown in Figure 1. A generalized schematic representation is used to emphasize that similar results have been obtained on PET and other polymers by a variety of methods. The general features which they[10] noted were:

1) The crossover showing T_G to be less intense than $T < T_G$ for $0.5 < \chi_c < 1.0$.

2) A finite value of the extrapolated loss function for $T < T_G$ at $\chi_c = 1$.

These results were interpreted in terms of a three-component model: perfect crystal, pure amorphous, and defect regions in the crystalline phase. It was assumed that the local mode, or $T < T_G$ relaxation, involved such short segments of the polymer chain that this motion, and the resultant loss, could occur both

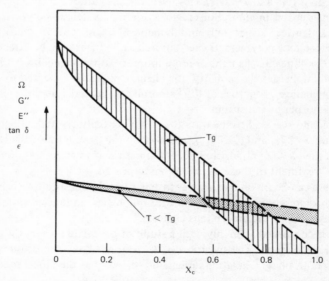

FIGURE 1   Schematic representation of data in Table I showing how several measures of mechanical or dielectric loss strength decrease with increasing fractional crystallinity, X_C.

in the amorphous phase and in defect regions. Hence, E'' for $T < T_G$ should not extrapolate to zero at $\chi_c = 1$. This work was both an inspiration and a guide for the present study. But first, additional literature indicating similar behavior is collected in Table I.

TABLE I
Polymer systems showing cross-over in strengths of T_G and $T < T_G$ relaxations[a]

Polymer	Method	Function Plotted	Frequency Hz	Reference
PET	Rheovibron	E'' max.	100	10
PET	Torsion pendulum	G'' max.	1	11
		$\int G'' (1/T)$[b]		
PET	Dielectric	ϵ''	100	12
Polypropylene[c]	Torsion pendulum	$\Lambda = \pi \tan \delta$	~1	13
$PCIF_3E$	Torsion pendulum	δ[d] $= $ log decrement	1	14, 15
$PCIF_3E$	Dielectric[e]	ϵ''	Range	16
Polypentenamer[f] (hydrogenated to various levels)	Rheovibron	$\tan \delta$	138	17

[a]This tabulation is not based on an exhaustive literature search. Its purpose is to show that different polymers and different techniques tend to exhibit the common behavior pattern shown in Figure 1.

[b]While this function is preferred on theoretical grounds,[1,11] it indicates that the strength of T_G increases with χ_c (see Figure 12 of Ref. 11). This anomalous behavior presumably is a direct result of the broadening of the T_G loss peak by crystallinity (see $G'' - 1/T$ plots in Figure 2 of Ref. 11).

[c]Variation in χ_c is a result of differences in tacticity (as measured by heptane insolubles). (See footnote g.)

[d]In a subsequent paper,[15] it is shown that different measures of loss strength can lead to contradictory conclusions about the variation of loss strength with χ_c. See also Section 4.3 of Ref. 1.

[e]It is not too clear that these data actually support Figure 1, in part because the $T < T_G$ relaxation seems to have two components.

[f]Change in χ_c results from changes in chemistry. χ_c is zero for the starting polymer and increases with degree of hydrogenation.

[g]Passaglia and Martin [*J. Res. Nat. Bur. Stds.* **68A**, 519 (1964)] reported dynamic loss measurements on a specimen of polypropylene whose χ_c was varied from 0.44 to 0.70 by thermal treatment. They plotted Δ, G'' and J'' as a function of temperature. They concluded that a change in χ_c may increase or decrease the magnitude of a given relaxation, depending on the mode of comparison employed. This paper is discussed in (15). A plot of Δ vs χ_c made by us from their data shows Figure 1-type behavior for $T < T_G$. However, T_G indicates an extrapolated finite intercept at $\chi_c = 1.0$. This may result in part because effective frequency increases slightly with χ_c.

We are aware that a serious problem exists concerning a proper definition, and hence measure, of the strength of a mechanical or dielectric relaxation, especially for a semi-crystalline polymer. This has been discussed in some detail by McCrum et al.[18] Footnotes (b) and (d) of Table I concern this problem. Gray and McCrum[15] have presented a critical analysis of different methods of presenting loss data, and have referred to earlier literature. Heijboer[19] has also discussed this problem. We had no choice but to use the original data in the literature. Gray and McCrum[15] calculated numerous functions of loss for T_G in $PCIF_3E$, but not for $T < T_G$.

One sees from Table I that the results are general to several polymers, several methods, several loss functions, and various frequencies. Inspection of the original data, or in some instances Figure 1-type plots made from the original data, showed the following:

The strength of the $T < T_G$ relaxation either crossed over, or extrapolated to cross over, the strength of the glass transition at high χ_c, either because the $T < T_G$ curve falls above the line for simple proportionality with $1 - \chi_c$, or because the curve for T_G falls below simple proportionality with $1 - \chi_c$, or both. A series of models related theories for such effects will now be discussed.

MODELS FOR A SEMI-CRYSTALLINE POLYMER

Any calculations of relaxation strengths in semi-crystalline polymers must consider one or more models for this two-phase system and related changes with χ_c. Available models are:[3-6]

1) Fringed micelle;
2) Irregularly chain folded as in bulk crystallized polymers;
3) Combinations of the two, especially in bulk crystallized specimens;
4) Chain folded crystals with:
   a) Regular tight loops, or
   b) Irregular loose loops, either from single crystals, or well annealed bulk crystals, and, of course,
5) Defects in the crystalline regions.

Some rather specific models have been discussed,[20-22] generally for the purpose of calculating crystallization kinetics, or equilibrium values of crystallinity or both. Several models specific to the purpose of this paper will be considered. Our calculations should be, and indeed are, rather insensitive to the model selected.

THEORETICAL REMARKS

In the following, we assume that the glass transition is associated with the conformational motion of relatively large chain segments (e.g., 40 or 50 chain atoms), whereas the $(T < T_G)$ transition is associated with the motion of much shorter chain segments (4 to 6 chain atoms). We shall consider various simplified models of the crystalline–amorphous morphology, and attempt to deduce the dependence of the amorphous transition "strengths" upon χ_c for each model. The basis for the estimation will be the simple assumption that a relaxation strength is proportional to the *number* of chain segments of the required size remaining in the amorphous region, and also proportional to the segment size.

Model A

Interrupted spherulitic growth. Consider a morphology composed of large, highly crystalline, spherulites in a continuum of amorphous polymer. Restriction of the amorphous chains is a minimum in this model. Neglecting the interactions at the relatively small surface area of the large spherulites, we expect the strengths of the glass transition and the $(T < T_G)$ transition to be simply proportional to the fraction of amorphous material $- (1 - \chi_c)$. This is plotted in Figure 3(a). The strength of the $(T < T_G)$ transition in completely amorphous polymer is taken to be 6/50 that of the glass transition.

Model B

Fixed number of crystalline zones, regularly spaced, without chain-folding. This model is shown schematically in Figure 2(b). In a repeating unit, n_a is the number of chain atoms in the amorphous zone, and n_c the number of chain atoms in the neighboring crystalline zone. $(n_a + n_c)$ is held fixed at the value $(n_a + n_c)^*$. If 50-sequences are responsible for the glass transition, and 6-sequences for the $(T < T_G)$ transition,

$$\text{No. of 6-sequences} = (n_a + n_c)^* (1 - \chi_c) - 5 \tag{1}$$

$$\text{No. of 50-sequences} = (n_a + n_c)^* (1 - \chi_c) - 49 \tag{2}$$

and the *relative strengths* (compared with completely amorphous material) of the two transitions are:

$$\text{Rel. Str. } (T < T_G) = (1 - \chi_c) - \frac{5}{(n_a + n_c)^*} \tag{3}$$

$$\text{Rel. Str. (glass)} = (1 - \chi_c) - \frac{49}{(n_a + n_c)^*}. \tag{4}$$

These relations are plotted in Figure 3(b), for an assumed value of $(n_a + n_c)^* = 200$. Again, the assumed strength of the $(T < T_G)$ transition in amorphous polymer is taken as 6/50 that of the glass transition.

Model C

Fixed size of crystallites, regularly spaced, without chain-folding. This model is indicated in Figure 2(c). n_c is fixed at n_c^*.

$$n_a = \frac{n_c^*(1 - \chi_c)}{\chi_c} \tag{5}$$

In n_a,

$$\text{no. of 6-sequences} = \frac{n_c^*(1 - \chi_c)}{\chi_c} - 5 \tag{6}$$

$$\text{no. of 50-sequences} = \frac{n_c^*(1 - \chi_c)}{\chi_c} - 49 \tag{7}$$

$$\text{Rel. Str. } (T < T_G) = (1 - \chi_c) - \frac{5 \chi_c}{n_c^*} \tag{8}$$

$$\text{Rel. Str. (glass)} = (1 - \chi_c) - \frac{49 \chi_c}{n_c^*} \tag{9}$$

This is plotted as Figure 3(c), setting $n_c^* = 200$.

Model D

Random spacing of equal-sized crystallites, without chain-folding. This model is indicated in Figure 2(d). All n_c are equal to the fixed value of n_c^*. The average value of n_a is:

$$\bar{n}_a = \frac{n^*(1 - \chi_c)}{\chi_c}. \tag{10}$$

We assume the distribution:

$$N(n_a) = \left(\frac{1}{\bar{n}_a}\right) \exp(-n_a/\bar{n}_a). \tag{11}$$

Averaging over these distributions by the integrals:

$$\int_5^\infty (n_a - 5)N(n_a)\,dn_a \text{ and} \int_{49}^\infty (n_a - 49)N(n_a)\,dn_a, \tag{12}$$

yields the estimated relative strengths:

$$\text{Rel. Str. } (T < T_G) = (1 - \chi_c) \cdot \exp\left[\frac{-5 \chi_c}{n_c^*(1 - \chi_c)}\right] \tag{13}$$

$$\text{Rel. Str. (glass)} = (1 - \chi_c) \cdot \exp\left[\frac{-49 \chi_c}{n_c^*(1 - \chi_c)}\right]. \tag{14}$$

These curves are plotted in Figure 3(d).

Model E

Chain folding. This model is shown in Figure 2(e). The chain folds f times before entering the amorphous region. The results of the earlier models (B through E) can all be repeated here, with $n_c = f \cdot l_c$. This merely results in *large values* of n_c for a given crystallite dimension. Large values of n_c in turn lead to relative strength curves which approach the simple expression: Re. Str. $= (1 - \chi_c)$.

2(b): MODEL B. (Regular)

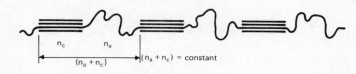

2(c): MODEL C. (Regular)

2(d): MODEL D. (Random)

$$n_c = \text{constant; average } n_a = \frac{n_c(1-X_c)}{X_c}$$

2(e): MODEL E. (Chain-Folding)

f = # folds/chain

FIGURE 2   Models.

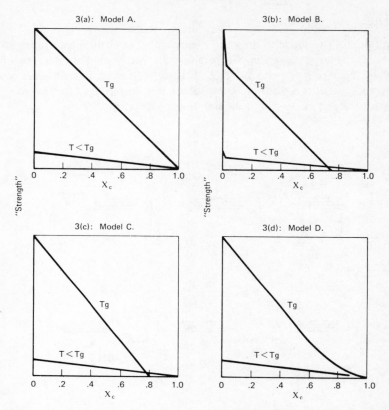

FIGURE 3   Calculated dependence of loss strength for T_g and $(T < T_g)$, for several models.

DISCUSSION OF RESULTS

The sharp initial drop in calculated strength of T_G shown in Figure 3(b) is a consequence of the discontinuity in structure on going from no crystallinity to any crystallinity, however small in amount. A plot of G'' vs χ_c in Figure 14 of Ref. 7 shows this initial rapid drop in strength.

The assumptions used in calculating the strength of the $(T < T_G)$ process require it going to zero at $\chi_c = 1$. That this contrasts with experimental observations is a direct consequence of definitions and/or physical measurements of χ_c. A term could be added to Eqs. (3), (8) or (11) which would be zero at $\chi_c = 0$, and increase, not necessarily linearly, to give a finite intercept at $\chi_c = 1$. Such an exercise is not germane to our purpose.

SUMMARY

The results of the above crude calculations can be summarized as follows:

1) The strengths of amorphous transitions in semi-crystalline polymers should decrease monotonically with increasing χ_c.

2) With some types of morphology, the curves can be expected to fall somewhat below the straight lines representing simple proportionality with $(1 - \chi_c)$.

3) Such downward deviations from linearity would be expected to be more pronounced for the glass transition than for the $(T < T_G)$ transition, and more pronounced in fine-textured morphologies than in coarse-textured morphologies.

4) Predicted patterns are rather *insensitive* to details of the model. We would hesitate to draw *any* structural conclusions based on a deviation from simple proportionality.

5) At the same time, these calculations suggest that the type of behavior shown in Figure 1 should be general to all semi-crystalline polymers, regardless of their morphologies, except for defects at high χ_c which were not considered in our models.

6) The related problem of extensive confusion in the literature over interpretation of amorphous loss peaks based only on intensities (tan δ_{max}, E''_{max}, G''_{max}, etc.) has been discussed elsewhere.[7]

7) The general question of the most significant measure of the "strength" of an amorphous transition is recognized, but is beyond the scope of this note.

8) The *ad hoc* assumption that relaxation strength is proportional to the number of chain atoms involved in a relaxation and also to the number of (amorphous) segments of that length yields predicted trends which are qualitatively similar to observed results.

References

1. N. G. McCrum, B. E. Read and G. Williams, *Anelastic and Dielectric Effects in Polymeric Solids* (Wiley, New York, 1967).
2. L. E. Nielsen, *Mechanical Properties of Polymers* (Reinhold, New York, 1962; 2nd Ed. 1974).
3. P. H. Geil, *Polymer Single Crystals* (Wiley, New York 1963).
4. R. L. Miller, in *Encyclopedia of Polymer Science and Technology*, 4, edited by N. Bikales (Wiley–Interscience, New York, 1966), p. 449ff.
5. A. Keller, *Rept. Progr. Phys.*, 31, 623 (1968).
6. J. D. Hoffman, G. T. Davis and J. I. Lauritzen, Jr., *Treatise on Solid State Chemistry*, 3, edited by N. B. Hannay, Chapt. 7 (Plenum, New York, 1975).
7. R. F. Boyer, *J. Polymer Sci.* C 50, 189 (1975), especially pp. 212–218.
8. R. F. Boyer, *J. Polym. Sci.*, C-50, 195–198 (1975).

9. M. Goldstein, *J. Chem. Phys.*, **51**, 3728 (1969).
10. M. Takayanagi, M. Yoshino and S. Minami, *J. Polymer Sci.*, **61**, S-7 (1962).
11. K. H. Illers and H. Breuer, *J. Coll. Sci.*, **18**, 1 (1963).
12. Y. Ishida, *J. Poly. Sci.*, *A-2*, 7, 1835 (1969).
13. H. A. Flöcke, *Kolloid.-Z. Z. Polym.*, **180**, 118 (1962).
14. N. G. McCrum, *J. Polymer Sci.*, **60**, S-3 (1962).
15. R. W. Gray and N. G. McCrum, *Polymer Letters*, 4, 639 (1966).
16. A. H. Scott, D. J. Scheiber, A. J. Curtis, J. I. Lauritzen, Jr. and J. D. Hoffman, *J. Res. Natl. Bur. Stand.*, 66A, 269 (1962).
17. K. Sanui, W. J. MacKnight and R. W. Lenz, *Macromolecules*, 7, 101 (1974).
18. N. G. McCrum, B. E. Read and G. Williams, *Anelastic and Dielectric Effects in Polymeric Solids* (Wiley, New York, 1967), pp. 136–140.
19. J. Heijboer, *Mechanical Properties of Glassy Polymers – Containing Saturated Rings*, Doctoral Dissertation (Vitgeverij Waltman, Delft, 1972) especially Chapter 2, Section 2.3, p. 18ff.
20. F. Gornick and J. L. Jackson, *J. Chem. Phys.*, 38, 1150 (1963).
21. J. I. Lauritzen, Jr. and R. Zwanzig, *J. Chem. Phys.*, 52, 3740 (1970).
22. I. C. Sanchez, *J. Macromol. Sci., Rev. Macromol. Chem. C-10*, 113 (1974).

DISCUSSION

Dr. H. G. McCrum (*Oxford*): You have used different measures of relaxation magnitude in your plot against crystallinity. The α relaxation of polychloro-trifluoroethylene is certainly associated with the glass transition ($T_G = 52°C$). But for the α relaxation a plot of G''_{max} against crystallinity is of positive slope (Gray and McCrum, *Polymer Letters*, 4, 639 (1966). Plots of J''_{max} and Λ''_{max} are of negative slope. What is your rationale for selecting the measures of relaxation strength that you have quoted?

T. Alfrey and R. F. Boyer: We used data supplied by the original authors, as shown in Table I. Because of Dr. McCrum's important question, our text has been modified to include a brief discussion and some pertinent literature references. See also the answer to a related question by Dr. McCrum appearing in the discussion portion of the paper by Stadnicki, Gillham and Boyer, this volume, page 362.

Molecular Motions and Configurational Thermodynamic Properties in Amorphous Polymers

ROBERT SIMHA

Department of Macromolecular Science, Case Western Reserve University, Cleveland, Ohio 44106, U.S.A.

1 INTRODUCTION

Relaxation processes in amorphous and semicrystalline systems have been explored by a variety of dynamic and thermodynamic methods, and have formed a basis for attempts at identifying the actual molecular processes involved in these multiple relaxations.

Changes in thermodynamic quantities, primarily heat capacities for the glass transition, and thermal expansivities for sub-glass relaxations as well, have served as diagnostic tools for the location of these processes. On the other hand these quantities characterize the equilibrium (or quasi-equilibrium) state of a system and these are the quantities which a molecular theory of the state aims to predict. The states of concern to us here are the melt, the corresponding glass, and transition zones between these and within the glass. The primary emphasis is to be on the glass and its relationship to the corresponding melt.

For a number of years a dichotomy has existed between two groups of investigators. On one side are the students of dynamic processes and on the other the phenomenological thermodynamicists who seek relationships between changes in different quantities at the transitions (the Ehrenfest-type relations), by postulating the freeze-in of structural parameters. The physical thought behind this postulate is that below T_g structure and configurational contributions are effectively frozen, that is, they no longer vary with pressure and temperature. For example, the thermal expansivity is due entirely to a vibration mechanism; no temperature dependent structural changes are available. On the other hand, it has

203

been amply demonstrated, that amorphous relaxations in both amorphous and semicrystalline systems are clearly resolved by measurements of the thermal expansivity. It will be recognized that this is a purely intermolecular property in contrast to the heat capacity. Thus this type of experimentation clearly indicates a temperature and pressure dependent gradual freeze-in.

It is important then to develop a theory which starts out with the melt, traverses the transition zone, and goes as far into the glass as possible. In particular, if a quantitatively satisfactory description of the equilibrium melt can be formulated, what modifications are required for the glassy state? The fact that in the latter the laws of thermodynamic equilibrium do not apply, places demands not only on the theoretical formulation, but equally on the design of meaningful experiments, which can be subjected to theoretical analysis.

2 THERMAL EXPANSION AND RELAXATIONS IN THE GLASSY STATE

We confine ourselves here to an illustration of the dilatometric results mentioned in the Introduction. In Figure 1 are displayed 3 sets of quantities for poly(cyclohexylene dimethylene terephthalate(*cis*)) as a function of temperature between *ca* 230 and 10 K, namely the linear expansivity, its temperature coefficient[1] and the logarithmic decrement.[2] We note first the breadth of the loss peak; second the corresponding double peak in $d\alpha'/dT$ and the closeness of T_Δ, the position of the former corrected to a frequency of $10^{-3.8}$ Hz,[1] to a dilatometric peak; and finally the precipitous reduction in α' below about 40 K. It is of interest to compare PCHDMT with poly(cyclohexylmethacrylate) in Figure 2. The loss modulus was determined by Heijboer.[3] In addition to qualitatively similar features, we observe the significantly narrower, but enhanced peak in $d\alpha'/dT$, whereas the dynamic loss peak increases considerably less. Since the latter has been ascribed to a chair—chair conformational transition in the cyclohexyl ring, one can conclude that relaxation processes arising from side-chain motions affect the thermal expansion more than the shear modulus. We will revert to the low temperature region displayed in Figures 1 and 2 in connection with the configurational thermodynamics of the glass.

3 THE LIQUID STATE

It will suffice for our purposes to briefly recapitulate the essentials of the theory to be employed here.[4] Recall the well known approximation due to Lennard-Jones and Devonshire which replaces the fluctuating field acting on a given

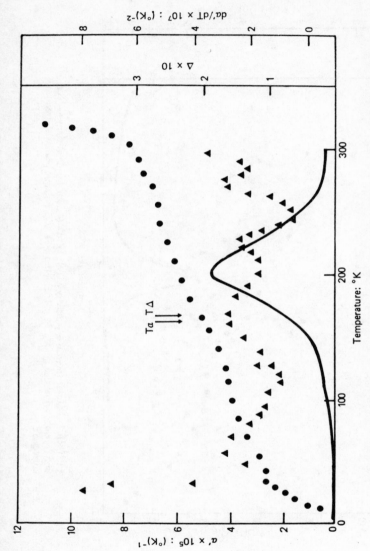

FIGURE 1 Linear thermal expansivity α' (○); $d\alpha'/dT$ (△); and log decrement (—) of poly(cyclohexylene dimethylene terephthalate-cis) T_α, dilatometric peak; $T_\triangle$, dynamic peak with frequency correction.

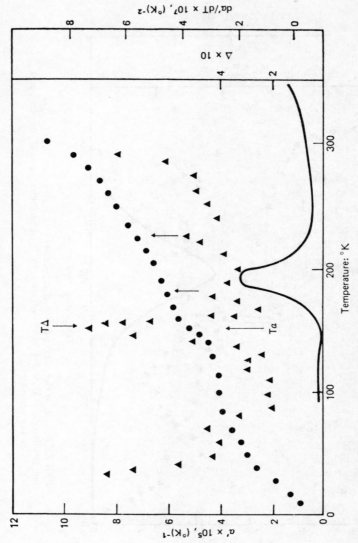

FIGURE 2 Linear thermal expansivity α' (○); $d\alpha'/dT$ (▲); and log decrement (—) of poly(cyclohexyl methacrylate). T_α, dilatometric peak; T_Δ, dynamic peak with frequency correction.

constituent particle by an average, which in turn is defined by placing the surroundings on a lattice. This approximation was extended by Prigogine and his
colleagues to fluids composed of chain molecules.[5] The defects of the original
theory in ascribing too much order to the system, carry over to amorphous
polymers,[6] albeit to a lesser degree than in the rare gas fluids. One reason for this
is the range of comparatively low (reduced) temperatures pertinent for the
former. Nonetheless, the deviations are significant. In order to retain the mathematical advantages of a lattice model, lattice vacancies or holes may be introduced. The number of these varies with volume and temperature in a manner to
be given by the theory through a maximization of the configurational partition
function.

The explicit results have been presented and discussed in detail on various
occasions. Hence it will suffice to summarize the working relations derivable
from the configurational partition function in symbolic form, with the tildes
indicating reduced variables and h the fraction of unoccupied sites, as follows:

Equation of state:

$$\tilde{P} = f[\tilde{V}, \tilde{T}; h(\tilde{V}, \tilde{T})]. \tag{1}$$

Minimum condition:

$$(\partial \tilde{F}/\partial h)_{\tilde{V}, \tilde{T}} = \phi(\tilde{V}, \tilde{T}; h; c/s) = 0 \tag{2}$$

with the solution

$$h = h(\tilde{V}, \tilde{T}; c/s). \tag{2a}$$

The combination of Eqs. (1) and (2a) then yields the explicit equation of
state. Here c/s represents the number of volume dependent degrees of freedom
per segment, with the ratio being of the order of unity for fluids of flexible
chains.[4] No further discussion of this factor is required for our purposes. Finally
we obtain the reduced internal energy $\tilde{U}[\tilde{V}; h(\tilde{V}, \tilde{T})]$ and entropy $\tilde{S}[\tilde{V}; h(\tilde{V}, \tilde{T})]$.
The scaling factors for the variables of state are explicitly defined in the theory
and depend on the details of molecular structure in a manner which need not
occupy our attention here. In practice they are derived from a superposition of
the experimental P–V–T surface of a given system onto the theoretical reduced
surface. This procedure then automatically determines the scaling parameters for
energy and entropy. We note from Eqs. (1) and (2) the validity of a practical
principle of corresponding states, provided the functional relationships resulting
from different values of the characteristic ratio c/s are practically superimposable.
This turns indeed out to be the case.[4]

Extensive comparisons between experiment, originating in our laboratory or
elsewhere, with theory are available,[7-11] and only a few typical results are shown
here. In Figure 3 a reduced atmospheric pressure isobar is plotted on a double

logarithmic scale. On this curve are superimposed experimental curves for several polymers and an oligomeric system.[8] This construction then yields the volume and temperature scaling parameters. The characterization of the melt is completed by means of pressure data. Figure 4 depicts the reduced compressibility factor $\tilde{P}\tilde{V}/\tilde{T}$ as a function of the reduced density at a series of temperatures for polymers of styrene and o-methyl styrene between their respective glass temperatures (100 and 130°C) and 200°C under a maximum pressure of 2 kbar. These figures illustrate the quantitative success of the theoretical equation of state and the existence of a corresponding states principle. We complete this section by exhibiting in Figure 5 the compressional energy for diverse polymers.[11] An explicit display of compressional entropies can be omitted, since they are immediately derived from the preceding graph and the equation of state.

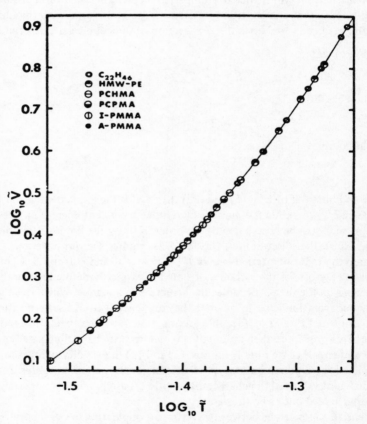

FIGURE 3   Thermal expansion of polymer melts and an oligomer fluid at atmospheric pressure. Solid line, theory, Eqs. (1) and (2a).

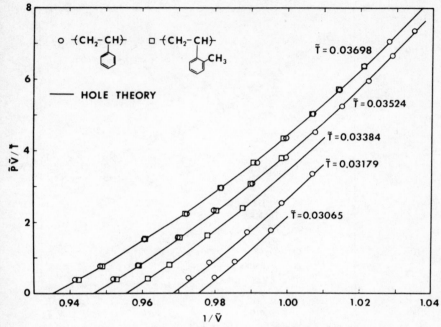

FIGURE 4   Reduced compressibility factor as a function of reduced density. Points, polystyrene and poly(o-methyl styrene). Solid line, theory, Eqs. (1) and (2a).

The information deduced from the equation of state studies of a broad spectrum of polymers and n-paraffins has provided a correlation between the specific
compressibility or specific entropy scaling factor and the scaling temperature.[11]
This proves useful for the prediction of high from low pressure data in addition
to structural implications.

The results illustrated give us some confidence in the computation or estimation of quantities which are not readily accessible to direct measurement. Specifically, we make use of the equilibrium theory in comparisons of the supercooled
melt with the corresponding glass.

4   THE LIQUID–GLASS TRANSITION AND THE GLASSY STATE

The point of view to be taken here is that under the conditions of the experiment
thermal equilibrium is established, but viscoelastic relaxation is negligible.
Consider then the modes of formation of and pressure cycles in the glass,[12] as
shown schematically in Figure 6. First we may cool under a fixed pressure, at-

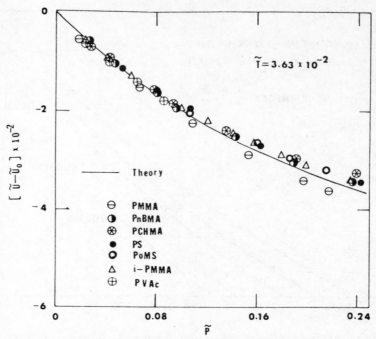

FIGURE 5   Compressional energy as a function of pressure for polymer melts in reduced coordinates. Solid line, theory.

mospheric or otherwise, starting out in the melt at point C. The glass may be subjected to various pressures, say from A to G, and temperatures, which ultimately return the system to a point F. In this manner a transition line B F is defined. Moreover, for a specified cooling rate r and formation pressure P' we can uniquely specify any (quasi)-equilibrium quantity Q as $Q(P, T, P', r)$ and a transition temperature $T_g^\dagger$ (P, P', r).[13] Thus by keeping P' and r fixed, we can define a constant formation glass. The rate r will usually be of the order of 30 to 5 K/h. It becomes then important to compare the state and other physical properties as well, of different constant formation glasses. A simple example of this is the well known densification effect.[14] Alternatively, a series of structurally different glasses may be prepared by different isobaric cooling paths, on pressurizing the liquid to different points D. This will yield a transition line B E. All possible variations with partial compression taking place in the melt, as exemplified by point I, are of course possible. We note the considerable difference between the two pressure coefficients of the transition arising from the differences in the compressibilities of liquid and glass. These have been considered for styrene and various methacrylate polymers in Refs. 7, 9, 11 and 15, with pressures extending up to 2 kbar. However in this work, a single constant forma-tion system with P' held at 1 bar was examined. A detailed study of polyvinyl acetate[10,16] extends only to 800 bar, but encompasses data for $P' = 1$ and 800 bar. Figure 7 depicts the transition lines for these two glasses, with transition temperatures indicated as $T_g^\dagger$, and for the variable formation system as T_g. All coordinates are in reduced form and the scale factors are: $P^* = 9380$ bar, $T^* = 9419$K, $V^* = 0.8141$ cm^3/g.[10] The lines must, of course intersect at 1 and 800 bar respectively.

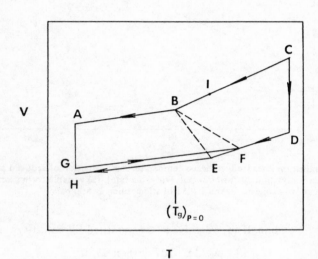

FIGURE 6   Alternative paths for glass formation. For explanation see text.

It is the relationships between these three lines which we wish to investigate now by means of the theory, specifically in terms of the structural parameter h. For this purpose we consider the pressure coefficients of T_g and $T_g^\dagger$. We have immediately:

$$dT_g/dP = (\partial T/\partial P)_h + (\partial T/\partial h)_p \times dh/dP. \tag{3}$$

On the other hand there obtains:

$$dT_g^\dagger/dP = \Delta\kappa/\Delta\alpha, \tag{3'}$$

where $\Delta\kappa$ and $\Delta\alpha$ are the respective changes in compressibility and thermal ex-pansivity at $T_g^\dagger$. Equation (3') is an immediate consequence of the intersection of two surfaces $V = V(T, P)$.[13,17] We now make the tentative assumption that for a constant formation system, the structural parameter should remain constant

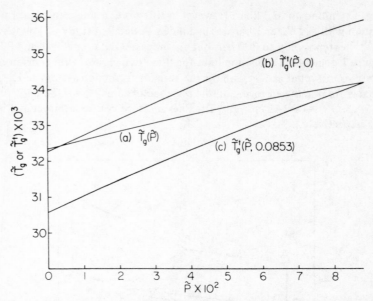

FIGURE 7   Transition lines (reduced glass temperature as a function of reduced pressure) for different thermodynamic histories of poly(vinyl acetate). (a) variable formation; (b) isobaric cooling at atmospheric pressure; (c) isobaric cooling at 800 bar.

along the transition line. If so, we arrive at the modified Ehrenfest-type equation

$$dT_g/dP = \Delta\kappa/\Delta\alpha + (\partial T/\partial h)_P \times dh/dP \tag{4}$$

In order to test these relations, we compute first the function $h(T, P)$ by eliminating in Eq. (2a) the volume with the aid of the equation of state. Figure 8 illustrates the behavior of h along the three transition lines. The T_g-curve bridges the two $T_g^{\ddagger}$-lines with a near constancy observed for the latter two, compared with the pronounced variation in the former. The analysis of poly(vinyl acetate) is continued in terms of Eq. (3′), where the partial derivative $(\partial T/\partial P)_h$ is substituted for the left hand side. Table I shows the numerical results.

TABLE I

Application of Eqs. (3), (3′) and (4)

	$\Delta\tilde{\kappa}/\Delta\tilde{\alpha}$	$(\partial\tilde{T}/\partial\tilde{P})_{\tilde{h}}$	$(\partial\tilde{T}/\partial\tilde{h})_p$	dh/dP	$d\tilde{T}_g/d\tilde{P}$	Eq. (4)
$P' = 0$	0.0480	0.0453	0.1746	−0.1036	0.0264	0.0299
$P' = 0.0853$	0.0373	0.0376	0.2080	−0.1093	0.0148	0.0146

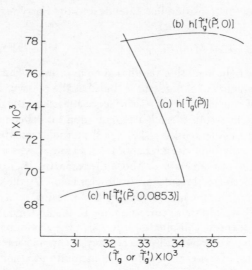

FIGURE 8   Theoretical hole fraction h as a function of T_g or $T_g^{\dagger}$, for variable, line (a), and two constant formation poly(vinyl acetate) glasses, as in Figure 7. For explanation, see text.

The agreement between the first two and last two columns respectively is gratifying indeed. For the constant formation glasses it is possible to predict the pressure coefficient of the transition temperature without a knowledge of the changes in compressibility and thermal expansivity. However the scaling parameters must be obtained from appropriate melt studies. A further prediction of the densification coefficients, i.e. the derivative $(\partial V/\partial P')_{P,T}$ has been presented recently.[18] We conclude moreover that the relationship between the transition lines of the two types of glasses is indeed regulated by the behavior of the hole function h and described by the modified Ehrenfest Eq. (4). A final observation is the non-constancy of h along the $T_g - P$ line, implying that no iso-condition for the onset of the glass transition can be formulated in the frame of this equilibrium theory. We can surmise this to be a general result irrespective of the particular theoretical formulation.

This is as far as an equilibrium theory can take us. When we turn now to the glassy state and wish to retain as a working hypothesis the formalism of the model, then at least the equilibrium condition imposed on the hole function, Eq. (2), must be abandoned and replaced by another condition to be introduced into the pertinent expressions for the thermodynamic functions.

A discussion in thermodynamic terms is appropriate only for glasses formed under constant conditions. Here we may test the assumption that the constancy of h observed along the transition line extends into the glassy region. In the relationship

$$-P = (\partial F / \partial V)_T = (\partial F / \partial V)_{T, h} + (\partial F / \partial h)_{T, V} \times (\partial h / \partial V)_T \qquad (5)$$

the product on the right hand side vanishes at equilibrium, when the first factor disappears. It also vanishes for constant h, and thus the pressure equation remains formally intact. Of course, the $P-V-T$ surface is altered by the different behavior of the h-function. The results of such a computation for different polymers have been surveyed first by Somcynsky and Simha,[19] and subsequently extended.[10,11,15] An illustration for poly(vinyl acetate) appears in Figure 9, where the volume is plotted as a function of temperature. Above the respective T_g's of the two glasses theoretical and experimental results agree closely. Below T_g the observed volumes deviate significantly from the lines computed by "freezing" h at T_g and are intermediate between these and those corresponding to the undercooled melt. The latter indicate the theoretical limits towards which the glass should relax upon prolonged annealing. To remove this discrepancy, an appreciable temperature and pressure dependence of h, i.e. a "structure" contribution to the thermal expan-

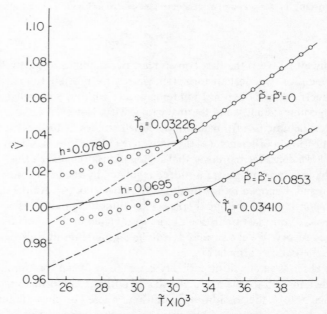

FIGURE 9   Volume–temperature relation for poly(vinyl acetate) system in reduced coordinates for the two cases $P = P'$. Circles experimental. Solid and dashed lines, equilibrium equation of state, Eqs. (1) and (2a). Solid lines below T_g, Eq. (1) subject to restraint:

$$h = h[T_g(P')] = \text{const.}$$

sivity must be maintained. While the ultimate goal is the derivation of this contribution from theory, we will consider h here as a disposable parameter.[15]

When the mathematical form of Eq. (1) is retained, but condition (2) disregarded and h instead derived from experiment, the results are illustrated in Figure 10 for the two glasses of poly(vinylacetate). Again the hole fraction is intermediate between the extrapolated equilibrium liquid and the value obtaining at T_g. Thus an intermediate "degree of freeze-in" of the unoccupied volume ensues and a characteristic ratio[8]

$$F.F. = 1 - \left[\left(\frac{\partial h}{\partial T}\right)_{P, \text{glass}} \Big/ \left(\frac{\partial h}{\partial T}\right)_{P, \text{liquid}} \right] \tag{6}$$

with limits zero and unity for the completely frozen glass and melt respectively, can be defined as another structural parameter. A comparison is exhibited in Figure 11 for different glasses, all formed at atmospheric pressure. Ignoring the loop, a systematic decrease of $F.F.$ with increasing reduced glass temperature is observed, with values ranging between 0.9 for poly(dimethylsiloxane) and less than 0.5 for poly(α-methyl styrene). We note once more the well known fact that the glass temperature is not a corresponding temperature in the equation of state sense, although there is a clustering around $\widetilde{T}_g$-values of 0.26–0.33. The correlation between $F.F.$ and the actual glass temperature can be expressed in first approximation by the linear relationship:[8]

$$F.F. = 1.107 - 1.346 \times T_g. \tag{7}$$

That is, high T_g-glasses are more "liquid"-like and less "frozen" at their respective T_g's than their opposites. Correspondingly, the unoccupied volume fraction $1 - y_g$ is larger for the former. The loop in Figure 11 is defined by a homologous series of n-alkyl methacrylates, starting with the methyl member. It reflects the well established initial effect of flexible side chains in increasing the thermal expansivity of glasses.[20] This effect is characterized here quantitatively in terms of the comparatively reduced frozen fraction for a given $\widetilde{T}_g$, followed by an increase beyond the n-hexyl compound.

The results are indicative of the gradual freeze-in of molecular motions in the glassy state and consistent with the results of the dilatometric studies illustrated earlier. Clearly, the concept of the glass as a completely frozen state, invoked in the past, is untenable. Moreover we expect the hole fraction to reflect the onset of sub-glass relaxations and this has been directly demonstrated for the β-relaxation region.[8,11] An indirect indication of a practically complete freeze-in results from the dilatometric studies at liquid He temperatures[1] and their theoretical interpretation.[21] In Figure 12 are shown on a reduced scale the thermal expansivities of several structurally diverse polymers as a function of temperature. The actual range displayed here extends from abou 14 to 190 K. The solid line is derived from the cell theory, that is the quasi-lattice model

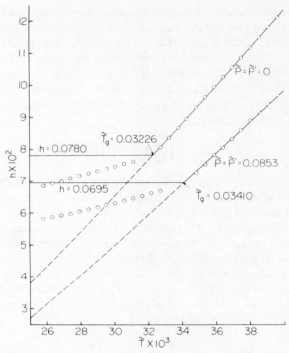

FIGURE 10   Hole fraction h as a function of temperature for poly(vinyl acetate). Circles from Eq. (1). Solid lines for $T \geqslant T_g$ and dashed lines, from equilibrium equation of state, Eqs. (1) and (2a). Solid lines for $T \leqslant T_g$, $h = const$.

alluded to in Section 3. Because of the low temperatures involved two factors must be taken into consideration. First, the cell potential given by the Lennard Jones–Devonshire theory is expanded to quadratic terms resulting in a force constant which is an explicit function of volume. Secondly, a fourth parameter in addition to the scaling factors appears, namely the limiting quantum temperature at $T = 0$. The scaling temperatures and volumes were determined by superposition of the volume-temperature curves, all polymers taken at atmospheric pressure, with the additional assumption of a *universal* reduced quantum temperature. The agreement between experiment and theory at low temperatures is noteworthy by itself. Important for our point of view however is the success of a theory which disregards "free volume" contributions to the thermal expansion below 50 K or so, depending on the polymer, in any case far below the glass temperature, and its systematic failure thereafter.

Having obtained the structural parameter h from a combination of the equation of state (1) with experimental $P–V–T$ data, one may proceed to the thermo-dynamic functions. Before doing this however, we reconsider Eq. (5) and the

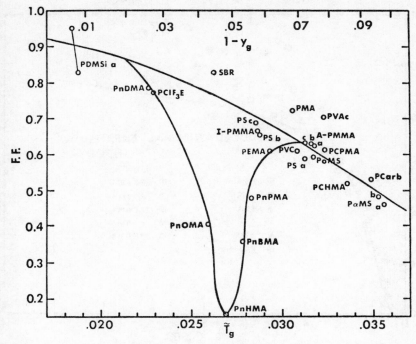

FIGURE 11 Correlation between frozen fraction, Eq. (6), $\tilde{T}_g$ and hole fraction $h = 1 - y$ at T_g.

analogous temperature derivative of the free energy. For the equilibrium fluid, Eq. (2a), combined with the appropriate derivatives of the configurational partition function Z, yields the expressions for the internal energy and the entropy.[4] This involves the use of the extremum condition $(\partial Z/\partial h)_{V,T} = 0$. Hence, it is inconsistent to evaluate the thermodynamic functions by employing the expressions applicable to the equilibrium liquid. By recalling the underlying quasi-crystalline model and considering the expression for U as simply a lattice sum, the use of the equilibrium expression may be defended. However, no analogous rationalization for an entropy relation exists. In any case a consistent procedure requires the use of the configurational partition function to obtain, so far from experiment, the function $h(V, T)$. The method adopted[22] retains the original mathematical form of the partition function,[4] treats Eq. (5) as a partial differential equation in h, where the derivatives of the free energy are known functions of h, and employs the experimental P–V–T surface. We illustrate in Figure 13 the results for the compressional energy ΔU of glassy poly(vinylacetate) formed under atmospheric pressure. The dashed lines represent the experimental values derived by integration of the empirical Tait equation. The adjustment procedure

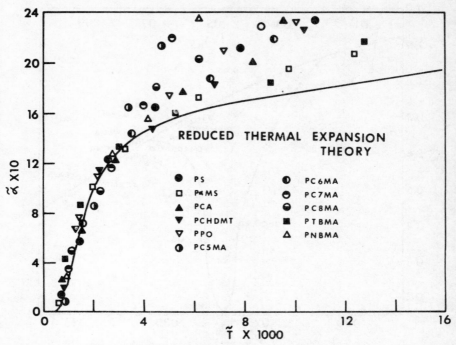

FIGURE 12   Low temperature thermal expansivity as a function of temperature in reduced
coordinates for different polymers. Solid line, theory. For specification of polymers, see
Figure 6, Ref. 21.

(A.P.) yields quite good estimates but an improvement results from the consistent use of the partition function (P.F.). Analogous findings in respect to the entropy are of course to be anticipated. A pronounced improvement however, ensues for the internal pressure.[22] In this manner we have obtained a consistent thermodynamic description based on the concept of partial freeze-in, at the expense however, of a greater numerical effort than in the simpler adjustment procedure. It is gratifying then to note that the numerical changes in h and F_T are relatively small. For example, F_T for the glass depicted in Figure 13 changes from 0.74 to 0.82 at $T = T_g$ and atmospheric pressure.[22] The comparative conclusions for different glasses discussed above remain valid.

Earlier we have compared the freezing fractions of different polymer systems at their respective T_g's and atmospheric pressure. It is of interest now to compare the two constant formation glasses of PVAc. At $P' = P = 1$ bar and the corresponding T_g, $F_T = 0.82$. At $P' = 800$ bar, but the same T and P, $F_T = 0.81$.[22] The previous values, obtained by the adjustment procedure were 0.74 and 0.75 respectively. The point here is the near equality of the two values, notwithstanding the differences in the densities, i.e. the well known densification effect. This could be indicative of a common thermodynamic characteristic of the two glasses. Or it may simply mean that F_T does not represent a sufficiently sensitive parameter to reflect other physical property changes.

The discussion has so far dealt with the pressure effects on the thermodynamic functions, which are derivable from the equation of state. A complete analysis requires calorimetric data to obtain the differences between melt and glasses at a given T and P. McKinney and Simha have offered some theoretical values[22] and comparisons with calorimetric data on the same sample.[23] As an example, at 0°C and atmospheric pressure, i.e., about 30°C below T_g, the enthalpy differences between low pressure glass and melt and depressurized glass and melt are about 7.4 and 3.4 J/g respectively. Thus the difference between the two glasses is predicted to be 4 J/g, with the corresponding entropy difference of 0.014 J/(gK). These are indeed large differences which should be experimentally established. Next we can separate the entropy into a lattice and a hole contribution, by setting

$$S(y) = S(1) + S_{hole}.$$

Here $S(y)$ is the expression derived from the hole theory with y determined by Eq. (2a) for the melt and by the procedure described earlier for the two glasses. $S(1)$ is obtained by setting $y = 1$ throughout. At 0°C and atmospheric pressure, the hole contribution in the undercooled liquid amounts to 24% of the total

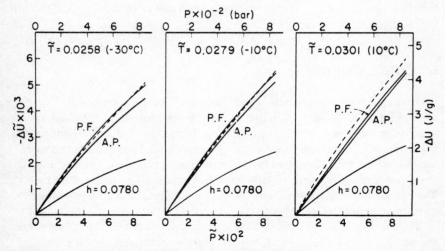

FIGURE 13   Compressional energy of poly(vinylacetate) glass, formed by isobaric cooling at atmospheric pressure, as a function of pressure. Dashed lines, from experimental equation of state. Lines marked P.F. and A.P., partition function and adjusted parameter procedure respectively. Other lines, $h = const$. For explanation, see text.

computed intermolecular entropy, whereas it is 29 and 26% for the $P' = 0$ and 800 bar glasses respectively. Thus a significant residue of hole entropy remains in the glass, but the differences between the three fractions are not large. Upon lowering the temperature to $-30°C$, the corresponding numbers become 17, 26 and 23 respectively. As is to be expected, differences between undercooled liquid and glass are enhanced but there is little variation in the relationship between the two glasses. The differences between the total intermolecular entropies of the two glasses at both temperatures amount to about 0.01 J/(gK), certainly, a significant and calorimetrically detectable amount, with the pressurized glass having of course the lower entropy. The corresponding enthalpy difference equals 3.8 J/g.

Most experimentation available so far concerns the difference between the liquid and glass at the transition temperature and specifically the heat capacity C_p. Consider first the liquid. The intermolecular contributions as computed by means of our theory, amount to about $11-13\%$ of the measured C_p for polystyrene and atactic poly(methylmethacrylate), varying slightly with temperature. For poly(vinylacetate) this ratio is 17.5%, based on unpublished data[23] yielding 1.92 J/(gK), whereas a value of 0.337 J(/gK) is derived from the theoretical entropy and the relation

$$C_p = T(\partial S/\partial T)_p.$$

The same set of measurements yields for the difference at $T = T_g = 304$ K, $\Delta C_p = 0.59$ J/(gK). The computed difference on the other hand is 0.234, a considerably smaller value.[22] The corresponding entropy differences are 0.079 and 0.0232 J/(gK).

5 CONCLUSIONS

We have shown that quasi-static measurements of the thermal expansivity are very useful tools for the exploration of relaxational motions in the glassy state, in contrast with the results of the heat capacity method. Full correspondence in the results of conventional low frequency dynamic methods and the enhanced resolution have been illustrated here for amorphous polymers. Detailed and analogous findings for semicrystalline systems have been presented elsewhere.[24-27] The bearing of thermal expansivity data at low temperatures on free volume questions[28] is not discussed here.

The theory of the liquid state outlined is of interest per se, since it describes successfully the equation of state and the changes of the volume dependent thermodynamic properties with pressure. We have utilized it to establish the connection with the corresponding glasses obtained by different formation histories and the relationship between the latter. This has been accomplished so far in a semiempirical manner by treating the hole fraction in the glassy state as an adjustable parameter, to be obtained from the experimental equation of state. It has allowed us to demonstrate the constancy of h along the transition lines of constant formation glasses and to predict the pressure dependence of the thermo-dynamic functions without any further parametric adjustments. The relaxation processes revealed by the dilatometric studies, the behavior of the hole function h in the glass, and the results of the low temperature theory all indicate that the freeze-in of structure in polymeric glasses, rather than being solely characteristic of and completed at the liquid–glass transition, is gradual over the temperature range of the glass and practically completed only far below T_g.

Significant differences between experiment and theory have emerged from an analysis of the differences in heat capacity and entropy between liquid and glass. This is so irrespective of measurement and extrapolation errors which might reduce ΔC_p from the value of 0.59 for poly(vinyl acetate) quoted on p. 220 to 0.48 J/(gK).[29] Underlying this and the calculation of the other thermodynamic quantities has been the basic assumption that the differences between the liquid and the glass can be interpreted solely in terms of the different behavior of the hole function h. It is possible however that a change in the number of volume dependent degrees of freedom must be allowed for in addition. If so, this will tend to increase the computed differences between liquid and glass. Such an effect would have to be insensitive to pressure, at least in our range of observations, i.e. up to 2 kbar. Clearly a more detailed analysis of thermal data is required. An independent separation of the intermolecular contribution to the heat capacity is desirable in this context.

Within the frame of the theory employed, structure in the liquid or glass is characterized by the h-function. For a given polymer glass, a characteristic quantity is the frozen fraction, Eq. (6), which differs significantly, as does h at T_g, from system to system. The question then arises whether corresponding differences in respect to properties other than the density are observable.

We have left open the question of the size distribution or in other words, the clustering of holes in the model. Considering the relatively large values of h in the high T_g polymers, see Figure 11, clustering can be anticipated.[11] This may form a basis for a quantitative interpretation of the reduced variation of h, see Figure 10, and thus of the equation of state properties below T_g. Answers to this as to the other questions raised remain to be found.

In conclusion, it is a pleasure indeed to acknowledge frequent and fruitful discussions of the questions considered with Raymond F. Boyer.

The research described was supported by grants GH-36124 and DMR 75-15401 from the National Science Foundation.

References

1. J. M. Roe and R. Simha, *Int. J. Polymeric Mater.* 3, 193 (1974).

222   R. SIMHA

2. A. Hiltner and E. Baer, *J. Macromol. Sci. Phys B* **6**, 545 (1972).
3. J. Heijboer, *J. Polymer Sci. C* **16**, 3412 (1968).
4. R. Simha and T. Somcynsky, *Macromolecules* **2**, 342 (1969).
5. I. Prigogine, N. Trappeniers and V. Mathot, *Faraday Soc. Disc.* **15**, 93 (1953).
6. V. S. Nanda, R. Simha and T. Somcynsky, *J. Polymer Sci. C* **12**, 277 (1966), where further references are given.
7. A. Quach and R. Simha, *J. Appl. Phys.* **42**, 4592 (1969).
8. R. Simha and P. S. Wilson, *Macromolecules* **6**, 908 (1973).
9. A. Quach, P. S. Wilson and R. Simha, *J. Macromol. Sci. Phys. B* **9**, 533 (1974).
10. J. E. McKinney and R. Simha, *Macromolecules* **7**, 894 (1974).
11. O. Olabisi and R. Simha, *Macromolecules* **8**, 211 (1975).
12. U. Bianchi, A Turturo and G. Basile, *J. Phys. Chem.* **71**, 3555 (1967).
13. M. Goldstein, *J. Phys. Chem.* **77**, 667 (1973).
14. N. I. Shishkin, *Sov. Phys.-Solid State* **2**, 322 (1960).
15. A. Quach and R. Simha, *J. Phys. Chem.* **76**, 416 (1972).
16. J. E. McKinney and M. Goldstein, *J. Res. Natl. Bur. Stand. A* **78**, 331 (1974).
17. E. Passaglia and J. M. Martin, *J. Res. Natl. Bur. Stands. A* **68**, 273 (1964).
18. J. E. McKinney and R. Simha, *Bull. Amer. Phys. Soc.* **20**, 310 (1975); *J. Res. Natl. Bur. Stands. A* **81**, 283 (1977).
19. T. Somcynsky and R. Simha, *J. Appl. Phys.* **42**, 4545 (1971).
20. R. Simha and R. F. Boyer, *J. Chem. Phys.* **37**, 1003 (1962).
21. R. Simha, J. M. Roe and V. S. Nanda, *J. Appl. Phys.* **43**, 4312 (1972).
22. R. Simha and J. E. McKinney, *Macromolecules* **9**, 430 (1976).
23. C. A. Angell, private communication to J. E. McKinney.
24. P. S. Wilson, S. Lee and R. F. Boyer, *Macromolecules* **6**, 914 (1973).
25. S. Lee and R. Simha, *Macromolecules* **7**, 909 (1974).
26. J. B. Enns, M. S. Thesis, Case Western Reserve University, June 1975, *J. Macromol. Sci. Phys. B* **13**, 11 (1977); ibid. B **13**, 25 (1977).
27. S. Lee, R. F. Boyer and R. Simha, in preparation.
28. See R. Simha and C. E. Weil, *J. Macromol. Sci. Phys. B* **4**, 215 (1970).
29. B. Wunderlich and L. D. Jones, *J. Macromol. Sci. Phys. B* **3**, 67 (1969).

DISCUSSION

D. J. Meier (*MMI*): Your data indicate that a single ordering parameter is not adequate. Does this mean that the distribution of "holes" should be studied?

R. Simha: As we have seen, in the equilibrium melt it suffices to treat the *total* hole fraction as the structural parameter. In the glass, the partial freeze-in illustrated in Figure 10 suggests indeed a consideration of the size distribution or, in a lattice model the clustering of holes, and a replacement of the equilibrium restraint, Eq. (12a), by a freeze-in condition for, say the isolated holes only. As is seen in Figure 11, at T_g the fraction of holes and hence the number average cluster size are comparatively large in high T_g (low frozen fraction) systems. For the polymers in Figure 11, the mean size varies between 2.5 and 1.1 units.[11] Therefore, a significant hole contribution to the thermal expansion and other thermodynamic properties of the high temperature glass can, at this point, be qualitatively understood. However, one should not conclude that a *single, frozen* ordering parameter is necessarily inadequate. A maximum frozen hole size, for example, could represent such a parameter.

M. Shinohara (*Dow Corning Corp.*): I have what may be a very naive question. Would you please explain why one can use thermodynamic equilibrium theory to treat polymeric glasses which may be super-cooled or super-heated?

R. Simha: Equilibrium thermodynamics applies to the description of the liquid—glass transition lines. As for the glassy state itself, there are several prerequisites to be examined: Compared with the rate of experimentation, thermal and pressure equilibrium is rapidly established, but viscoelastic relaxation is slow. Thus time dependent processes do not need to be considered and for a given P and $T < T_g$, a stationary value of V obtains. In the experiments analyzed here, these conditions are satisfied. For the purposes of our discussion, the fact that the glass is not a true equilibrium system, is manifested by the dependence of properties on the thermodynamic history. That is, any extensive quantity Q is a function not only of the ambient pressure and temperature, *after* the glass has been formed, but depends also on the cooling rate and the formation pressure P', as discussed on p. 210. However, once the glass has been generated under a specified constant set of formation conditions, we are dealing with a thermodynamically reversible system, i.e. the results obtained are independent of the path and we can define, for example, a unique equation of state. This is not the case for the variable formation histories illustrated in Figure 6.

T_g of Polymers by the Electron Spin Probe Technique[1]

PHILIP L. KUMLER†

Department of Chemistry, Saginaw Valley State College, University Center, Michigan 48710, U.S.A.

and

RAYMOND F. BOYER‡

The Dow Chemical Company, Midland, Michigan 48640, U.S.A.

Since the pioneering studies by McConnell[2] in the early sixties, electron spin resonance (ESR) has been used extensively in biological systems to study molecular motion. Since that time the use of "spin-labels" — stable free radicals that can be attached to a specific site on a molecule in a complex system and whose ESR spectrum would contain information about the environment of the label — has become quite useful and widespread within the biological sciences.[3] In view of the extensive use of this technique by biological scientists it is somewhat surprising to find very limited extension of ESR techniques to macromolecular science.

The work to be described today is an attempt to determine the potential utility of ESR — specifically a "spin-probe" technique — as a tool for the study of molecular motion in polymeric materials.

The theoretical basis for the ESR phenomenon is quite similar to that of the NMR phenomenon. ESR spectroscopy is, however, only applicable to a system that contains at least one unpaired electron. A free electron possesses a spin (S) of $\frac{1}{2}$, and can (in the absence of an external magnetic field) exist in either of two equal energy or degenerate spin states ($M_s = +\frac{1}{2}$ or $M_s = -\frac{1}{2}$). This degeneracy is removed in the presence of an external magnetic field (see Figure 1) causing one Zeeman level to increase in energy while the other decreases. The lower energy state ($M_s = -\frac{1}{2}$) corresponds to the parallel alignment of the magnetic moment

---

†Present address: State University of New York, College at Fredonia.
‡Present address: Midland Macromolecular Institute.

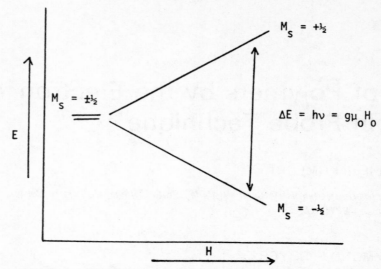

FIGURE 1   Splitting of the electron Zeeman levels in a magnetic field.

of the electron to that of the external field while the higher energy state
($M_s = +\frac{1}{2}$) corresponds to the antiparallel situation.

If electromagnetic radiation of frequency ν is present which satisfies the
resonance condition, then transitions between these Zeeman levels can occur.
The conditions for resonance to occur are defined by Eq. (1):

$$\Delta E = h\nu = g\mu_0 H_0 \tag{1}$$

The g factor in Eq. (1) is a universal constant related to the magnetic field *at the
electron*; for a completely "free" electron the value is 2.00232. When an
external magnetic field is applied, however, an internal magnetic field is
generated in the sample which will either add to or subtract from the external
field. This will cause a deviation in the value of g (from 2.00232) for electrons in
real systems and the g value may thus be considered as a quantity characteristic
of the environment in which the unpaired electron is located. It should be noted
that the g value for a radical corresponds to the chemical shift in an NMR
experiment. μ_0 is the electronic Bohr magneton and H_0 is the strength of the
applied magnetic field. Typical of the conditions necessary to satisfy the
resonance condition for radicals are an applied magnetic field of 3400 gauss and
a frequency in the microwave region around 9500 MHz.

A block diagram of a typical ESR spectrometer is shown in Figure 2. The
essential components of an ESR spectrometer are: (1) a source of microwave
radiation — typically a klystron, (2) a sample cell or cavity connected to the
klystron by a waveguide, (3) a direct-current magnetic field, (4) a detection

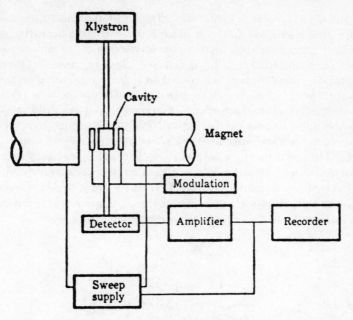

FIGURE 2   Block diagram of an ESR spectrometer.

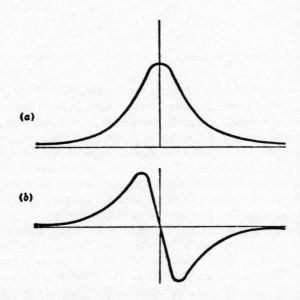

FIGURE 3   (a) Absorption signal (b) First derivative of absorption signal.

system, and (5) a recorder or oscilloscope. The klystron oscillator produces monochromatic microwave radiation which is transmitted to the cavity by a waveguide of dimensions appropriate to the wavelength of the microwave radiation. The transmitted radiation is detected (diode rectifier) and the signal, after appropriate amplification, is recorded on a chart or displayed on an oscilloscope. Instead of displaying the signal as the absorption curve (typical of NMR experiments), phase-sensitive detection is employed and the displayed signal closely approximates the first derivative of the absorption signal (Figure 3).

In the present work we have used a class of stable free radicals called nitroxides (Figure 4a) that have been extensively used in spin-label studies of biological molecules. Utilization of nitroxides that possess gem dimethyl substituents in both alpha positions (Figure 4b) confers a high degree of stability to the radical and also results in a quite simple three line spectrum. Specifically we have used nitroxides of the general formula shown in Figure 4c.

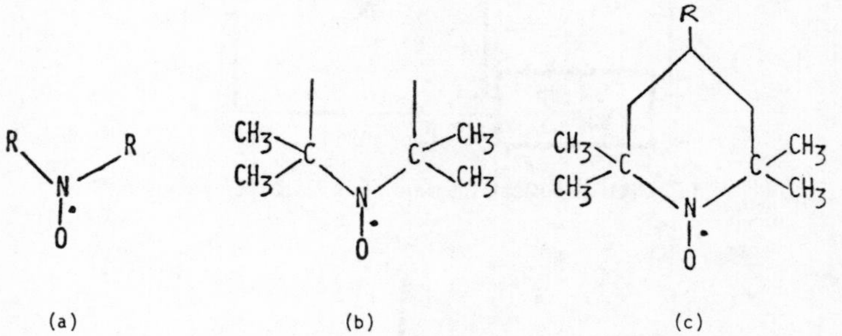

FIGURE 4   General molecular formulas for nitroxides.

It is instructive to briefly consider the reasons why the ESR spectra of nitroxides of type 4c appear as a three-line spectrum. In most free radicals the unpaired electron is present in a molecular orbital that encompasses more than one atom. If one (or more) of these atoms possesses a nuclear magnetic moment then, because of the interaction of this nuclear magnetic moment with the electron spin, the Zeeman levels suffer additional splitting and hyperfine structure (hfs) appears in the ESR spectrum. The nuclear interaction which causes this splitting is termed hyperfine interaction and the corresponding splitting of the spectral lines is called hyperfine splitting. For the particular case of interest here — that is nitroxides of type 4c — the unpaired electron interacts with the nuclear spin of nitrogen ($S = 1$). In a static magnetic field the nuclear spin of nitrogen can assume $2S + 1$ values, that is +1, 0, or −1. The magnetic moment of the electron will interact with each of these three levels and the Zeeman levels are thus split into three sublevels (see Figure 5). When the resonance conditions of field strength and frequency are satisfied, the electron spins change their orientation and absorb the energy of this field, while the nuclear spins do not change their orientation during the time required for the electronic transition. Thus, electronic transitions occur between levels with the same projections of nuclear spin. The selection rule corresponding to this

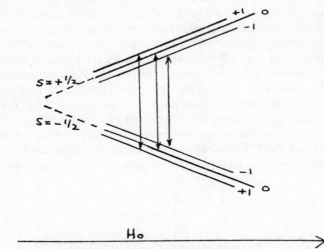

FIGURE 5   Hyperfine splitting of the energy levels of an unpaired electron with a nitrogen nucleus.

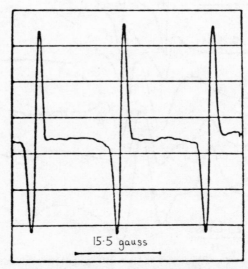

FIGURE 6   ESR spectrum of nitroxide 4c (R = OH) in ether solution.

situation is of the form $\Delta m = 0$, and thus the hyperfine structure of the ESR spectrum will consist of three equidistant lines, all of equal intensity (see Figure 6). As previously stated the presence of gem dimethyl substituents "insures" the simplicity of the spectrum; if there were hydrogen atoms (rather than methyl groups) at any of the alpha positions additional hyperfine splitting would be expected from interaction with the nuclei of the protons.

The ESR spectrum shown in Figure 6 is characteristic of a nitroxide of type 4c undergoing very rapid tumbling and rotational reorientation, such as a dilute solution of this nitroxide in a solvent of very low viscosity; the hyperfine splitting (hfs) is 15.5 gauss. The observed spectra of nitroxides are, however, quite sensitive to a number of parameters such as the polarity of the environment, the actual molecular motion of the nitroxide spin-probe, and the orientation of the spin-probe molecule with respect to the applied magnetic field. As an example of this type of effect Figure 7 shows the ESR spectra of nitroxide 4c (R = OH) in aqueous solutions of varying glycerol concentrations. As the glycerol concentration is increased (A to F) two related phenomena are seen to occur. The individual lines of the three-line spectrum broaden (and not at the same rate)

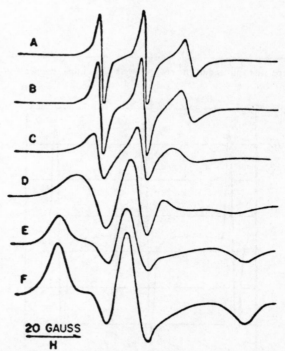

FIGURE 7   ESR spectra of nitroxide 4c (R = OH) in aqueous solutions of varying glycerol concentrations. Taken from: J. Hsia and L. Piette, *Arch. Biochim. Biophys.* **129**, 296 (1969).

and the separation between the outermost lines (or any two lines) increases. Both of these results are due to the fact that the nitroxide's rate of tumbling is decreasing as the viscosity of the solvent increases; the rotational correlation time, τ, increases as the viscosity of the solvent increases. The spectrum in Figure 6 is characteristic of a rapidly tumbling nitroxide ($\tau \simeq 10^{-10}$ s) while a spectrum such as Figure 7F is characteristic of a slowly tumbling nitroxide ($\tau \simeq 10^{-7}$ s); spectra such as 7F are often referred to as that of a "strongly immobilized nitroxide". It should be noted that even in spectrum 7A there is some asymmetry and line-broadening present; at this viscosity there is already some observable decrease in the rate of rotational reorientation. It should also be noted that the separation of the two outermost peaks in spectrum 7F (62 gauss) is approximately twice that of the same separation (31 gauss) in spectrum 6.

Although nitroxide spin-probes had been extensively used to study the molecular motion and structure of liquids and biological systems, the use of spin-probes to study motion in non-biological polymers was pioneered by Rabold[4] and others[5-7] in the late sixties. In his pioneering work Rabold doped a series of polymers with low (less than 0.01% by weight) concentrations of nitroxide 4c (R = —O—CO—Ph). The ESR spectra of the doped polymers were recorded at a series of temperatures. Rabold observed that at low temperatures the spectra were quite similar to figure 7F while at high temperatures they were quite similar to Figure 7A. He plotted the separation (in gauss) between the outermost lines of the spectrum as a function of temperature and sigmoidal curves as shown in Figure 8 resulted for a series of polymers. He

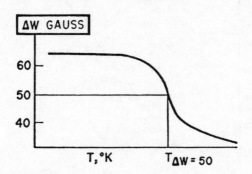

FIGURE 8   Extrema separation versus temperature plot.

selected as a parameter characteristic of the line shape the temperature at which the extrema separation was equal to 50 gauss; this parameter was called $T_{\Delta W = 50}$ (we have now renamed this parameter T_{50G}). For a series of four polymers [polyethylene, polypropylene, poly(vinyl chloride) and polystyrene] Rabold observed that there was a linear correlation of T_{50G} with either the glass temperature (T_g) or the melting temperature (T_m). This early work, as noted by Rabold, suggested the potential utility of the ESR spin-probe technique to gain information about molecular motion in polymeric materials.

The utility of the spin-probe technique to study molecular motion in polymers was extended by Boyer in 1973.[8] Using published and unpublished data of Rabold, he was able to correlate the T_{50G} values for a series of nine polymers with accepted values of the glass temperature (T_g). It was observed that when T_{50G} values for the nine polymers were plotted versus T_g values, the points appeared to lie on a straight line (Figure 9). Accepting this linear correlation of T_{50G} with T_g, and extrapolating Rabold's ESR data for a series of ethylene copolymers to amorphous polyethylene, it was argued (on the basis of ESR and other pertinent data) that the T_g of amorphous polyethylene was 195 ± 10 K. This was the first attempt to use ESR spin-probe data to resolve conflicting opinion about the T_g of polymers for which the temperature of the glass transition was a matter of controversy.

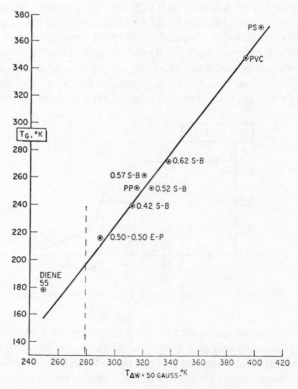

FIGURE 9   Correlation of T_{50G} values with T_g values.

On the basis of the above pioneering work we were interested in the plausibility of utilizing the ESR spin-probe technique to study the glass transition in polymeric materials. A number of potential applications of the technique were obvious. Perhaps this method could be useful for determining the T_g of polymers for which there was controversy about the glass transition temperature; the method could potentially be used as a rapid and reliable method for determining T_g of polymers for which limited (or no) T_g data was available in the literature. The primary goal of the work we have initiated, whose preliminary results will be described here, was to determine the validity, scope, and utility of this technique for a wide-ranging series of polymeric materials.

There are a number of questions that we feel need to be resolved and/or answered to substantiate the validity and utility of the technique.

1) Does the apparent linear correlation between T_{50G} and T_g, as suggested by Rabold and Boyer, hold over an even wider temperature range than they have investigated? The polymers studied in the earlier works had T_g's varying from 179 K (polybutadiene) to 373 K (polystyrene). Would the linear correlation be valid for polymers with both higher and lower T_g's.

2) Would the linear correlation be valid and would the method be useful for polymers other than the vinyl type? The earlier works were limited to polymers of this type and it was of obvious interest whether the technique would be valid for other type polymers and copolymers.

3) Was the method really useful for resolving conflicting opinion about the actual T_g in cases where controversy exists?

4) What would be the effect of molecular weight on the observed T_{50G}? Would in fact the T_{50G} vary in a predictable and reproducible fashion as a function of chain length as the T_g does?

5) How is the ESR spin-probe method affected by the detailed nature of the probe? How do variations in the size, shape, polarity, and functionality of the nitroxide spin-probe affect the observed T_{50G}?

6) What would be the effect of covalently bonding the nitroxide to the polymer chain rather than mechanical incorporation of the nitroxide into the polymer matrix? Could this technique be utilized as a spin-labelling technique, similar to the extensive spin-labelling studies that have been done with bio-polymers?

7) Is the method limited to the study of glass transitions or could it in fact be used to gain information about other molecular transitions?

The first question that we attempted to resolve was whether or not the apparent linear correlation of T_{50G} with T_g would be valid over a much wider temperature range. We have determined the T_{50G} values for a series of polymers that both duplicate some of the earlier data and includes polymers with both higher and lower T_g values than previously studied. During this study it became apparent that with some polymers (for example polycarbonate) the extrema separation changed quite slowly over a wide temperature range; that is the plot of extrema separation versus temperature gave a quite broad sigmoidal curve (Figure 10). With other polymers, for example poly(dimethylsiloxane), the extrema separation changed quite rapidly over a narrow temperature range and a "narrow" sigmoidal curve resulted (Figure 11). There are, however, a number of important similarities between the "broad" and the "narrow" curves. For either type of curve the upper limit of extrema separation falls between 60 and 65G at low temperature and the spectrum is characteristic of a "highly immobilized" nitroxide; also with either type of curve the lower limit of extrema separation

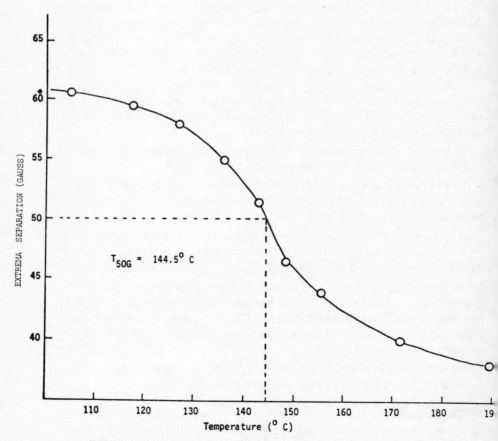

FIGURE 10   Extrema separation versus temperature for polycarbonate.

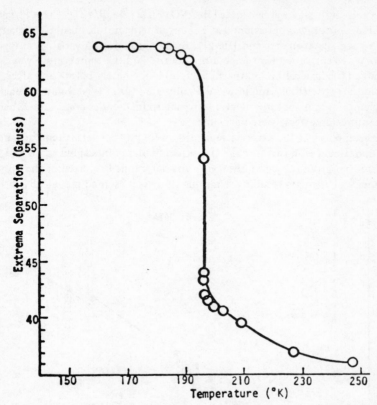

FIGURE 11   Extrema separation versus temperature for poly(dimethyl siloxane).

falls between 35 and 40G at high temperature and the observed spectrum approaches that of a freely-tumbling nitroxide.[9] A freely-tumbling nitroxide would be expected to have an extrema separation of 31G (see Figure 6). In addition it should be noted that with either the "broad" or "narrow" sigmoidal curve the T_{50G} corresponds very closely to an inflection point on the curve; therefore, at T_{50G} the extrema separation is changing most rapidly and therefore the rate of rotational reorientation of the probe molecule is undergoing its most dramatic change. It is for these reasons that we feel that the observed T_{50G} is a quite reasonable parameter to correlate with the glass temperature.

For the study of the correlation of T_g with T_{50G} we chose a series of 15 polymers and copolymers of differing structural type covering a wide range to T_g's. Only polymers for which there was little or no controversy about the T_g were included in this phase of the study. Each of the polymers was doped with low concentrations (less than 0.01% by weight) of 2,2,6,6-tetramethyl-4-

hydroxypiperidine-1-oxyl benzoate (BzONO, Figure 4c, R = O—CO—Ph) and
the ESR spectrum was determined at a series of temperatures using a Varian
4502 X-band spectrometer and 100 kHz field modulation. A variable temperature
accessory was employed for temperature control and the temperature was
determined (using a calibrated thermocouple) immediately before and after
recording the spectrum. Samples were equilibrated for at least 3 min for each
10°C change in temperature; spectra were determined under conditions where
power saturation effects were negligible.

For each polymer the extrema separation were plotted versus temperature
and a smooth curve drawn through the experimental points (Figures 10 and 11,
for example). From this curve the T_{50G} was determined for each of the polymers.
A summary of the data resulting from this study is presented graphically in

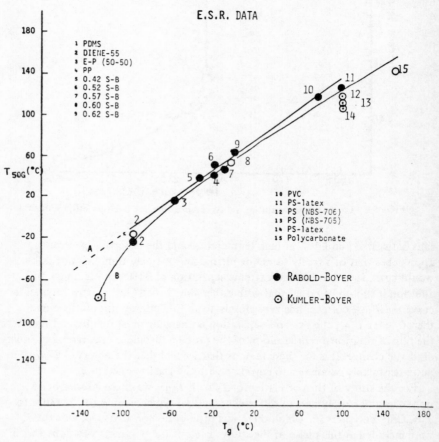

FIGURE 12   Correlation of T_g with T_{50G}.

Figure 12. There are a number of pertinent features to the data. Line A in Figure 12 is the linear correlation previously suggested by Boyer,[8] line B is the smooth curve that we have drawn through the experimental data points. It is obvious from Figure 12 that there is significant deviation from linearity in the correlation, especially at low T_g values. At high T_g values the correlation line B seems to be linear and agrees quite well with the previously suggested correlation line A). It should also be noted that at low T_g values the T_{50G} is significantly higher than the T_g; however, at high T_g values the T_{50G} is approximately equal to T_g. The apparent non-linearity of the T_{50G} versus T_g correlation was initially quite surprising to us and it was to this apparent anomaly that we first directed out attention.

It has long been established, and widely accepted, that an *observed* transition temperature is not necessarily equal to the *actual* transition temperature. The *observed* transition temperature is, in fact, dependent upon the frequency of the test method utilized. In general, the *observed* transition temperature increases as the frequency of the test method increases. In addition, the apparent activation energy (ΔH_A) for the glass transition is *not constant*, but increases sharply as T_g is approached.[10-12] Utilizing these ideas an elementary treatment of the kinetics involved was performed. For a generalized Arrhenius treatment, Eq. (2) is expected to be valid where f_i is an appropriate frequency, ΔH_A is the apparent activation energy for the glass transition, and T_i is the temperature corresponding to f_i:

$$f_i = A \exp\left[-\Delta H_a/(RT_i)\right]. \tag{2}$$

For the ESR spin-probe method described here, Eq. (3) and (4) are therefore appropriate where f_{50G} and f_{T_g} are the frequencies at T_{50G} and T_g respectively:

$$\log(f_{50G}) = \log A - \frac{\Delta H_a}{(2.3)(R)(T_{50G})} \tag{3}$$

$$\log(f_{T_g}) = \log A - \frac{\Delta H_a}{(2.3)(R)(T_g)}. \tag{4}$$

Assuming identical values of A (reasonable for any one polymer) leads to:

$$\log\left(\frac{f_{50G}}{f_{T_g}}\right) = \frac{\Delta H_a}{2.3R}\left(\frac{T_{50G} - T_g}{T_g \cdot T_{50G}}\right). \tag{5}$$

Substitution into Eq. (5) of appropriate values of f_{50G} (10^7 Hz) and f_{T_g} (1 Hz),[13] and solution of the equation for T_{50G} leads to Eq. (6):

$$T_{50G} = \frac{T_g}{\left[1 - (0.03)\left(\dfrac{T_g}{\Delta H_a}\right)\right]}. \tag{6}$$

As a test of this derived equation we have calculated expected T_{50G} values for a series of polymers of widely varying activation energies and glass temperatures (ΔH_A values from 18 to 115 kcal/mol and T_g values from −125 to 150°C) using values quoted by Boyer, except for that for poly(dimethylsiloxane). The resultant data is graphically summarized in Figure 13 and includes a *theoretical* correlation line. A comparison of this theoretical correlation line of Figure 13 with the data and correlation line B of Figure 12 shows very good agreement. The surprisingly good agreement (both position and shape) of the two correlation

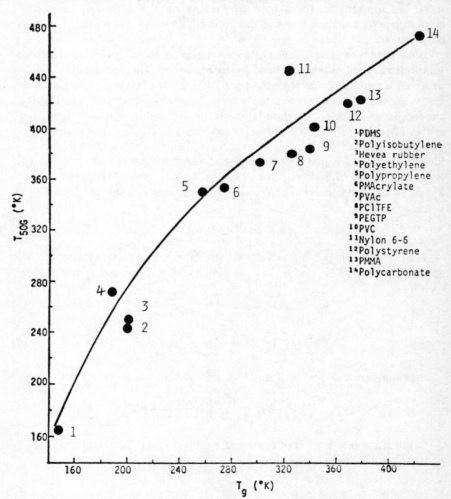

FIGURE 13   Calculated correlation of T_g with T_{50G} from Eq. (6) using known values of T_g and ΔH_A.

lines reinforces our belief that the actual correlation of T_{50G} with T_g is not, in fact, linear as previously suggested. It should also be noted that Eq. (6) predicts that T_{50G} values would be significantly higher than T_g for polymers with low ΔH_A values whereas T_{50G} values would be approximately equal to T_g for polymers with high ΔH_A values.

The non-linear nature of the correlation of T_{50G} with T_g should not, however, preclude the use of this technique to determine the T_g for polymers which have unknown and/or controversial values for this temperature. If we accept the validity of the correlation line shown in Figure 12 then it would be relatively easy to determine T_g values for any polymer whose T_g lies between the temperature limits of Figure 12.

In order to demonstrate the utility of the spin-probe method for studying the glass transitions of polymers with unknown/controversial glass temperatures, we have studied a series of polymers of this type. Our results are summarized in Table I.

TABLE I

Glass temperatures by the ESR spin-probe method

Polymer	T_{50G} K	T_g K[a]	T_g K[b]
Ethylene—vinyl acetate copolymers			
0.82 wt fraction ethylene	289	210	208[14]
0.72 wt fraction ethylene	293	216	215[14]
0.55 wt fraction ethylene	296	220	229[14]
Ethylene—propylene copolymers			
0.84 mole fraction ethylene	283	205	201[15]
0.75 mole fraction ethylene	277	200	204[15]
0.63 mole fraction ethylene	279	202	210[15]
Poly(vinyl fluoride)	321	251	250[25]
			314[16]
Poly(vinylidene fluoride)			
PVF$_2$ film	347	287	228[16]
Kynar 301 (latex)	347	287	286[17]
Kynar 881 (suspension)	354	296	300[18]
Kynar 7201			
PVF$_2$ + 30% tetrafluoroethylene	340	276	–
Tefzel			
(50/50 ethylene—			
tetrafluoroethylene)	362	309	–
Poly(ethylene oxide)	272	196	206[16]
Poly(methylethylsiloxane)	197	141	148[19]

[a]Estimated from Figure 12.
[b]Literature values.

Ethylene copolymers

The values of T_g (ESR) listed in Table I for the ethylene–vinyl acetate co-polymers were read from correlation line B of Figure 12 using the observed value of T_{50G}. These values agree quite well with the extrapolated curve of Illers for high ethylene content, semi-crystalline polymers, extrapolated using the Gordon–Taylor copolymer equation.[14]

The values of T_g (ESR) for the ethylene–propylene copolymers (determined as above) agree very well with the observed values of Manaresi and Giannela[15] who also fitted their data with the Gordon–Taylor equation.

Both of these results lend additional support to the proposition that the glass temperature of polyethylene is 195 ± 10 K. Figure 14 is a graphical super-position of our data and that of Illers[14] for ethylene copolymers. The value of 195 ± 10 K, supported by our present ESR-spin-probe data, is in contrast with values of T_g for amorphous polyethylene occasionally reported in the literature as 145 ± 5 K and with values of 240 ± 10 K reported as the T_g for crystalline polyethylene. This result seems to suggest that the nitroxide spin-probe in ethylene–vinyl acetate copolymers and ethylene–propylene copolymers is responding to the glass transition of an isolated amorphous phase, essentially free from restraints by crystallites. It would obviously be of value to study, by the ESR spin-probe method, samples of polyethylene with varying thermal histories and resultant varying degrees of crystallinity; such studies are currently underway in our laboratories.

Fluorinated polymers and copolymers

A series of fluorinated polymers and copolymers have also been studied by the ESR spin-probe technique. The value of T_g for poly(vinyl fluoride) determined in the current study (251 K) is seen to be in very good agreement with one of the two values most often cited for T_g while at considerable discrepancy from the other often quoted value. The current data may help to resolve this controversy.

Three different samples of poly(vinylidene fluoride) have been examined in the current study. All samples gave T_g values at approximately 290 K. Even though this value is quite different from the generally accepted value of 228 K[16] it is quite consistent with values of 286 K and 300 K reported by Japanese[17] and American workers respectively.[18] It should also be noted that the Japanese investigators — using thermal expansion data — did detect another transition at 221 K. The current data, coupled with the data mentioned above, may suggest that further study is needed of the transitions occurring in PVF_2.

In addition to the above homopolymers, two fluorine-containing copolymers have been studied by the current method. A copolymer of PVF_2 with 30% tetra-fluorethylene gave a T_g of 276 K, and a 50/50 ethylene–tetrafluorethylene

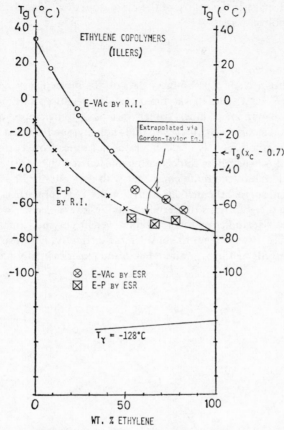

FIGURE 14 T_g data for ethylene copolymers.

copolymer gave a T_g of 309 K. For both of these copolymers there is very limited data available on transition temperatures and the T_g is, at best, unclear.

Other polymers

Two other examples of polymers are included in Table I. The T_g of 196 K for polyethylene oxide is in reasonable agreement with the generally accepted value of 206 K for the T_g of this polymer,[16] and suggests the validity of the spin-probe method for determining the T_g in polymers of the polyoxide type.

A value of 141 K has been determined for the T_g of poly(methylethyl) siloxane. This value, which is slightly lower than the 148 K value for poly-(dimethylsiloxane) is in good agreement with the T_g determined by other methods,[19] and suggests the utility of the current method in studying transitions in a variety of polysiloxanes.

Polystyrene

We have also obtained some preliminary data on the molecular weight dependence of T_g for a series of varying molecular weight polystyrenes. A series including styrene monomer, dimer, trimer, various oligomers up to and including high molecular weight (*ca* 180,000) polystyrene has been studied by the spin-probe method. The T_g's calculated for this series from correlation line B of Figure 12 fall off sharply with decreasing molecular weight (see Figure 15). The results of this study are quite consistent with data obtained by other methods[20, 21] and suggest the obvious utility of the ESR spin-probe technique for determining the glass transition temperature of low molecular weight polymeric materials; studies on low molecular weight polymeric materials *via* other methods are often subject to considerable difficulty. Our results thus far suggest that the ESR method is quite reliable and experimentally facile for low

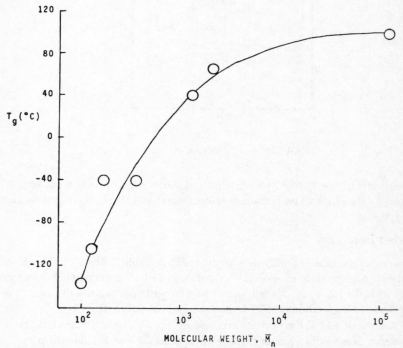

FIGURE 15 T_g data (from ESR) for varying molecular weight polystyrenes.

molecular weight materials. A similar study of the effect of chain length on T_g for a series of poly(dimethylsiloxanes) is currently in progress and will be reported at a later date.

Discussion

From the data reported in the current study, it appears as if the ESR spin-probe technique is a valuable addition to the arsenal of methods available for the study of molecular motions in polymeric materials. The method seems to be especially useful for studying polymers for which the T_g is either controversial or unknown and may also facilitate the determination of T_g for low molecular weight polymers.

Another potentially useful and interesting application of the ESR spin-probe method would be for the estimation of apparent activation energies, utilizing Eq. (6), of any polymer for which the glass temperature was non-controversial.

Of the seven questions posed at the start of this presentation we have presented data giving answers, or partial answers, to some of the questions. The correlation of T_{50G} with T_g does apparently hold over a wide temperature range, but the correlation is not linear. The method seems to be applicable to polymers other than vinyl type. The method seems to be appropriate for co-polymers as well as homopolymers; it seems to be helpful in resolving controversy about the actual T_g where controversy exists. The T_g does vary in a predictable and reproducible fashion as a function of chain length. We have not as yet reported any data pertinent to questions (5), (6), and (7). We have only very limited data at the present time pertinent to these questions.[22]

Since the current method depends upon the paramagnetic properties of an added "foreign body" to the polymer matrix, it would not be surprising if the observed transition temperature (T_{50G}) was dependent on the detailed nature (size, shape, polarity, etc.) of the particular spin-probe utilized as well as being dependent on the nature of the polymeric material as we have already sufficiently demonstrated. Rabold has presented some data relevant to this question[4] – using two different spin-probes to study the same polymer – but a more extensive study of these types of parameters of spin-probe structure is currently underway in our laboratories. Our very preliminary data, along with that of Rabold, do in fact suggest that the observed transition temperature is decidedly dependent on the molecular properties of the nitroxide chosen as the spin-probe.

It should be noted that wide-line NMR experiments empirically exhibit the same type of temperature dependency as the ESR spin-probe experiments (see Ref. 23 for example). It has also been established that NMR line-width parameters do correlate with glass temperatures but no direct comparison of the two methods has been made. Both the NMR and ESR methods are high-frequency test methods but we expect the results from these two types of experiments to be complementary. Whereas the NMR method "observes" the host polymer directly (signals from protons or other nuclei covalently bonded to the polymer itself) the ESR method "observes" the dynamic properties of a "guest" paramagnetic molecule (this would not be true if the spin-probe were covalently bonded to the polymer as mentioned above).

We would also like to note that we have been unsuccessful thus far in attempts to observe and detect multiple transitions in any of the polymers or copolymers that we have studied thus far by the ESR spin-probe method. As has previously been pointed out by Boyer[24] low-frequency test methods should have the highest resolving power when scanning a polymer for multiple transitions; transitions having different activation energies will tend to "merge" as the frequency of the test method is increased. It is perhaps not surprising that we have failed to detect these other transitions which have typical activation energies much lower than the glass transition. This may well, however, tend to make the ESR spin-probe method selective for the glass transition — a quite favorable situation for resolving conflicting opinions in polymers exhibiting multiple transitions.

In addition to areas mentioned above a number of other areas are in various stages of investigation in our laboratories. We are evaluating the method for applicability to other types of homopolymers (for example polyesters, polyurethanes, polyamides, polyacrylates, and polyanhydrides). We are investigating, in addition, the potential application of the ESR spin-probe method to a study of the glass transition(s) in block copolymers and what variations are brought about by converting the experiment into a spin-label experiment rather than a spin-probe experiment by covalently bonding the paramagnetic probe to the polymer itself.

Acknowledgments

We sincerely acknowledge valuable discussions with Dr. Gary Rabold (Dow Chemical) and the invaluable technical assistance of the following Saginaw Valley State College students: George Bousfield, Michael Gebler, and Steven Keinath. The work was made possible by generous donations (ESR spectrometer, operating supplies, and financial support) to Saginaw Valley State College by the Dow Chemical Company.

References and footnotes

1. Presented in part at the 169th National Meeting of the American Chemical Society, April 1975; see *Polymer Preprints* 16, 1, 572 (1975).
2. S. Ohnishi and H. McConnell, *J. Amer. Chem. Soc.* 87, 2293 (1965).

3. Ian C. P. Smith, in *Biological Applications of Electron Spin Resonance*, Chap. 11, edited by H. M. Swartz, J. R. Bolton, and D. C. Borg (Wiley-Interscience, New York, New York, 1972).
4. G. P. Rabold, *J. Polymer Sci. A-1* 7, 1203 (1969).
5. A Savolainen and P. Tormälä, *J. Polymer Sci. A-2* 12, 1251 (1974), and references therein.
6. N. Kusumoto, M. Yonezawa, and Y. Motozato, *Polymer [London]* 15, 793 (1974), and references therein.
7. A. L. Buchachenko, A. L. Kovarskii and A. M. Vasserman, in *Advances in Polymer Science*, edited by Z. A. Rogovin (Wiley and Sons, New York, 1974) pp. 37ff.
8. R. F. Boyer, *Macromolecules* 6, 288 (1973).
9. On the basis of the two curves presented here it is tempting to speculate that "narrow" curves are typical of polymers with low T_g's and "broad" curves are typical of polymers with high T_g's but such is not the case. Within the series of 15 polymers described here and other work in our laboratories there is no obvious correlation of "broadness" or "narrowness" with T_g. Work is however in progress to attempt to correlate "broadness" with some parameter(s) relating to the structure of the polymer.
10. M. L. Williams, R. F. Landel, and J. D. Ferry, *J. Amer. Chem. Soc.* 77, 3701 (1955).
11. N. B. McCrum, B. E. Read, and G. Williams, *Anelastic and Dielectric Effects in Polymeric Solids* (Wiley, New York, 1967).
12. D. W. McCall, in *Molecular Dynamics and Structure of Solids*, edited by R. S. Carter and J. J. Rush (National Bureau of Standards, Washington D.C., Special Publication 301, June 1969).
13. A value of 2×10^7 Hz has previously been estimated as appropriate for f_{50G}.[4] The value of 1 Hz was selected for f_{Tg} from a consideration of the frequency dependency of the glass transition; the value of 1 Hz would be in the linear portion of the plot of log f vs $1/T$[12] but would not be in the region where deviations from linearity (at very low frequencies) have been observed.[10]
14. K.-H. Illers, *Kolloid-Z. Z. Polym.* 190, 16 (1963).
15. P. Manaresi and V. Giannella, *J. Appl. Polymer Sci.* 4, 251 (1960).
16. J. Brandrup and E. H. Immergut, Eds., *Polymer Handbook* (Interscience, New York, 1966).
17. H. Sasabe, S. Saito, M. Asahina, and H. Kakutani, *J. Polymer Sci. A-2* 7, 1405 (1969).
18. A. Peterlin and J. Elwell, *J. Materials Sci.* 2, 1 (1967).
19. K. Polmanteer, personal communication.
20. T. G. Fox and P. J. Flory, *J. Polymer Sci.* 14, 315 (1954).
21. K. Ueberreiter and G. Kanig, *Z. Naturforschung* 6a, 551 (1951).
22. Data pertaining to these questions will be presented elsewhere.
23. See Ref. 12.
24. R. F. Boyer, *Rubber Chem. Technol.* 36, 1303 (1963).
25. M. Görlik, R. Minke, M. Trautvetter and G. Weisgerber, *Angew. Makromol. Chem.* 29/30, 137 (1973).

DISCUSSIONS

N. Kusumoto (*Case Western Reserve Univ.*): How did you determine the narrowing temperature of the ESR line separation? In some cases, you found spectra that showed sub-splitting around T_{50G}, and this could cause some difficulty in estimating T_{50G}.

P. L. Kumler and R. F. Boyer: In some cases we did, in fact, observe other peaks during the temperature runs but this did not present any severe problems in determining the empirical parameter, T_{50G}. In cases where this sub-splitting occurred, perhaps due to isotropic motion of the nitroxide probe, we simply plotted the extrema separation of the *outermost* peaks as a function of temperature. When this situation did occur, it was generally easier to identify the outermost peaks in the low-temperature spectra and then "follow" the separation of these two selected peaks as the temperature was increased in small increments.

N. Kusumoto: Have you considered the possibility of independent motion of the probe within the polymer matrix? In other words, is the motion of the probe, which is not chemically bonded to the polymer chains, truly representative of the motions within the polymer matrix?

P. L. Kumler and R. F. Boyer: Although we have not attempted, as yet, to define in any detail how the T_{50g} relates specifically to motion within the polymer matrix we do feel that the observed correlation of T_{50g} with T_G for a wide variety of both polymer type and molecular weight suggests quite strongly that the motion of the nitroxide probe is, in fact, representative of molecular motion of the polymer chain. We expect that work currently in progress involving a series of probes of differing size and molecular functionality will help us to elucidate in more detail the molecular basis of the correlation.

N. Kusumoto (*Case Western Reserve Univ.*): In my opinion, the motion of the probe is frozen at very low temperature, and the probe begins rotational oscillation with increasing temperature, in some cases around a certain molecular axis causing incomplete narrowing; subsequently, isotropic tumbling occurs to complete the line narrowing. This process may occur in the so-called "cavity" model. However, near T_g the probe (as well as the polymer segments) may jump into an adjacent hole (free volume) as was suggested previously by us (*Polymer*, **15**, 793 (1974)). Regarding the effect of probe size, as the probe becomes larger, the probability of its finding a larger hole will become smaller, and therefore T_{50g} will increase. An effort has been made by us to correlate T_g and T_{50g} to the volume ratio of the probe to segments according to the free volume theory (*Preprint of ESR Symposium*, Matsuyama, Japan, 1973):

$$T_{50g} - T_g = 52 \left[2.9 \left(\ln \frac{1}{f} - 1 \right) f - 1 \right],$$

where f is V_p^m / V_m^m, V_p^m is the molar volume of the probe, and V_m^m is the molar volume of a segment. From the known volume of the probe and the measured value of T_{50g}, V_m^m could be estimated.

A. Eisenberg (*McGill Univ.*): Have you checked the ESR Correlation times for your systems?

P. L. Kumler and R. F. Boyer: We have not determined rotational correlation times in the present work, as yet. It should be noted, however, that in Rabold's earlier work he did obtain linear plots of log (rotational frequency) *vs* $1/T$ for a series of polymers and was thus able to determine rotational activation energies of nitroxide 4c (R = O—C—Ph). We expect that our data will yield similar

                                              ‖

                                              O

results.

J. F. Jansson (*MIT*): How large must the polymer motion be before it can be observed with ESR probes?

P. L. Kumler and R. F. Boyer: At the present time, we do not know. Again here we expect that our study of various probes will help elucidate this question.

Nature of Biopolymer–Water Association by Dynamic Mechanical Spectroscopy

A. HILTNER, H. SHIRAISHI, S. NOMURA and E. BAER

Department of Macromolecular Science, Case Western Reserve University, Cleveland, Ohio 44106, U.S.A.

Poly-L-hydroxyproline has dispersion processes below physiological temperature which are influenced by moisture in a manner very similar to that observed in the structural protein collagen. Both materials sorb a large percentage of water which does not have the properties of bulk water. This non-freezable water is associated with two dispersion peaks, β_2 at 150 K (1 Hz) and β_1 at 200 K (1 Hz). Proposed mechanisms involve diffusion of bound water molecules and water plasticized side group or ring motions respectively. From changes in temperature location and intensity of the β peaks with water content at least three types of water are distinguished in poly-L-hydroxyproline and three in collagen.

INTRODUCTION

The relation between water and biological macromolecules is one of the most important phenomena of all life processes. Although the human body is 50–60% water, only 10% of this is found in the vascular system and is considered to have fluid properties. The rest is bound by poorly understood mechanisms in the cytoplasm of cells and in gels of extracellular connective tissue. The physical and mechanical properties of hard and soft connective tissue, including bone, ligament and tendon, are altered appreciably by changes in the bound water content. The most important protein in connective tissue is collagen. Despite the obvious significance, many aspects of water–collagen interactions remain a mystery. In large part this is due to the complex chemical composition of the collagen molecule, which includes more than 19 different amino acids, and the equally complex hierarchies of structural organization which exist in connective tissue. On a molecular level, water is probably associated through hydrogen bonds with the peptide linkages and with polar side groups, e.g., the hydroxyl groups of serine, tyrosine and hydroxyproline; it can also form hydration spheres around ionized residues such as lysine and glutamic acid. On a structural level, water

249

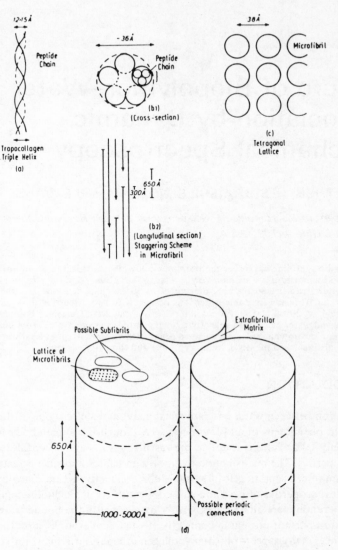

FIGURE 1   Scheme of structural hierarchies from molecular dimensions upwards.
(a) Tropocollagen: perpective view. (b) Microfibril (b₁) cross sectional view, (b₂) longitudinal view. The arrow headed lines are the tropocollagen triple helices and correspond to the
large circles in (b₁). The quarter stagger is indicated but not the mild twist of the whole
assembly. (c) The lattice formed by microfibrils. (d) The collagen fibril. Perspective view.
The fibrils are drawn as isolated in the extrafibrillar matrix. The 65 nm longitudinal period
and the tetragonal lattice of microfibrils is indicated. The possibility of a subfibril of
approximately tens of nanometers within the fibril is indicated in the sketch. Also, there is
a possible connectedness between fibrils across matrix (drawn dotted). (Reference 1).

molecules can be accommodated between the structural units of each hierarchical level of organization (Figure 1). In general, each type of association will differ as regards the energy of the interaction and the effect on the physical properties.

Molecular models are essential if the individual molecular interactions are to be separated and characterized. The most convenient and versatile models are the synthetic poly-α-amino acids. They have the peptide backbone, can assume collagen-like conformations, and can be synthesized as homopolymers or with more than one amino acid in a known sequence. Two synthetic polypeptides which assume a helical conformation closely resembling the collagen helix are poly-L-hydroxyproline (PHP) and poly-L-proline II (PPII). Factors which make this the stable conformation include the absence of a peptide hydrogen for hydrogen-bonding and the restrictions in flexibility imposed by the ring. Hydrogen bonding as it occurs in PHP between hydroxyl and peptide carbonyl is intermolecular. It is not a coincidence that hydroxyproline and proline are important constituents of collagen and combined account for about 220 per 1000 residues. The presence of proline and hydroxyproline with the ring structure as part of the peptide bond, has a stiffening effect on the collagen molecule and locks the chain into its helical conformation.

In the present study, the interaction of water with a collagenous tissue, human dura-mater, and a synthetic polypeptide, poly-L-hydroxyproline, is compared. The dynamic mechanical technique was chosen for the investigation (1) because of the known sensitivity of this method to interaction of water with macromolecular substances and (2) because it has application to problems involving the relationships between structural and mechanical change.

EXPERIMENTAL

Dynamic mechanical measurements were made with a free-oscillating, inverted torsional pendulum at about 1 Hz over the temperature range 80–300 K.[2] Because poly-L-hydroxyproline formed an intractable film, the torsional braid technique was used.[3] In this method the polymer is cast from solution onto a viscoelastically inert glass braid support.

Heat of fusion of the sorbed water was measured by differential scanning calorimetry. The instrument used was a Perkin-Elmer Model DSC-IB. Heating scans were made over the temperature range 200–300 K under a nitrogen gas stream (30 ml/min) at a heating rate of 10°C/min.

Poly-L-hydroxyproline (MW 10,700) was obtained from Miles Laboratory and was cast from aqueous solution. The infrared spectrum of the cast film showed the polymer to be in the poly-L-proline II conformation, a left-handed 3/1 helix with *trans* configuration of the ring.

Water content was varied by equilibrating specimens over saturated salt solutions. The water content is expressed as weight percent gain relative to the vacuum dried polymer. The hydrated specimens were quenched in liquid nitrogen immediately after weighing to retard water loss. When reweighed at the end of an experiment the weight loss was always less than 5% of the total water content.

RESULTS

Morphologically the poly-L-hydroxyproline (PHP) films showed no organization above the molecular helix except for some poorly developed crystallinity. This polymer does not form a triple helix nor the fibrilar hierarchy characteristic of native collagen; and comparisons in the relaxation behavior must be made on the molecular level. This does not alter significantly the value as a model for relaxation behavior of collagen since by and large low temperature relaxation behavior is interpreted in terms of molecular structure with molecular organization having a secondary effect.

The driest PHP obtainable under existing experimental conditions showed broad, weak maxima at 135 (γ) and 220 K (β_2). The presence of water altered both the temperature and intensity of the β_2, and at higher water content a β_1 peak is also apparent (Figure 2). The effect of water, particularly on the β_2, shows several regimes of behavior (Figure 3). The intensity of the β_2 initially increases with water content reaching a maximum at 16% (1 water/peptide). In this regime I, the β_2 temperature is almost constant, decreasing only slightly

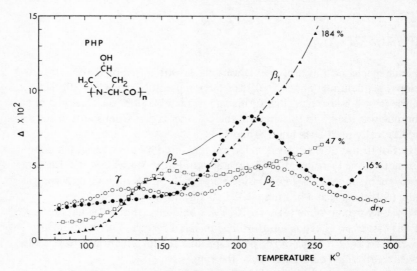

FIGURE 2   Mechanical loss spectra of poly-L-hydroxyporline.

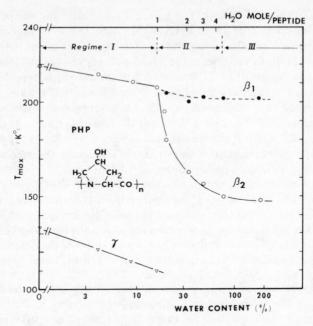

FIGURE 3   Effect of water on the temperature of the γ, β_1 and β_2 losses in poly-L-hydroxyproline.

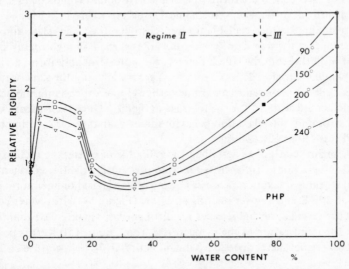

FIGURE 4   Rigidity of poly-L-hydroxyproline as a function of water content at various temperatures.

from 220 to 208 K at 16% water. The γ intensity decreases and this peak could not be detected above 16% water. In regime II (16 to 75% water) the intensity of the β_2 decreased, and the temperature also decreased rapidly at first then leveled off at 150 K. In this regime a second β peak (β_1) is also indicated at about 220 K. In regime III ($> 75\%$ or about 5 water molecules/peptide) there are no significant changes in either β peak. Freezable water, i.e., water exhibiting a melting endotherm in DSC measurements, was observed only when water content exceeded 75%. The heat of fusion of the freezable water was the same as bulk water, 80 cal/g, but the temperature of fusion, 260 K was suppressed. Melting point suppressions of this magnitude are frequently observed in the water associated with hydrophilic biopolymers.[4, 5]

Although absolute values of G' could not be obtained, the frequency in free oscillation ($f^2 \propto G'$) was measured for one specimen at several water contents (Figure 4). The rigidity initially increases, reaches a maximum and begins to decrease at 16% water and finally increases again with the formation of ice. The shape of the rigidity curve of PHP does not change over the temperature range investigated, so apparently the antiplasticizing effect of the initial 16% water is not associated with either of the β processes.

Hydrated specimens of poly-L-proline II (PPII) also show both the β_2 and β_1 peaks, but only the two regimes in β_2 behavior corresponding to II (0–50% water) and III ($> 50\%$) described for PHP.[6] In fact, the β_2 temperature curve is superimposable with that of PHP in regimes II and III. The rigidity curve of PPII does not show a maximum, but is initially flat before the increase due to freezing water. The helical conformation of the PHP and PPII molecules is the same, the significant difference is in chemical structure where PHP has a hydroxyl group on the ring available for hydrogen-bonding. The relaxation data would indicate that the initial 16% water in PHP is sorbed on high energy sites not present in PPII. From the stoichiometry, the water probably forms a 1:1 hydrogen-bonded complex with the hydroxyl group. Possibly the same water molecule is hydrogen-bonded to a peptide carbonyl to create an intermolecular bridge, which could account for the increase in rigidity. Once the hydroxyl groups are saturated, the relaxation behavior indicates that water sorption by PHP and PPII is essentially identical.

With the regime concept as described for PHP, it is of interest to reconsider the relaxation behavior of collagen as reported previously.[7] Native collagenous tissue exhibits three relaxation processes below physiological temperature: β_2 at 150 K, β_1 at 200 K and a water melting at 270 K (Figure 5). When water content is taken as the variable, several regimes are distinguished (Figure 6). From 0† to 25% water the temperature of the β_2 decreases from 260 to 150 K. It is possible to define two regimes here, divided by the inflection in β_2 temperature at 3%

---

†There is undoubtedly some residual water in the vacuum dried collagen which can not be removed without destroying the structure.

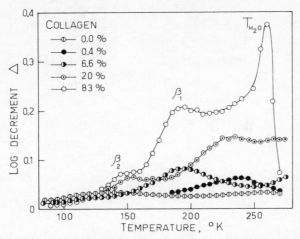

FIGURE 5   Mechanical loss spectra of collagen (human dura-mater).

water: regime I (0–3%) and regime II (3–25%). Above 25%, in regime III, temperature of the β_2 is constant. A β_1 peak is first observed at 10% water at about 290 K, the limit of the experiment. Temperature of the β_1 also decreases with increasing water contact and a third regime III' (25–50% water) might be distinguished from the point at which temperature of the β_1 becomes constant and a water melting peak is detectable.

The regimes described for collagen are also associated with changes in rigidity curve (Figure 7). Considering first the rigidity at 90 K, which is below the temperature at which any relaxation processes (β_2, β_1) are activated, the rigidity

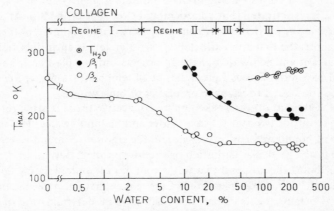

FIGURE 6   Effect of water on the temperature of loss peaks in collagen.

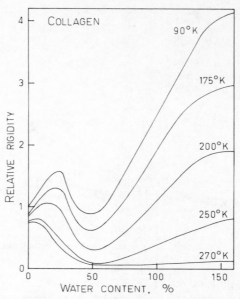

FIGURE 7   Rigidity of collagen as a function of water content at various temperatures.

initially increases reaching a maximum at 25% water, then decreases with a minimum at 50% water whereupon the rigidity increases again with the formation of ice. For three higher temperatures, 175, 200 and 250 K, there is a water content, described by Figure 6, at which the β_2 is activated. This of course is a consequence of the shift in β_2 temperature with water content. The maximum in rigidity at these temperatures is observed at progressively lower water content, the water content corresponding to the maximum is slightly above that at which the β_2 process is activated (but considerably below that of the β_1). At 270 K, which is above the β_2 at all water contents, there is no maximum in the rigidity curve. In short, the first approximately 25% water acts as an antiplasticizer at temperatures at and below that of the β_2 process and as a plasticizer above. An additional 25% water acts as a plasticizer at all temperatures. Water in excess of this can be considered to have the properties of bulk water.

The large amount of water taken up by connective tissue is often refered to as belonging to one of two types, "free" water and "bound" water. The actual definition of bound water is very vague and the amount of bound water varies according to the particular definition or experiment used. Several studies, however however, support the proposal that water contents of approximately 25 and 50% are significant.[5, 8] For example, three of the regimes have also been identified in calorimetric measurements of Haly and Snaith:[8] DSC determination of the heat of fusion of sorbed water in rat tail tendon distinguished unfreezable water

(0–30%), water with an unusually low heat of fusion (30–60%) and freezable water with the bulk heat of fusion ($>$ 60%) and freezable water with the bulk heat of fusion ($>$ 60%). Low temperature x-ray measurements in this laboratory[9] also indicate that some freezable water is present in collagen at water contents as low as 40%. This water does not have the identical properties of bulk water. When a specimen is quenched rapidly the x-ray pattern observed is that of the unusual cubic form of ice, evidence that this initial freezable water also interacts with the collagen substrate. An irreversible transformation of the cubic ice to the normal hexagonal form occurs at about 220 K. This transformation is not associated with either β process as these are completely reversible.

DISCUSSION

There are some strong similarities in the relaxation behavior of PHP and collagen. Both sorb a considerable amount of water which does not have the properties of bulk water as evidenced by the absence of melting. The β_2 and β_1 relaxation maxima occur repeatedly at approximately the same temperatures. The behavior of the β_2 temperature with respect to water content indicates that several types of water may be interacting with the macromolecule.

Before speculating on mechanistic interpretations of the β peaks, it is instructive to consider an example from the literature on commercial synthetic polymers. It is well known that the relaxation behavior of nylon 6, a linear polyamide, is altered by the presence of water. Of particular interest is the β peak observed in wet specimens.[10] With the first 3% water the β temperature shifts (from 230 to 190 K) with a maximum in intensity at 3%. Further changes with additional water, up to 8 or 9%, are slight. The model for the sorption of water by nylon 6 is proposed by Puffr and Sebenda:[11] a total of three molecules of water are sorbed on two neighboring amide groups in the accessible or amorphous regions of the polymer. The first water molecule (0–3% water) forms a hydrogen-bonded bridge between carbonyl groups, and is termed tightly bound water. The other two molecules interact less strongly with the amide groups and are referred to as loosely bound water. The existence of a relaxation in ice at the same temperature has led to a suggested mechanism for the nylon β process.[12] The mechanism for the ice transition is considered to involve the diffusion of a water molecule in a Ll or Dl defect with the breaking and reformation of two hydrogen bonds; in nylon 6 the analogous process would involve diffusion of the tightly bound water molecules which are in a hydrogen-bonded configuration similar to the Ll defect.

Obviously some qualitative analogies exist between the nylon β peak and the β_2 of PHP, PPII and collagen, and also other hydrophilic poly-α-amino acids:[7, 13] (1) All exhibit a β peak in the 150–250 K region which is associated with the presence of water, and (2) The peak does not show a smooth change in either temperature or intensity with water content, the former reaches a limiting value and the latter goes through a maximum. Differences are primarily in stoichiometry and can not be resolved without a molecular model of water sorption by the polypeptides. That the analogies exist despite dissimilar chemical structures argues that the kinetic unit must be the water molecules themselves which are associated through hydrogen-bonding with the macromolecule.

The activation energy of the nylon β process at the limiting temperature is equal to the energy of two hydrogen-bonds. Activation energies were not obtained for the polypeptides; the apparent decrease in activation energy associated with the characteristic decrease in β_2 temperature may reflect simply the decrease in energy of the binding sites available as water content increases, or may indicate a cooperative aspect to the water diffusion.

No analogy to the β_1 exists in nylon 6. Possibly at higher water contents more complex motions of both water and macromolecule are activated to which the β_2 is a precursor. The appearance of the β_1 peak in PHP and PPII does not alter the interpretation proposed previously that this is a water-plasticized side group peak. In the case of PHP and PPII torsional deformation of the five-membered ring might be involved.

CONCLUSION

Synthetic polypeptides have found use as molecular models in the study of water–biopolymer interactions. The biopolymers sorb a considerable amount of bound water, i.e. water which does not have the properties of bulk water. The temperature location and intensity of dispersion processes below physiological temperature (β peaks) are found to depend on the amount and type of bound water. Changes in the dynamic mechanical parameters are not continuous with water content but exhibit regimes in behavior. The regimes observed in the structural protein collagen and the synthetic model polypeptide, poly-L-hydroxy-proline, are analogous in many respects. It is concluded that the regimes are associated with specific types of interactions between water and biopolymer. From this work and that on other synthetic polymers, some speculations are presented concerning the nature of these interactions and their significance as regards the mechanisms responsible for the β processes. Thus the β_2 (150 K, 1 Hz) is associated with diffusion of tightly bound water molecules accompanied by breaking and reforming hydrogen bonds. This may be a precursor to the β_1 (200 K, 1 Hz) process. The latter is postulated to involve local motion of the water–biopolymer complex.

The nature of biopolymer–water interactions is a difficult and complex question whose study will be facilitated by application of new methods and model systems. As progress is made, it is apparent that the two-state (bound and free) model for water in biopolymers will be refined. The general designation bound water (and possibly free water as well) will encompass a spectrum of bound water types, the nature and function of which are determined by the chemical and organizational composition of the biopolymer.

References

1. E. Baer, L. J. Gathercole and A. Keller, *Colston Papers* **26**, 189 (1975).
2. C. D. Armeniades, I. Kuriyama, J. M. Roe and E. Baer, *J. Macromol. Sci. Phys. B 1*, 777 (1967).
3. J. K. Gillham, *Crit. Revs. Macromol. Sci.* **1**, 83 (1972).
4. A. R. Haly and J. W. Snaith, *Biopolymers* 7, 459 (1969).
5. A. R. Haly and J. W. Snaith, *Biopolymers* **10**, 1681 (1971).
6. H. Shiraishi and E. Baer, in preparation.
7. A. Hiltner, S. Nomura and E. Baer, *Peptides, Polypeptides and Proteins*, edited by Blout, Bovey, Goodman and Lotan (John Wiley, N.Y. 1974) p. 485.
8. M. Luescher, M. Ruegg and P. Schindler, *Biopolymers* **13**, 2489 (1974).
9. S. Nomura, A. Hiltner, J. B. Lando and E. Baer, *Biopolymers* **16**, 231 (1977)
10. Y. S. Papir, S. Kapur, C. E. Rogers and E. Baer, *J. Polymer Sci. A-2* **10**, 1305 (1972).
11. R. Puffr and J. Sebenda, *J. Polymer Sci. C* **16**, 79 (1967).
12. S. Kapur, C. E. Rogers and E. Baer, *J. Polymer Sci., Polymer Phys. Ed.* **10**, 2297 (1972).
13. A. Hiltner, J. M. Anderson and E. Baer, *J. Macromol. Sci. Phys. B* 8, 448 (1973).

Dynamic Properties of Ionomers

A. EISENBERG

McGill University, Montreal, Quebec, Canada

The dynamic mechanical properties of ionomers reflect the complex structural features of these materials, notably the presence of various types of ionic aggregates, which were also found by a range of other techniques. Transient rheological data, small angle x-ray scattering, water uptake experiments below T_g, as well as glass transition studies, all suggest that below a certain ion concentration the ions exist as small multiplets (of a few ion pairs) which behave like transient crosslinks. Above that concentration, much more extensive clustering is observed; these clusters contain not only a large number of ions, but probably also some non-ionic material. The critical ion concentration varies from polymer to polymer occuring at <1 mol % in polyethylene, at *ca* 5–6% in polystyrene, at 10–15% in ethyl acrylate, and not at all in the phosphates.

In the glass transition region, the dynamic mechanical properties of several series of clustered ionomers exhibit two peaks; the one at lower temperature is related to the glass transition of the regions from which the ionic groups have been excluded, while the one at high temperatures reflects the presence of the clusters. The lower temperature peak decreases in height with increasing ion concentration while the higher temperature peak exhibits exactly the opposite behavior. Both shift to higher temperature as the ion content increases. This behavior was observed in all the clustered ionomers which have been studied to date by that technique.

In general, the clustered materials behave very much like phase separated block copolymer systems. For example, a styrene ionomer containing *ca* 10 mol % sodium methacrylate behaves like a block copolymer containing *ca* 60–70% of a high temperature block, the T_g of which is *ca* 200°C. This illustrates the strong amplification effect of the clusters.

The Determination of Activation Energies from Experiments at a Single Frequency. I: Theory

T. F. SCHATZKI

Western Regional Research Laboratory, Agricultural Research Service, Albany, CA 94710, U.S.A.

A derivation is made of the slopes of dynamic modulus contours in the usual Arhenius plot [ln ω vs $\beta = 1/kT$] by use of the relaxation spectrum. Defining the slopes as $S_i(T) = -(\partial \ln \omega/\partial \beta)_{E_i}$, where $i = 1$ refers to the storage, $i = 2$ to the loss modulus, relations are derived between $S_i(T)$ and the usual activation energy $q = -(\partial \ln \omega/\partial \beta)$ at max E_2. One finds $S_1(T) = S_2(T) = q$, but only when there exists a distribution of entropies of activation. When a distribution of energies of activation is present, additional terms appear in S_1 and S_2 which are discussed. A method of obtaining $S_1(T)$ for a temperature-varying, single-frequency experiment is developed (essentially by evaluating the ratio $dE_1/d\beta$ to E_2 point by point) which allows one to investigate q or its distribution by using solely the data from this common type of experiment. The present paper develops the mathematics involved without any resort to molecular models. In a paper to follow $S_1(T)$ and $E_2(T)$ spectra are displayed for a number of homologous series of polymers and molecular descriptions are derived therefrom.

INTRODUCTION

Measurement of the dynamic moduli of materials as a function of temperature and frequency has long been accepted as a sensitive method for elucidating the molecular mechanisms which cause relaxation processes. In the case of dielectric relaxation it is relatively simple to excite a single sample over a considerable range of frequency and temperature and thus map out an appreciable part of the relaxation spectrum. Because of the inefficiency of mechanical transducers mechanical relaxation experiments are limited in the frequency which is accessible for a given excitation mode. In order to map out the frequency domain in mechanical work one is commonly forced either to combine results from several modes or to extract as much as possible from the temperature scan alone. It is the purpose of the present communication to review the assumptions involved in

263

using limited-frequency range data to describe molecular relaxations. We shall then present a novel method of extracting activation energies from storage and loss moduli of a sample oscillating in a single mode over a range of temperatures. The method will be illustrated in the publication to follow,[1] referred to as II here-forth.

The common technique in mechanical modulus experiments is to measure the loss modulus $E_2(\omega, T)$ for at least two frequencies, ω_1 and ω_2, over a range of T encompassing, if possible, a peak in E_2. The two resulting curves are shifted along the $\beta = 1/kT$ axis to "overlap" and the activation energy q is computed from $-\ln(\omega_1/\omega_2)/(\beta_1 - \beta_2)$. If the two curves are indeed of the same shape and intensity the amount of shift is unique. Otherwise one usually matches the maxima, although it must be realized that the shift (and hence q) strictly speaking differs at the high and low temperature side of the peak. We note that when ω_1 and ω_2 are kept constant over the entire temperature range, $\ln(\omega_1/\omega_2)$ is constant as well. In the case of free oscillation, however, the two frequencies correspond to the fundamental and an overtone or to two samples of differing shapes and ω will vary with temperature as the resonance frequency changes with T because the storage modulus E_1 changes as the loss peak is traversed. Nevertheless, the ratio ω_1/ω_2 does not vary much. We may consider what happens in terms of a contour map in the ln ω vs. β plane. In Figure 1 we show constant E_1 contours and in Figure 2 constant E_2 contours. Figures 1 and 2 are consistent, although purely hypothetical, and are meant for illustrative purposes only. They were drawn using the known relaxation maps of a number of polymers (see e.g. McCrum et al.[2]). We note the glass transition (KGK), and two low temperature processes of essentially constant activation energy (DCD and ABA), the one at the lower temperature having the lower activation energy. Also shown are $S_1(T)$ contours taken from data presented in II. The suggestion of a high temperature process is indicated at H. The arrowed lines represent a free oscillation traverse, dropping in frequency as the temperature is raised (β is lowered). Fixed frequency experiments would correspond to a horizontal line. In drawing the contours we have further assumed that the relaxation peak maxima decrease as the frequency is lowered, a situation which is commonly observed, at least for dielectric loss spectra. Full contour maps, such as are shown in the figures, are almost never reported since data over a large frequency and temperature range would be required. An early exception is the work of Deutsch, Hoff and Reddish.[3] In general only the position of the E_2 maxima are reported, usually because data are drawn from a number of experimentors.

Although the location of the ridge proper does indeed describe the relaxation map in a general way, it is not, by itself, adequate to describe the map fully. For this one would like to have the entire shape of the ridge and its variation with temperature, information which is contained in the complete $E_1(\omega,\beta)$ or $E_2(\omega,\beta)$ function. A partial description would at least involve the slope of the contours

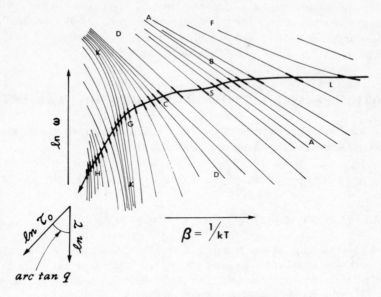

FIGURE 1   Contours of constant storage modulus, E_1.

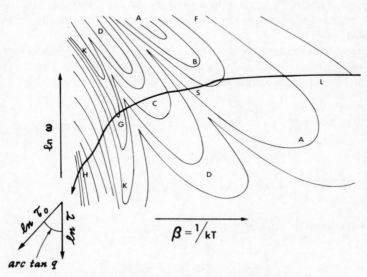

FIGURE 2   Contours of constant loss modulus, E_2.

of Figures 1 or 2 $(d \ln \omega/d\beta)_{E_i}$, $i = 1$ or 2, at all ω and β. We shall have considerable use for these slopes and shall designate them by $S_1 = -(d \ln \omega/d\beta)_{E_i}$, $i = 1$ or 2. In an experiment in which T is varied but where ω is constant or essentially so, as shown by the arrowed lines in Figures 1 and 2, we shall refer to $S_1(T)$ as the "contour spectrum" and to $E_2(T)$ as the "loss spectrum."

RELATION BETWEEN CONTOUR AND RELAXATION SPECTRA

As is well known, if only a single relaxation process is active (exponential stress decay) the frequency dependence of E_1 and E_2 is given by

$$E_1(\omega, T) = E_1(T) - \frac{\Delta E}{1 + \omega^2 \tau^2(T)} \qquad E_2(\omega, T) = \frac{\omega \tau(T)}{1 + \omega^2 \tau^2(T)} \Delta E \qquad (1)$$

$$\Delta E = E_1(\infty, T) - E_1(0, T) \qquad E^\infty(T) = E_1(\infty, T).$$

Here τ plays simply the role of a parameter. If the temperature is changed Eq. (1) may still represent the E_1 and E_2 frequency dependence but different parameters E^∞ and τ may be needed. This is specifically indicated in (1) by writing E^∞ and τ as functions of T. A common form is

$$\tau(T) = \tau_0 e^{\beta q}, \qquad \beta = 1/kT. \qquad (2)$$

In the present report Eq. (2) is to be viewed simply as a transformation from τ to τ_0, with q serving the role of a temperature shift factor. No molecular implication is involved.

Linear systems cannot have more rapid frequency dependence than that shown in (1). They commonly do have a more slowly varying response. This is expressed by use of a relaxation spectrum H for which (1) becomes

$$E_1(\omega, T) = E^\infty(T) - \int_{-\infty}^{\infty} H(\ln \tau(T), T) \frac{1}{1 + \omega^2 \tau^2(T)} d \ln \tau(T) \qquad (3)$$

$$E_2(\omega, T) = \int_{-\infty}^{\infty} H(\ln \tau(T), T) \frac{\tau(T)}{1 + \omega^2 \tau^2(T)} d \ln \tau(T).$$

The distribution function obeys

$$\int_{-\infty}^{\infty} H(\ln \tau(T), T) d \ln \tau(T) = \Delta E$$

and can be derived exactly if the entire frequency dependence of either E_1 or E_2 is known, i.e. Eq. (3) is invertable. H may depend directly on T as well as through $\tau(T)$, as indicated by $H(\ln \tau(T), T)$. By the $\tau(T)$ dependence we describe the shift of H along the $\ln \tau$ axis as T is varied. This shift may (or may not) be described by Eq. (2) for constant q, but it can always be so described if q is allowed to vary with T. However, if the *shape* of $H(\ln \tau(T))$ varies as well with T, then we must write $H(\ln \tau(T), T)$ and we are dealing with a distribution of activation energies.

It should be clearly understood that if such a distribution of activation energies is present no unique way exists on phenomenological grounds to separate the shift of H from its change of shape, i.e. the T dependence of H which arises through τ and that which is direct. As a consequence no unique T dependence of $\tau(T)$ and hence no activation energy can be derived, not even one which is T dependent. It follows that distributions of activation energies and entropies can only be obtained using molecular models and cannot be derived from dynamic moduli (or dielectric constants) alone. In accordance with the usual custom we divide the $\tau(T)$ and direct T dependence of H in such a way as to *minimize* (in some sense) the change in shape of H with T and derive q accordingly. This minimization of change of shape is done conveniently by assigning to $\ln \tau(T)$ the shift of the maximum H. Since $H(-\ln \omega)$ is approximately proportional to $E_2(\omega)$ (see below) the maximum in $H(\tau)$ corresponds to $(\partial E_2/\partial \ln \omega)_\beta = 0$ and very nearly to $(\partial E_2/\partial \beta)_\omega = 0$, i.e. the top of the ridge in E_2 space in Figure 2. This is the common choice for establishing q, as we saw above.

We now proceed in the usual fashion. We make the transformation (2), with the proviso that q may be an explicit function of T, in the integrals of Eq. (3). We obtain for a given ω and β

$$E_1(\omega, \beta) = E^\infty(\beta) - \int_{-\infty}^{\infty} H(\ln \tau_0, \beta) K_1(\omega \tau_0 e^{\beta q}) \, d \ln \tau_0$$

(4)

$$E_2(\omega, \beta) = \int_{-\infty}^{\infty} H(\ln \tau_0, \beta) K_2(\omega \tau_0 e^{\beta q}) \, d \ln \tau_0$$

with the kernels

$$K_1(z) = (1 + z^2)^{-1} \text{ and } K_2(z) = z(1 + z^2)^{-1}.$$

In the integrals of Eq. (4) the entire T dependence now appears explicitly through β and implicitly through any T dependence of τ. We may thus proceed to differentiate.† We obtain

$$(\partial E_1/\partial \beta)_\omega = -kT^2 \, dE^\infty/dT - \int_{-\infty}^{\infty} \left(\frac{\partial H}{\partial \beta}\right)_0 K_1(z) \, d \ln \tau_0$$

(5a)

$$+ 2q' \int_{-\infty}^{\infty} H(\ln \tau_0, \beta) K_2^2(z) \, d \ln \tau_0$$

$$(\partial E_2/\partial \beta)_\omega = \int_{-\infty}^{\infty} \left(\frac{\partial H}{\partial \beta}\right)_0 K_2(z) \, d \ln \tau_0 + q' \int_{-\infty}^{\infty} H(\ln \tau_0, \beta) K_3(z) \, d \ln \tau_0 \qquad (5b)$$

$$(\partial E_2/\partial \ln \omega)_\beta = 2 \int_{-\infty}^{\infty} H(\ln \tau_0, \beta) K_2^2(z) \, d \ln \tau_0 \qquad (5c)$$

$$(\partial E_1/\partial \ln \omega)_\beta = \int_{-\infty}^{\infty} H(\ln \tau_0, \beta) K_3(z) \, d \ln \tau_0 \qquad (5d)$$

with

$$K_3(z) = z(1 - z^2)/(1 + z^2)^2$$

and where

$$q' = q \left[1 + \frac{d \ln q}{d \ln \beta}\right]$$

and

$$z = \omega \tau_0 e^{\beta q(T)}.$$

By division we obtain the contour slopes

$$S_1(\omega, \beta) = -\left(\frac{d \ln \omega}{d\beta}\right)_{E_1} = \frac{(\partial E_1/\partial \beta)_\omega}{(\partial E_1/\partial \ln \omega)_\beta} =$$

---

† As usual a subscript on a partial derivative indicates the variables to be kept constant. For conciseness of notation we use 0 instead of τ_0, and 1 instead of τ_1. Further, the subscript z on a function indicates evaluation at $z = \omega \tau = 1$, the subscript x indicates constant experimental geometry (see Appendix).

$$\frac{-kT^2\,dE^\infty/dT}{2\displaystyle\int_{-\infty}^{\infty} H(\ln \tau_0, \beta)K_2^2(z)\,d\ln \tau_0} - \frac{\displaystyle\int_{-\infty}^{\infty} \left(\frac{\partial H}{\partial \beta}\right)_0 K_1(z)\,d\ln \tau_0}{2\displaystyle\int_{-\infty}^{\infty} H(\ln \tau_0, \beta)K_2^2(z)\,d\ln \tau_0} + q + \beta\,\frac{dq}{d\beta} \qquad (6a)$$

$$S_2(\omega, \beta) = -\left(\frac{d\ln \omega}{d\beta}\right)_{E_2} = \frac{(\partial E_2/\partial \beta)_\omega}{(\partial E_2/\partial \ln \omega)_\beta} = \frac{\displaystyle\int_{-\infty}^{\infty} \left(\frac{\partial H}{\partial \beta}\right)_0 K_2(z)\,d\ln \tau_0}{\displaystyle\int_{-\infty}^{\infty} H(\ln \tau_0, \beta)K_3(z)\,d\ln \tau_0} + q + \beta\,\frac{dq}{d\beta}. \qquad (6b)$$

Before discussing Eq. (6) in detail it is useful to make an approximation, valid for virtually all polymeric relaxations; i.e. we note that $H(\ln \tau_0)$ and $(\partial H/\partial \beta)_0$ vary slowly compared to the kernels in the region near $z = 1$. Far from $z = 1$ the kernels are either very small or, in the case of K_1, essentially constant. This allows us to approximate the integrals by expanding H or $(\partial H/\partial \beta)_0$ by Taylor series around $z = 1$ from which we obtain to first order†

$$\int_{-\infty}^{\infty} fK_1(z)\,d\ln \tau_0 = \int_{-\infty}^{z=1} f\,d\ln \tau_0 \qquad (7a)$$

$$\int_{-\infty}^{\infty} fK_2(z)\,d\ln \tau_0 = f_z \int_{-\infty}^{\infty} K_2(z)\,d\ln \tau_0 = (\pi/2)f_z \qquad (7b)$$

$$\int_{-\infty}^{\infty} fK_2^2(z)\,d\ln \tau_0 = f_z \int_{-\infty}^{\infty} K_2^2(z)\,d\ln \tau_0 = (1/2)f_z \qquad (7c)$$

$$\int_{-\infty}^{\infty} fK_3(z)\,d\ln \tau_0 = (\partial f_z/\partial \ln \tau_0)_\beta \int_{-\infty}^{\infty} \ln \tau_0\,K_3(z)\,d\ln \tau_0 = (-\pi/2)(\partial f_z/\partial \ln \tau_0)_\beta, \qquad (7d)$$

where $f = H(\ln \tau_0)$ or $(\partial H/\partial \beta)_0$. Substitution of Eq. (7b) into Eq. (3b) yields $E_2(\omega, \beta) = \pi/2(H(-\ln \omega, \beta))$, while substituting (7c) into (5c) yields $(\partial E_1/\partial \ln \omega)_\beta = H(-\ln \omega, \beta)$ from which we get the well-known approximation

---

† Schwarzl and Staverman,[4] among others, have given higher approximations, but the above suffices here.

$2E_2/\pi = (\partial \ln E_1/\partial \ln \omega)_\beta$. Substituting the latter as well as Eq. (7a) into Eq. (6) allows us to write Eq. (8).

$$S_1(\omega, \beta) = \frac{dE^\infty/d\beta}{2E_2/\pi} - \frac{\displaystyle\int_{-\infty}^{z=1} (\partial H/\partial \beta)_0 \, d \ln \tau_0}{2E_2/\pi} + q + \beta \frac{dq}{d\beta} \qquad (8a)$$

$$S_2(\omega, \beta) = - \frac{(\partial H/\partial \beta)_0}{(\partial H/\partial \ln \tau_0)_\beta} + q + \beta \frac{dq}{d\beta}. \qquad (8b)$$

Eq. (8) is more convenient than Eq. (6) and the approximations, as noted, are not serious for most polymer systems.

Of the two expressions, (8a) and (8b), the former is more useful. The first term of the r.h.s. of Eq. (8b) contains the denominator $(\partial H/\partial \ln \tau_0)_\beta$, which goes to zero *by definition* at the top of the ridge. Hence, unless $(\partial H/\partial \beta)_0$ vanishes *in the neighborhood* of the H maximum, S_2 will contain a singularity at the ridge. But this vanishing requires a spectrum of strictly constant shape and intensity, at least near the maximum in the $\ln \tau_0$ domain, a condition which does not seem to be achieved in polymers. The usual situation is then a singularity in S_2 which appears in Figure 2 as a turning around of the contours at the top of the ridge. Commonly one finds a shape which is temperature independent but an intensity which varies with β. While S_2 can then not be directly associated with q the following technique is commonly applied. One measures $E_2(\omega, \beta)$ for two (or more) frequencies and obtains an "average activation energy" by shifting the E_2 curves to lie above one another, matching the position, but not the height of the peaks. In actual fact, the true shift (and hence $1/S_2$) will be too large on one side of the peak and too small on the other. At the peak itself the shift will not be defined, since no amount of shifting will match E_2 values. If even the shape of the E_2 spectra change the value of q found becomes somewhat arbitrary and one reports "q for constant frequency", "q for constant temperature" or similar quantities. The important information regarding temperature variation of the spectrum is lost.

In the present report we concentrate on S_1 which does not suffer from the difficulties indicated. One can obtain S_1, just like S_2, by performing experiments at two (or more) frequencies over a range of temperature and derive S_1 from the spectrum shift. However, as we mentioned in the introduction, many experiments are conducted as a function of T but at a single frequency only. In this case $S_1(T)$, but not $S_2(T)$, can still be obtained. Using (8) in (6a) we have

$$S_1(T) = \frac{(\partial E_1/\partial \beta)}{2E_2/\pi} = - \frac{\pi k T^2}{2E_2} \left(\frac{\partial E_1}{\partial T}\right)_\omega \qquad (9a)$$

for the contour spectrum from a constant frequency experiment. Introduction of $(A-6)$ from the appendix yields

$$S_1(T) = -\frac{\pi k T^2}{2E_2} \left(\frac{\partial E_1}{\partial T}\right)_x \left[1 - \frac{E_2}{\pi E_1}\right] - \frac{kT^2 \lambda}{2} \tag{9b}$$

for a constant geometry experiment. All the quantities in (9a) and (9b) are available directly or from numerical differentiation. No corresponding expression for $S_2(T)$ is available.

DISCUSSION

One may now proceed in several ways. One may use Eq. (9) to evaluate the contour spectrum for a number of polymers of a homologous series and empirically compare the results to see what may be learned. Alternatively, one can start with a theoretical model of polymer relaxation, strictly H and E^∞, evaluate the r.h. terms of Eq. (6) or (8) individually and compare the sum with experiment. Finally, if both E_1 and E_2 contour spectra are available one could compare them through Eq. (6) or (8) to see what can be learned about the E^∞ term in which they differ. The homologous series approach is taken in II, while the theoretical technique is used in what follows. The final method, that of comparing the two contour spectra, we hope to make the subject of a future communication in which we shall analyze some dielectric results.

Returning now to Eq. (8), our master equation, we find four terms, a term involving the very high frequency (IR) response, a term involving the change of the shape of H with T, and two terms involving the shift of H (along $\ln \tau$) with T. Our interest in Eq. (8a) arises from its relation to the activation energy q and we wish to establish the relative importance of the various terms of the r.h.s. The last two terms may be viewed as arising from the transformation given in Eq. (2), which when applied to Eq. (2), results in Eq. (8). Since any function $q(\beta)$ may be used in Eq. (2), quite independent of any molecular basis for an activation energy, any value can be assigned to the last two terms of Eq. (8). Hence there is nothing unique about the relative size of the last three terms of (8a). It is only the sum of the last three terms which is an experimental quantity. Thus, on the face of it, Eq. (8a) tells us very little. The same applies to (8b).

Nevertheless, there exists a special case for which we have seen that Eq. (8a) and (8b) may be simplified considerably: when H shifts only with temperature but does not change shape. If $q(\beta)$ is so chosen in Eq. (2) to represent just this shift, call it $q_0(\beta)$, then $(\partial H/\partial \beta)_0 = 0$ and the corresponding terms of Eqs. (8a) and (8b) disappear. It follows that S_1 and S_2 will then be functions of β only and not of ω. Further, if $q_0(\beta)$ is of the form $c_1 + c_2 T$, where c_1 and c_2 are constants, $S_2(\beta)$ will be constant. Since S_1 and S_2 are experimentally observable, this result can be tested, at least in principle. We may turn the argument around and state, without proof, that if S_1 and S_2 are independent of frequency the shape of H is preserved; if S_2 = constant, $q_0 = c_1 + c_2 T$; and if $S_1 = S_2$ furthermore, $dE^\infty/d\beta = 0$. Of course, most of these results can be obtained directly from $E_2(\omega, \beta)$ since $E_2(\omega,\beta) = (\pi/2)H(-\ln \omega, \beta)$.

Suppose now that this special situation applies; in fact, suppose $H(\omega, \beta)$ $= H(\tau_0 e^{\beta q_0})$ with q_0 constant. Let us apply the transformation of Eq. (2) with a different q, say q_1, with $\ln \tau_1 = \beta q_1 + \ln \tau_0$. From what we have said above it follows that $S_1 = S_2 = q_0$, but now the term involving $(\partial H/\partial \beta)_1$ will not be zero, instead it will equal $q_1 - q_0$ in both (8a) and (8b). It is instructive to directly evaluate the integral term in (8a) and (8b). It is instructive to directly evaluate the integral term in (8a) for this "incorrect" choice of q; although we know the result, the evaluation will teach us about the meaning of the integral term in the more usual case. To do so we shall use the symbolic relaxation map of Figures 1 and 2 and we shall choose to evaluate the terms of S_1 at a point (ω, β) on the indicated traverse to the right (lower T) of the lowest temperature transition, say the point L. We transform the integral of Eq. (8a) by

$$-\frac{\pi}{2E_2} \int_{-\infty}^{z=1} \left(\frac{\partial H}{\partial \beta}\right)_1 d \ln \tau_1 = \frac{\pi}{2E_2} \int_{-\infty}^{z=1} \left(\frac{\partial \ln \tau_1}{\partial \beta}\right)_H \left(\frac{\partial \beta}{\partial \ln \tau_1}\right)_\beta d \ln \tau_1. \quad (10)$$

In the derivation of Eq. (7) we noted the approximate relation $2E_2(\omega, \beta) = \pi H(-\ln \omega, \beta)$. From this it follows that Figure 2 may be viewed not only as a contour map of E_2, but also as a contour map of $\pi/2H(-\ln \tau, \beta)$ simply by viewing the $\ln \tau$ coordinate as the reverse of the $\ln \omega$ coordinate as indicated in the Figures. Hence the integration is to be carried out in Figure 2 along a straight line in the τ_1 direction from the upper right to L. Since the $\ln \tau_1$ axis does not coincide with the $\ln \tau_0$ axis (the latter is perpendicular to the parallel H or E_2 contours) the factor $(\partial \ln \tau_1/\partial \beta)_H$ is not zero. However, it is constant, i.e., $q_0 - q_1$, and may thus be factored out of the integral. The integral remaining becomes $H_1 - H(\ln \tau_1 = -\infty)$ which becomes H_z (at L) since H must vanish in the limit. Since $H_z = 2E_2/\pi$ Eq. (10) reduces exactly to q_0 under our assumptions, regardless of the value of q used in (2).

When the spectrum varies with temperature for any choice of q and hence the contours are not parallel, the assumption of constant $(\partial \ln \tau_1/\partial \beta)_H$ (i.e. constant angle between $\ln \tau_1$ axis and contour) fails and the integral in (10) contributes the effect of changing spectrum. The exact sign and size of this contribution will depend on the variation of H with β. An interesting case arises if there exists more than one relaxation ridge, as is shown in Figure 2. While the ridge DCD does not affect the situation at B the converse is not true since the integration path in (10) must pass through B to reach C. This is to be expected, of course, for the relaxation which has already occurred (at the frequency considered) will affect E_1 and hence S_1. For a simplistic approximation assume that the contours to the right of the valley at S in Figure 2 are all parallel to ABA with slope $-q_B$, those of the left of S parallel to DCD with slope $-q_C$. For this model the inelastic terms of Eq. (8a), would amount to q_B to the right of S while to the left of S we would obtain

$$\left(\frac{E_2 - E_{2s}}{E_2}\right) q_C + \frac{E_{2s}}{E_2} q_B, \tag{11}$$

where E_{2s} is the value of E_2 in the valley corresponding to S. Hence as T increases the inelastic contribution to S_1 would remain at q_B until the first peak is passed and then rise towards the q_C level as E_2 rises above E_{2s} In II we find some step behavior in actual systems although Eq. (11) must be considered a simplification, at best.

The need for interpreting single-frequency, variable temperature data has been recognized previously, particularly by Read and Williams.[5] They point out that under certain assumptions, principally the invariance of the relaxation intensity ΔE with temperature, an average activation energy can be obtained from the area of the loss spectrum

$$\int_0^\infty E_2(T)\, d\beta = \frac{\pi}{2} \Delta E \langle q^{-1} \rangle. \tag{12}$$

Here $\langle q^{-1} \rangle$ is essentially given by

$$\langle q^{-1} \rangle = \int_0^\infty H(\ln \tau, \beta) q^{-1}\, dq, \quad \tau = \tau_0 e^{\beta q}, \quad q \text{ fixed but distributed} \tag{13}$$

i.e. the relaxation distribution averaged inverse of the activation energy. Details are given in McCrum, Read and Williams.[2]

Expression (12) has been relatively rarely applied in practice. The difficulty is that the integral is required over the entire relaxation region and this will be available only if relaxation peaks are well separated. Serious overlap between ridges having different slopes at the ridge (as in Figure 2) is very common, however. Moreover, as we will find in II, $S_1(T)$, which plays the role of the distributed q in (12) above, varies extensively over the range of even a single peak and thus an averaged $S_1(T)$ or $q(T)$ gives little information about a process. For this reason we prefer to display $S_1(T)$ as a spectrum, avoiding the integration of (10) above.

Appendix

In the main text calculation, the expression $(\partial E_1/\partial\beta)_\omega$ is used in the deviation of the contour $(-d\ln\omega/d\beta)_{E_1}$. The value of $(\partial E_1/\partial\beta)_\omega$ is available directly from experiment only if the latter is carried out at truly constant ω as is the case in forced oscillation (as with the Vibron instrument) or in dielectric work. Mechanical transducers achieve their maximum efficiency at resonance modes of the sample, however. For this reason it is common in modulus measurements to vary the temperature while operating throughout at sample resonance. Since the sample geometry remains constant (except for thermal expansion, a very small effect) and the modulus changes, the frequency will change through the experiment. It is thus useful to derive the expressions of the text for constant sample geometry rather than constant frequency.

A homogeneous, unsupported sample at mechanical resonance will obey

$$E_1 = c\omega^2, \tag{A-1}$$

where c is a constant which will depend on the mode of oscillation and the sample geometry and ω is the resonance frequency. Differentiating the logarithm of (A-1) with respect to β yields,

$$\frac{1}{E_1}\left(\frac{\partial E_1}{\partial\beta}\right)_x = \frac{1}{c}\left(\frac{\partial c}{\partial\beta}\right)_x + 2\left(\frac{\partial\ln\omega}{\partial\beta}\right)_x, \tag{A-2}$$

with the subscript x indicating that the resonance mode has been kept constant. From (A-1) the dimension of c is kg/m and hence the $(\partial c/\partial\beta)_x$ term involves the linear thermal expansion coefficient of the sample, λ, i.e.

$$\frac{1}{c}\left(\frac{\partial c}{\partial\beta}\right)_x = -kT^2\lambda \tag{A-3}$$

Since E_1 is solely a function of ω and β we have, by differentiation at constant x,

$$\left(\frac{\partial E_1}{\partial\beta}\right)_x = \left(\frac{\partial E_1}{\partial\ln\omega}\right)_\beta\left(\frac{\partial\ln\omega}{\partial\beta}\right)_x + \left(\frac{\partial E_1}{\partial\beta}\right)_\omega, \tag{A-4}$$

while in the main text we derived

$$\left(\frac{\partial E_1}{\partial\ln\omega}\right)_\beta = \frac{2}{\pi}E_2. \tag{A-5}$$

Substituting (A-2), (A-3) and (A-5) into (A-4) and rearranging we find the desired relation between $(\partial E_1/\partial\beta)_\omega$ and the experimentally accessible $(\partial E_1/\partial\beta)_x$

$$\frac{\pi}{2E_2}\left(\frac{\partial E_1}{\partial\beta}\right)_\omega = \frac{\pi}{2E_2}\left(\frac{\partial E_1}{\partial\beta}\right)_x\left[1 - \frac{E_2}{\pi E_1}\right]\frac{kT^2\lambda}{2}. \tag{A-6}$$

For a glassy polymer solid λ will be on the order of a few times 10^{-5} deg^{-1} hence the last term will be on the order of 1 cal m^{-1}. In the main text we noted that the l.h.s. of (A-6), which corresponds to the contour, amounts to 5 kcal m^{-1} or more. Hence the last term in (A-6) may be ignored in the usual case. In any event λ is easily obtainable to rough accuracy by placing a sample in two different baths (say liquid nitrogen and dry ice) and measuring its length with calipers.

The expression (A-6) is applicable to a homogeneous sample vibrating in a resonance mode. For a composite sample, consisting in part of an elastic structural support, some intermediate value between $(\partial E_1/\partial \beta)_\omega$ and $(\partial E_1/\partial \beta)_x$ will be obtained, while a lightly coated steel reed, e.g., would yield $(\partial E_1/\partial \beta)_\omega$ directly.

References

1. T. F. Schatzki, in preparation.
2. N. G. McCrum, B. E. Read and G. Williams, *Anelastic and Dielectric Effects in Polymeric Solids* (Wiley, London, 1967).
3. K. Deutsch, E. A. W. Hoff, and W. Reddish, *J. Polymer. Sci.*, **13**, 565 (1954).
4. F. Schwarzl and A. J. Staverman, *Appl. Sci. Res. Act.*, 127, 1953.
5. B. E. Read and G. Williams, *Trans. Faraday Soc.*, **57**, 1979 (1961).

Molecular Mechanics Calculations of the Energetics of Relaxations in Polymeric Materials

RICHARD H. BOYD

Department of Chemical Engineering, Department of Materials Science and Engineering, University of Utah, Salt Lake City, Utah 84112, U.S.A.

Empirical molecular force fields have reached the state of development where reliable predictions of the energetics of conformational change in organic molecules and polymers can be made. Furthermore, computational strategies have been developed that permit the calculation of the energetics of polymeric chain motions including the effects of the condensed phase surroundings providing the motions are reasonably localized. The relationships between the quantities resulting from molecular mechanics calculations and the experimental quantities usually observed (relaxation location, width and intensity) are also discussed here.

INTRODUCTION

The molecular interpretation of relaxations in polymeric materials has been hampered by a lack of tractible models. Many apparently worthwhile suggestions have been made concerning the molecular basis of relaxation, but it has been difficult to quantitatively assess the consequences of these mechanistic proposals in terms of experimentally measured quantities. However, recently the development of techniques that fall into an area that for want of a better term may be called "molecular mechanics" promises to rectify this situation at least in part.

The molecular mechanics approach seeks to describe the energy of a molecule that can change conformation or be distorted as the sum of the energies of distortion of its individual bonds (bond twisting, bending, etc.) and the energies of steric interactions between various parts of the molecule. The energies of the individual terms are described by functions characteristic of the kind of bond involved or the kinds of atoms involved in the steric interactions. These functions are of certain standard forms and differ from bond to bond or atom to atom by

277

the constants or parameters in them. The latter are determined largely by fitting experimental data on selected test molecules. Much of the work in selecting and applying these parameters (or "force fields") has been done not with polymers but rather with small and medium sized organic molecules. But the functions can be transferred readily to polymers containing the same kinds of bonds and atoms.

Many of the mechanistic proposals concerning polymer relaxations involve relatively localized molecular motions. The crankshaft motions[1,2] proposed for the γ relaxation in polyethylene, kink[3,4] and twist[5] crystal defect motions, and chain rotation[6-8] associated with the α relaxation in polyethylene, are all examples of motions that are localized. These have promise of their energetics being amenable to molecular mechanics calculations. Because these motions take place in a condensed phase and do depend on the matrix of surrounding material, the application to relaxations is inherently more difficult numerically than the application to isolated molecules. However, it appears that it is now possible to make such calculations. The motions associated with the glass rubber transition presumably require a still elusive statistical mechanical description of the highly cooperative processes before energetics calculations will be of value. In the present paper we will present a brief description of the molecular mechanics or force field method and the present state of its application to relaxation processes in polymers.

THE FORCE FIELD

Non-bonded (steric) interactions

For many years there has been interest in describing the stability of simple molecular solids in terms of the potential energy of interaction between the constituent molecules. It is known, for example, that the interaction potential energy between rare gas atoms (neon—neon, argon—argon, etc.) determined from experiments in the gas-phase (or recently from quantum mechanical calculations) also accounts fairly well for the properties of the solid such as heat of sublimation, lattice parameter and compressibility.[9] It has become common practice to attempt to express such interaction potentials in terms of an empirical function such as,

$$V(R) = A \exp(-BR) - C/R^6 \tag{1}$$

or

$$V(R) = A'/R^n - B'/R^6 \quad (R = \text{internuclear distance}) \tag{2}$$

in which the parameters A, B, C or A', B', n are selected to hopefully give a good representation of the true potential. It is possible to determine the three parameters directly from experimental data on the solid by fitting the heat of sublimation, lattice parameter and compressibility.

In polyatomic cases (a simple example would be methane, CH_4), it is tempting to express the interaction energy between molecules as the sum of interactions between atomic centers. In methane there would be $5^2 = 25$

interactions between two molecules. These would be divided into three types $C\cdots C$, $C\cdots H$, $H\cdots H$. Each function would require three constants (A_{cc}, B_{cc}, C_{cc}, etc.) or a total of nine parameters would be required to represent the interaction energy. We will set aside the question of the validity of the assumption of representing the interaction energy as the sum of these twenty-five *interatomic non-bonded interactions* and take up the question of determining the nine parameters. It is obviously not possible to determine them directly from experimental measurements on one compound of the three quantities used for the rare gases. Three methods have been used to circumvent this difficulty. First, theory has been used to determine some of them. For example, the London dispersion force theory has sometimes been used to determine the R^{-6} attraction constant. Secondly, the unlike pair constants ($C\cdots H$ for example) have been assumed to be determinable as certain combinations of the like constants ($C\cdots C$ and $H\cdots H$). Third, it can be assumed that the interaction potentials are independent of the specific kind of molecule they originate in and depend only on the kind of atoms. For example, the $C\cdots C$, $C\cdots H$ and $H\cdots H$ potentials should be the same in methane, ethane, butane, etc. Then experimental measurements on a variety of compounds can be used as a data base to be statistically fitted in order to arrive at the parameters. The latter method also carries the important implication that the potentials, once determined, can be used to calculate the properties of a great variety of compounds not included in the data base but containing the same kinds of atoms. In the past, there has been a great deal of disagreement in the potentials found and the properties predicted among different investigators depending on the particular mix of the above three approaches used. However, it has become clear in the past few years that, by minimizing the use of theory, judiciously using combining rules and depending mostly on a significant data base, potentials can be determined that (although not unique in the resolution into individual atom$\cdots$atom potentials) do lead to accurate predictions of the properties of compounds containing the same atoms. A notable example is the set of carbon hydrogen potential functions (of the form of Eq. 1) determined by Williams[10, 11] by fitting the lattice parameters compressibilities and heats of sublimation of a variety of hydrocarbons. The only theoretical quantity used was a quantum mechanical calculation of $B_{H—H}$. Data on graphite was used to determine $B_{C—C}$. Combining rules were used, taking $B_{C—H}$ as the arithmetic mean of $C\cdots C$ and $H\cdots H$ and $C_{C—H}$ as the geometric mean but allowing $A_{C—H}$ to vary independently. Table I shows the predicted crystal parameters of polyethylene using these potential parameters. In general, the William potential functions have proven very successful in representing the energetics and geometry of packing of hydrocarbons (i.e., *inter-molecular* non-bonded interactions). Further, it has been found that these potentials also do a good job of representing the steric interactions between non-bonded atoms within the same molecule (i.e., *intramolecular* non-bonded interactions).

Bond deformations

Torsion   One of the principal applications of molecular mechanics calculations lies in evaluating the differences in energies between conformational isomers of the same molecule. These conformers are two or more forms of the molecule that differ in shape by virtue of rotation about the bonds of the molecular skeleton. For example, cyclohexane can exist in two forms, the more stable chair form which has all of the skeletal torsional angles in the gauche conformation and the boat form which has two torsional angles near the ellipsed position (Figure 1).

C   B

FIGURE 1   Chair (C) and boat (B) conformations of cyclohexane.

There is apparently no simple physical interpretation for the origin of the barrier to bond rotation. A representation that has empirically been found to work well for C—C bonds in hydrocarbons is to assume that there is an inherent barrier to rotation involving orbital interaction,
This part of the barrier is taken to be of the form,

$$V = V_0/2 (1 + \cos 3\phi).$$

Part of the barrier and especially the effect of different substituents on the barrier is assumed to arise from non-bonded interactions. The same non-bonded

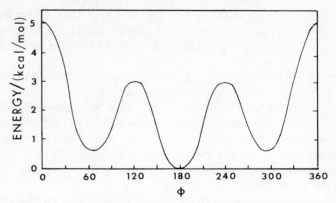

FIGURE 2   Calculated rotational energy diagram for polyethylene.

interactions as found from crystal structure fitting give a good representation of the barrier for rotation about C—C containing C and H substituents. Figure 2 shows the calculated barrier for bond rotation in polyethylene using a three-fold inherent barrier of 2.5 kcal/mol plus the non-bonded functions of Table I. It reproduces the features of the total barrier as well as they can be inferred from experiment. McCullough and Lindenmeyer[17] have argued that of the total

TABLE I

Calculation of crystal structure of polyethylene using modified William's potentials[a].

	Cell dimensions and setting angle			
	$a(A°)$	$b(A°)$	$\theta°$	ΔHs (kcal mol$-$CH$_2-$)
Calc.	7.166_b	4.914_b	47.1	1.850
exp.	7.155^b	4.899^b	$42°-48°C$	1.84

	$V(\text{kcal/mol}) = (A \exp (-BR) - C/R^6) \, 144.0$		
	A	$B(A^{°-1})$	C
C$\cdots$C	104.0	3.09	4.45
C$\cdots$H	30.0	3.415	0.96
H$\cdots$H	18.4	3.74	0.19

[a]Modified[12,13] to make C$\cdots$C repulsion slightly less steep than Williams[11] parameter set IV.

[b]At 77 K by Swan.[14]

[c]Refs. 15 and 16: The classical value has been 42° (Ref. 16), but there is now evidence that 48° may be a better value (Ref. 15).

barrier in ethane (2.8 kcal/mol) 75–90% must arise from the inherent barrier and 10–25% from the H· · · ·H non-bonded interactions. The Williams H· · · ·H non-bonded functions of Table I satisfy this bond (0.3 out of 2.8 kcal/mol).

Bond bending and stretching   Although variations in the energy from the torsional barrier and from the non-bonded interactions due to changes in the torsional angles are the most dominant in conformational change, distortions in valence angles and bond lengths are also important. In all but the most trivial situations they must also be included. Fortunately, a great deal is known about their energetics from vibrational spectroscopy. It is usually sufficient to consider these deformations as harmonic about natural bond angles Θ_0 and bond lengths R_0 and of the form,

$$V(\Theta) = \frac{k_{abc}}{2} (\Theta - \Theta^0_{abc})^2 \tag{3}$$

$$V(R) = \frac{k_{ab}}{2} (R - R^0_{ab})^2. \tag{4}$$

where Θ is the angle between the a—b—c bonds and R is the bond length of the a—b bond. Bonds are more easily bent than stretched and the distortions and energies due to bond bending are usually larger than from stretching.

Snyder and Schachtschneider[18, 19] have developed a set of force constants (k's) that are transferrable among alkanes and do an excellent job of reproducing the vibrational spectra of hydrocarbons. They have also included, as is the usual practice, potential terms for the interaction between different bending and stretching motions, $\Delta\Theta_1 \Delta\Theta_2$, $\Delta\Theta\Delta R$, etc. However, they did not include non-bonded interactions in their force field. Usually the energy terms for interaction between different bending and stretching motions are not significant for molecular mechanics calculations and are omitted. In our own work[12, 13] we have modified the Snyder and Schachtschneider work to allow for the effects of the non-bonded interactions and have omitted the $\Delta\Theta_1 \Delta\Theta_2$, $\Delta\Theta\Delta r$ cross terms. In Table II force fields which have included the internal degrees of freedom of stretching and bending as well as torsional and non-bonded interactions are listed.

APPLICATION TO MOLECULES

In some of the early applications[31] to polymers the molecule was considered to be rigid except for the torsional coordinates and the total conformational energy

TABLE II

Molecular force fields that include stretching and bending along with torsion and non-bonded interactions.

Force field	Consistent with vibrational frequencies[a]	Area of application (References)				
		alkanes	alkenes	aromatics	ketones	hahides
Boyd	+	12, 13	(b)	20		
Lifson	+	21	22			
Allinger	−	23	24	25	26	27
Bartell	−	28				
Hägele-Pechhold	−	29				
Schleyer	−	30				

[a]In the sense of including the fitting of vibrational frequencies in the data base and comparing calculated with experimental frequencies in other molecules.

[b]Developed by S. M. Breitling and furnished with computer program (MDLBD3). Available on request for molecular property calculations.

to be made up only of the torsional and non-bonded energy,

$$V = \sum \frac{V_0}{2} (1 + \cos 3 \, \phi_i) + \sum V_{NB}(R_{ij}). \tag{5}$$

In such "ball and stick" model calculations usually two adjacent torsional co-ordinates (5 consecutive skeletal atoms) were considered sufficient to generate all of the important steric interactions. From a contour map of the total energy *versus* these two torsional coordinates, the most stable conformation of the chain could be determined by locating the deepest valley. Such techniques sufficed for the simpler polymer chains. However, in polymers with serious steric interferences (polyisobutylene, polystyrene, polypropylene, etc.) the rigid skeleton adjacent bond pair approximation is inadequate for assessing the differences in energy between different conformations. The same problem has been encountered in strained cyclic organic molecules where valence angle distortion (and to a lesser extent bond length distortion) is significant and quite a number of torsional coordinates must be included. Thus, the total energy most conveniently includes *all* internal degrees of freedom,

$$V = \sum \frac{k_{ij}}{2} (R_{ij} - R_{ij}^0)^2 + \sum \frac{k_{ijk}}{2} (\Theta - \oplus\, {}^0_{ijk})^2$$

$$+ \sum \frac{V^0}{2} \, ijkl \, (1 + \cos 3 \, \phi_{ijkl}) + \sum V_{NB}(R_{ij}), \tag{6}$$

where the terms include all bond stretching, bond bending, torsional, and non-bonded interactions in the molecule. A polymer is treated by considering adequately long but finite homologs. The stable conformations can in principle be located by minimizing the total energy with respect to all of the internal degrees of the freedom of the molecules. An adequately long segment of a polymer or an interesting strained cyclic hydrocarbon can involve up to 50—60 atoms (150 to 180 degrees of freedom; x, y, z for each atom) and of the order of a thousand terms in Eq. (6). In including the surrounding condensed phase in relaxational models in polymers, up to ~1000 atoms could profitably be included. Thus, it is not a trivial problem minimizing the total energy function. However, efficient computer algorithms have been developed which permit this in quite reasonable computation times. We have demonstrated[32] that one algorithm is not only efficient but has the further advantage of being readily adaptable to computing the vibrational frequencies of the molecule after convergence to the minimum energy structure. This is important because it allows ready adjustment of the input energy function parameters to fit a data base of experimental vibrational frequencies as well as conformational energies. Also the thermodynamic functions (free energy, enthalpy, entropy and heat capacity) of the molecule can be calculated from the vibrational frequencies. Conformational free energies can be calculated in addition to conformational energies.

Force field calculations (with all internal degrees of freedom participating, Table II) have been applied to essentially every saturated hydrocarbon for which experimental data are available.[33] In general, the calculated structures and energies agree well with experiment, well enough that one can have some confidence in the method. It has also been applied to non-conjugated olefins and to some aromatics, ketones and chlorides. It has also been successfully applied to predicting the most stable conformation of sterically hindered polymers (polyisobutylene)[34] and the energy differences between various conformations having steric interferences (polypropylenes).[35]

Before taking up the application to relaxation processes in polymers let us consider a simpler illustrative example, the chair–boat conversion of cyclohexane.[36] From an examination of stick models (such as Fieser Prentice-Hall Framework Models), it is clear, that by changing the value of one of the torsional coordinates of the chair form from the *gauche* value progressively, through eclipsed and over to a *gauche* value of the opposite sense, one can "induce" the chair–boat conversion. This same transition can also be induced in molecular mechanics calculations by constraining the selected torsional coordinate to a series of preselected values. The energy is minimized subject to the fixed torsional angle constraint at each of the predetermined values of the selected torsional angle. Thus, if the transition has the proper topological behavior, a minimum energy path is traced out. This path is very much analogous to the "reaction coordinate" in absolute reaction rate theory.[37] In fact absolute reaction rate theory can be applied to the transition and ground state to arrive at a rate constant for the conversion. Since the energy minimization method allows calculation of the vibrational frequencies, the partition functions of the transition and ground states may be computed and absolute rate theory applied in the way it was originally intended but seldom can be. That is, the rate constant k is given in terms of the partition Q and energy barrier E_0 by

$$k = \frac{k_B T}{h} \frac{Q^{\ddagger}}{Q_0} \exp\left(-\Delta E_0^{\ddagger}/RT\right) \qquad (7)$$

$$k = \frac{k_B T}{h} \exp\left(\Delta S^{\ddagger}/R\right) \exp\left(-\Delta E^{\ddagger}/RT\right). \qquad (8)$$

Table III shows the application to cyclohexane chair–boat conversion and it may be noticed that the agreement is excellent.

There are some restrictions in the application of the method. The selected torsional coordinate must be single valued over the transition. One could have a split valley and go up the wrong chimney. However, with some judicious

TABLE III
Chair—boat conversion of cyclohexane[a]

	$\Delta H^{\ddagger}$ kcal/mol	$\Delta S^{\ddagger}$ cal K^{-1} mol^{-1}	ΔH kcal/mol
Chair—transition state	10.0 (calc) 10.8 (exp)	3.4 2.8	
Chair—boat (twist)			6.1 (calc) 5.7 (exp)

[a]Ref. 36, but recalculated on the basis of the inherent torsional barrier adjusted from 2.1 to 2.5 kcal/mol.

reference to models, it has been possible to apply it to a variety of conformational changes in polymer chains.

RELAXATION PROCESSES IN POLYMERS

Relaxation theory and molecular mechanics

As indicated in the introduction, molecular mechanics will be of use where well-defined conformational transitions take place between a limited number of states. Thus, site models are appropriate. A localized conformational situation is identified (such as a conformational defect in a crystalline polymer). A set of conformational changes (of the defect†) is isolated and the energies of the stable conformations and the energy paths between determined. These stable conformations can be called sites and the energy paths between them represent barriers between the sites. The dynamics of the defect can be treated as a rate process involving a group of sites and their barriers. The treatment will differ from that illustrated for the chair—boat conversion of cyclohexane mainly in that the processes take place in a condensed phase and the effects of the surroundings must be included. Also, in addition to the time—temperature location of the transition (the values and temperature dependence of the relaxation times), we shall want to know about the *intensity*. From the intensity contribution of each relaxation time, we can construct the shape or *width* of the relaxation. The relaxations between molecular mechanics calculations, site theory and experimentally determined quantities are shown in Table IV.

† For want of a better term, we will use the term defect here to mean any localized conformational situation under consideration. It could also be a crankshaft in an amorphous polymer or a side group motion.

The calculation of the relaxation times from the site energies and barriers is straightforward. The values of the relaxation times of course determine the time–temperature location of the transition. However, the shape or width and the center of the relaxation region as well as measures of total intensity will also depend on the intensity contribution of each relaxation time. The latter depends on the three factors listed in Table IV. The first, the defect concentration, can either be at thermal equilibrium or may be the result of freezing in the equilibrium at some higher temperature. In either case, the energy of creation of the defect will be the crucial determinant in the defect concentration. The thermal population of the sites above the lowest site of the defect will also affect the intensity. This population is at thermal equilibrium at times long compared to the longest relaxation time of the defect. It is determined by the Boltzmann factors resulting from the site energy differences. Finally, there is the inherent intensity of the defect motion.

TABLE IV

Relationships among molecular mechanics calculations, relaxation theory and experimentally observed quantities.

Experiment	Site theory	Molecular mechanics calculation
Relaxation location in time–temperature domain	(Absolute rate theory applied to the site system of the defect.)	
	(a) set of temperature dependent relaxation times	energies of sites and barriers between sites
Relaxation intensity	(b) temperature dependent intensity contribution of each relaxation time	(a) concentration of defects (from energy of inclusion of the defect)
Relaxation width		(b) energy differences between sites of one defect
		(c) inherent intensity factor of one defect (strains associated with each site of the defect)

The presence of a defect in a crystal lattice causes a strain in the lattice. Each site or state of an individual defect has its own strain field associated with it. Reorientation of the defect from one site to another causes the strain to change. When this occurs in the presence of a stress imposed on the specimen, work is done. The work done by the strain change accompanying defect reorientation in the presence of stress is in addition to the work of elastically deforming the lattice itself. If the number of defects reorienting is known then the total work of deformation can be computed. From this, the macroscopic strain can be deduced from the applied stress.

The number of defects reorienting on application of the stress can be calculated from the Boltzmann factors for the sites in the presence and absence of stress. The energy differences between sites in the presence of stress depends on the energy differences in the absence of stress plus the work of reorientation in the stress field. The latter is given by the *product of the stress times the strain change on reorientation. The strain change on reorientation is calculated in the absence of stress. It is not necessary to know the effect of stress on the energy differences between sites* in the sense of including an imposed stress or strain field in the calculation of the site energies. While it is formally correct to say "the relaxation is mechanically active because of a redistribution of defect orientations due to biasing by stress," the biasing is done by the work arising from the strain change (calculated in the absence of applied stress or strain) accompanying changing from one site to another (reorientation) in the presence of stress. The direct effect of applied stress or strain on the site energies leads to a *second order* effect, i.e., the non-linear dependence of strain on stress (the dependence of the modulus on stress). This situation is very much analogous to the dielectric relaxation case where the dielectric constant (polarization response to an electric field) is calculated from the average-square dipole moment in the absence of a field.

In summary, it is the task of molecular mechanics to provide the defect creation energy, the differences in energy among the sites (orientations) of the defect, the barriers between sites and the strains induced by each orientation. Below we will review some of the work that has been done in calculating these. As will be apparent rarely, if ever, have all of the above been calculated for a given postulated defect. Most of the studied have been confined to only some of them.

Application to polymers

Nearly all of the work on molecular mechanics calculations appropriate to relaxation has been done on polyethylene. This work is summarized in Table V. Rather than discuss all of this past work, it seems rather more appropriate here to discuss in detail some of our recent calculations that come fairly close to providing all of the information necessary for a comparison between model and experiment and also illustrate what can be done at present. These calculations refer to two closely related conformational phenomena, crankshaft motions and kink motions. In searching for particularly simple motions that could explain how polyethylene (and other polymers) could have such a low temperature

TABLE V

Molecular mechanics calculations on polyethylene.

Defect	Effect of condensed phase environment[a]	Creation energy	Reorientation or site energies	Barriers between sites	Site strains	Ref.
Reneker	1	+				39
	1	+	+			4
Chain fold	1	+	+			39
	1	+	+	+		4
Kink	1	+	+			4
	2, 3	+	+			40
	2, 3, 42	+	+	+	+	38, 42
Crankshaft	1, 3	+	+	+	+	38, 42
Chain rotation	2	+	+	+		8
Methyl branch in C$_{20}$H$_{42}$[b], 42	3	+	+		+	41

[a]1 = not included (isolated chain), 2 = included but considered rigid, 3 = included and allowed to deform.
[b]Chain internally rigid.

relaxation (γ region, $\sim-120°C$, 1 Hz), Schatzki[1] noticed that when the chain has a conformational sequence of — (T G T G T) —the end bonds are colinear. He further reasoned that motion about the stems should involve the barriers to two simultaneous C—C rotations and thus have an internal chain contribution to the activation energy that would leave room for some contribution from the matrix deformation. Boyer[2] proposed an alternative mechanism involving the motion about stems including three bonds. Because colinearity of the stems in this case requires appreciable valence angle distortion, it appeared that the motion about the stems would involve a higher intrachain contribution to the barrier. However the matrix inhibition might be less. Recently we have proposed[38] a generalization of the Boyer three-bond motion which involves the conformational sequence — (T G T) —between stems. In this case, the stems are parallel but not colinear. It has the advantage (along with the five-bond Schatzki motion) that the initial state is not sterically strained. Thus either of these motions (three-bond or five-bond) could take place from initial states that could be conformational sequences present in the chain in the liquid state and frozen in at the glass temperature (which we regard as associated with the β transition). We have modelled these motions and found that the generalized Boyer (three-bond) motion can move to a final internally unstrained state described as—(T G' T)— with the same stem positions. Furthermore, if one of the stem bonds is *gauche* of the opposite sense of the internal one G'←T G T→ then this motion involves two distinct barriers with an intermediate state between them. One of the barriers involves the conformational switch,

$$G'(T G T)\alpha \text{ (initial)} \longrightarrow G(T G' T)\alpha \text{ (intermediate)}.$$

The other one involves,

$$G(T G' T)\alpha \text{ (intermediate)} \longrightarrow T(T G' T)\beta \text{ (final)},$$

(where α is any conformation and β is obtained from it by $\sim120°$ rotation).

In order to model the effect of the surroundings, some model for the amorphous matrix had to be made. It was assumed to have the structure of the crystal but the amorphous volume. The details of the structure turned out to be unimportant since it was found that one of the barriers (the $G'(T G T)\alpha \rightarrow G(T G' T)\alpha$ switch) involved motion with very little swept out volume and was not influenced by the surroundings. The other barrier (motion about both stems) required considerable swept-out volume and was completely inhibited as far as contributing to the γ process. The Schatzki crankshaft motion requires even more swept-out volume and thus is completely inhibited.

It was found that the predicted (from the calculated relaxation times) location of the transition region in the time—temperature domain was in reasonable agreement with experiment. No attempt was made to estimate the intensity.

In the calculations modelling the surrounding, it was necessary in order to reduce computation time to approximate the —CH$_2$—groups as point force centers and not explicitly include the hydrogens. Since completing these calculations we have managed to effect improvements in the computer algorithm which now permit inclusion of the hydrogens. Calculations have now been made of the strains induced by the crankshaft motion for the purpose of estimating the intensity. We intend to publish these calculations elsewhere[42] but will summarize them here.

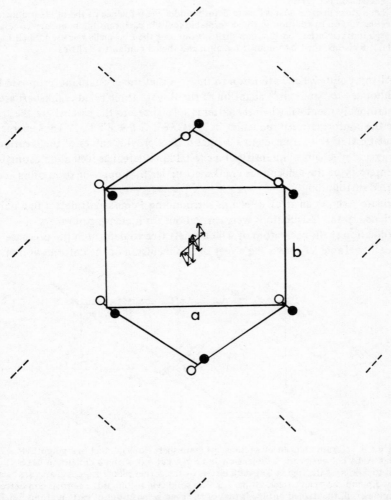

FIGURE 3   An array of chains surrounding a crankshaft defect in a polyethylene crystal.

TABLE VI
Effect of surroundings on defect energy
(T T T G T G′ T T T T T crankshaft or kink[a]).

	Number of coordination shells free to deform[b]		
	0	1	2
Static defect			
Energy (kcal/mol)	7.6	6.6	6.7

[a]Each chain is $C_{14}H_{30}$.
[b]Each atom in each shell is free to deform under the influence of the inter- and intra-molecular forces. In addition to the deformable shells in each case the boundary condition of a rigidly fixed outer coordination shell surrounding the free shells was used. Shell 1 contains 6 chains, shell 2 contains 12 chains, and shell 3 contains 18 chains.

At this point, we call attention to the fact that the crystal kink proposed by Pechhold et al.[3] and our[38] adaption of the Boyer[2] three bond crankshaft are conformationally identical when the latter is specialized to the case where the stems have the conformational sequence ... T T G′ ‒(‒T G T‒)‒T T T ... Thus energetics calculations of this sequence in a crystal matrix provide values of the creation energy and mechanical intensity of a crystal kink defect as well as an estimate of the intensity of the amorphous crankshaft under the approximation of an ordered surrounding.

Figure 3 shows an array of chains surrounding a crankshaft defect in a polyethylene crystal. Calculations were carried out for increasing numbers of coordination shells each atom of which is left free to deform in the presence of the defect. Table VI shows the effect of this inclusion of the deformation of the

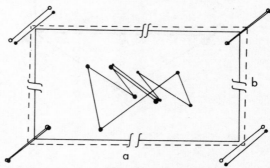

FIGURE 4   Strain induced by presence of crankshaft (kink) defect in a polyethylene crystal. A 4Å cut out of the a dimension and a 2 Å cut out of the b dimension has been taken to compress the figure. The dark circles of the corner planar zig-zag chains are carbon atoms 5,6 and the open circles, atoms 7,8. The solid cell outline is the normal non-defect containing cell, the dashed outline is that of the average positions of carbon atoms 5,6,7,8 of the corner planar zig-zag in the defect containing crystal.

surroundings on defect energy. It is worth noting that in the context of the size of the systems that can be computationally handled that the cases of 1 and 2 free coordination shells involve 308 and 836 free atoms, respectively. It is to be noted that the inclusion of one deformable shell has a significant but not crucial effect. Further, the energy has converged at the addition of two deformable shells. Figure 4 shows the strain induced by the presence of the defect. The principal deformation is a shear strain in a plane across the a b diagonal and parallel to the c direction. This shear is about 4% in each of the two orientations of the defect. This amounts to a change in strain of 0.08 on reorientation of the defect. The consequences of this strain in terms of relaxation intensity is still being pursued.

Acknowledgments

In the maturation of any area of physical science it undergoes a succession of development from qualitative conjecture through detailed model development which finally hopefully results in predictive power. The field of relaxation in polymeric materials appears to be progressing from the first stage towards the second. We all owe a debt to men like Ray Boyer who have contributed so importantly in this first stage to allow the second one to proceed.

The author is also indebted to the U.S. Army Research Office (Durham) and to the National Science Foundation for their financial support.

References

1. T. F. Schatzki, *J. Polymer Sci.* **57**, 496 (1962); T. F. Schatzki, *ACS Polymer Preprints* **6**, 646 (1965); T. F. Schatzki, *J. Polymer Sci. C* **14**, 139 (1966).
2. R. F. Boyer, *Rubber Chem. Tech.* **34**, 1303 (1963).
3. W. Pechhold, S. Blasenbrey and S. Woerner, *Kolloid-Z. Z. Poly.* **189**, 14 (1963).
4. P. E. McMahon, R. L. McCullough and A. A. Schlegel, *J. Appl. Phys.* **38**, 4123 (1967).
5. D. H. Reneker, *J. Polymer Sci.* **59**, 539 (1962).
6. J. D. Hoffman, G. Williams and E. Passaglia, *J. Polymer Sci. C* **14**, 173 (1966).
7. C. A. F. Tuijnman, *Polymer* **4**, 259 (1963).
8. R. L. McCullough, *J. Macromol. Sci. Phys. B* **9**, 97 (1974).
9. J. O. Hirschfelder, C. F. Curtis and R. B. Bird, *Molecular Theory of Gases and Liquids* (Wiley, New York, 1954).
10. D. E. Williams, *J. Chem. Phys.* **45**, 3770 (1966).
11. D. E. Williams, *J. Chem. Phys.* **47**, 4680 (1967).
12. C. F. Shieh, D. McNally and R. H. Boyd, *Tetrahedron* **25**, 3653 (1969).
13. S. J. Chang, D. McNally, S. Shary-Tehrany, M. J. Hickey and R. H. Boyd, *J. Amer. Chem. Soc.* **92**, 3109 (1970).
14. P. R. Swan, *J. Polymer Sci.* **56**, 403 (1962).
15. Kobayashi, N. Uyeda and A. Kawaguchi, U.S.—Japan Joint Seminar or the Polymeric Solid State, COSC Western Reserve Univ. Cleveland Ohio, (1973).
16. C. W. Bunn, *Trans. Faraday Soc.* **35**, 482 (1939).
17. R. L. McCullough and P. H. Lindenmeyer, *Kolloid-Z. Z. Polym.* **250**, 440 (1972).
18. R. G. Snyder and J. H. Schachtschneider, *Spectrochim. Acta.* **21**, 169 (1965).
19. J. H. Schachtschneider and R. G. Snyder, *Spectrochim. Acta.* **19**, 117 (1963).
20. R. H. Boyd, S. N. Sanwal, J. Shary-Tehrany and D. McNally, *J. Phys. Chem.* **75**, 1264 (1971).

21. S. Lifson and A. Warshel, *J. Chem. Phys.* **49**, 5116 (1968); A. Warshel and S. Lifson, *J. Chem. Phys.* **53**, 582 (1971).
22. O. Ermen and S. Lifson, *J. Amer. Chem. Soc.* **95**, 4121 (1973).
23. N. L. Allinger, M. A. Miller, F. A. Van-Catledge, and J. A. Hirsch, *J. Amer. Chem. Soc.* **89**, 4345 (1967); N. L. Allinger, J. A. Hirsch, M. A. Miller, I. J. Tyminski and F. A. Van-Catledge, *J. Amer. Chem. Soc.* **90**, 1199 (1968); N. L. Allinger, M. T. Tribble, M. A. Miller and D. H. Wertz, *J. Amer. Chem. Soc.* **93**, 1637 (1971).
24. N. L. Allinger and J. T. Sprague, *J. Amer. Chem. Soc.* **94**, 5734 (1972).
25. N. L. Allinger and J. T. Sprague, *J. Amer. Chem. Soc.* **95**, 3893 (1973).
26. N. L. Allinger, M. T. Tribble and M. A. Miller, *Tetrahedron* **28**, 1173 (1972).
27. N. L. Allinger, J. A. Hirsch, M. A. Miller and I. J. Tyminski, *J. Amer. Chem. Soc.* **91**, 3893 (1973).
28. E. J. Jacob, H. Thompson and L. S. Bartell, *J. Chem. Phys.* **47**, 3736 (1967).
29. P. C. Hagele and W. Pechhold, *Kolloid-Z. Z. Polym.* **241**, 977 (1970); G. Wobser and P. C. Hagele, *Ber. Bunsenges. Phys. Chem.* **74**, 896 (1970); P. C. Hagele, Ph.D. Thesis, Ulm (1972).
30. E. M. Engler, J. D. Andose and P. R. Schleyer, *J. Amer. Chem. Soc.* **95**, 8005 (1973).
31. For a summary see P. J. Flory, *Statistical Mechanics of Chain Molecules* (Interscience, New York, 1969).
32. R. H. Boyd, *J. Chem. Phys.* **49**, 2574 (1968).
33. N. L. Allinger, Calculations of molecular structure and energy by force-field methods, *Adv. Phys. Org. Chem.* **13**, 1 (1976).
34. R. H. Boyd and S. M. Breitling, *Macromolecules* **5**, 1 (1972).
35. R. H. Boyd and S. M. Breitling, *Macromolecules* **5**, 279 (1972).
36. K. B. Wiberg and R. H. Boyd, *J. Amer. Chem. Soc.* **94**, 8426 (1972).
37. S. Glasstone, K. J. Laidler and H. Eyring, *Theory of Rate Processes* (McGraw-Hill, New York, 1941).
38. R. H. Boyd and S. M. Breitling, *Macromolecules* **7**, 855 (1974).
39. V. Petraccone, G. Allegra and P. Corradini, *J. Polymer Sci. C* **38**, 419 (1972).
40. H. Scherr, P. C. Hägele, and H. P. Grossman, *Colloid Polymer Sci.* **252**, 871 (1974).
41. B. L. Farmer and R. K. Eby, *J. Appl. Phys.* **45**, 4229 (1974).
42. R. H. Boyd, *J. Polymer Sci-Polym. Phys. Ed.* **13**, 2345 (1975).

DISCUSSION

P. A. Hiltner (*Case Western Reserve Univ.*): We have observed mechanical relaxations (δ peaks) at very low temperatures in PE and other linear crystalline polymers which have an activation energy on the order of 2—4 kcal/mol. Would you care to speculate on possible crystalline defect mechanisms which would have an activation energy of this magnitude?

R. H. Boyd: Calculations show that this Reneker defect is quite flexible and that some reorientational motions (not the translation through the lattice) have quite low activation energy. That is one possibility.

Y. S. Papir (*Richmond College*): Do your elegant calculations predict a dependence of the intensity of the relaxation process on the number of Pechhold type kinks which are included in the crystalline matrix? I ask this question in light of the δ internal friction process reported by Papir and Baer for LPE and POM, where the experimental results imply that the relaxation process requires the presence of simple conformational kink defects but its strength appears to be independent of their number.

R. H. Boyd: Any defect mechanism would predict the intensity to be proportional to the number of defects.

R. Simha (*Case Western Reserve Univ.*): In computing the effect of the environment in a semi-crystalline, not to speak of an amorphous system, must not a disordered structure or defective lattice rather than a crystalline arrangement be assumed? This could exert a considerable influence on the result.

R. H. Boyd: Certainly the disordered structure is important, but in the case of crankshaft motion it seems to boil down to a situation where the flip-flop motion is unhindered by environment and the complete crankshaft motion is highly hindered. Thus the details of the packing in this case are not as important as one might have expected.

W. Wun (*Case Western Reserve Univ.*): How many neighboring chains have you considered in the potential energy calculation for polyethylene? How good is the energy calculation when only nearest neighbor chains are used? Also, how accurate is the activation energy as calculated by your method? It seems that a study of small molecules, such as cyclohexane, which may be handled by quantum mechanical calculations, would be useful, since one should be able to directly compare the activation energies obtained by your method and that from the quantum mechanical approach.

R. H. Boyd: These questions are answered pretty well in the text of the paper (Tables III and VI).

J. Heijboer (*TNO*): Would you please explain why you selected the lower energy barrier, and not the higher energy one, in your calculations as being the significant barrier?

R. H. Boyd: We observe that the lower energy barrier seems to predict a relaxation similar to the experimental one, but the higher one, if active, would lead to a relaxation entirely outside the temperature range of the γ process and would occur at much higher temperature.

Molecular Origin of the Cyclohexyl Loss Peak

J. HEIJBOER

Centraal Laboratorium TNO, Delft, The Netherlands

Most polymers containing cyclohexyl groups show a mechanical loss peak near 190 K at 1 Hz. The effects of chemical modifications (as well as comparison with NMR data) prove that this loss peak is due to the chair−chair transition of the cyclohexyl ring.

The location and height of the peak are remarkably insensitive to the molecular environment of the cyclohexyl group.

INTRODUCTION

Apart from the main mechanical loss peak, which is due to the glass transition, amorphous glassy polymers show smaller secondary loss peaks related to local motions of smaller groups of atoms. Although there often is an indication or even a proof of what group of atoms is responsible for a given secondary loss peak, the detailed molecular mechanism usually is not well-defined. An exception is the secondary loss peak found near 190 K at 1 Hz shown by polymers containing cyclohexyl rings. In this paper we will present arguments which in our opinion definitely show that this loss peak is caused by the chair−chair flipping of the saturated six-membered ring.

THE CONFORMATION OF THE CYCLOHEXYL RING[1]

As early as 1890 Sachse[2] pointed out that there are two forms of cyclohexane in which the C—C—C bonds are free from angle strain: the rigid or *chair* form and the flexible or *boat* form. Since in the chair form all hydrogen atoms are staggered with respect to one another, its energy is 5 to 6 kcal/mol lower than that of the most favourable flexible form (Eliel *et al.*[3]). At room temperature, therefore, the majority of the cyclohexane molecules assume the chair form.

The chair form is depicted schematically in Figure 1, which shows that it has two types of bonds connecting the subsituents to the carbon atoms of the ring: those pointing up and down from the median plane of the ring, called *axial* bonds, and those lying closer to the plane of the ring, called *equatorial* bonds. A mono-substituted cyclohexane therefore has two non-identical chair conformers.

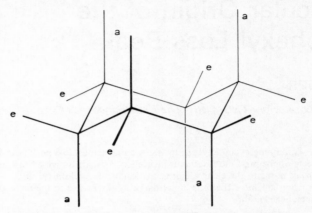

FIGURE 1   Schematic representation of the chair conformer of cyclohexane with axial (a) and equatorial (e) bonds.

In general, the axial position of a substituent is energetically less favourable than the equatorial position. This effect is mainly due to the interaction of the axial substituent with other axial substituents or with the hydrogen atoms in the 3-positions (two carbon atoms further on in the ring). A bulky substituent, such as *tert*-butyl, can only occupy the equatorial position, and only one of the chair conformers can exist.

The transition of one chair conformer to the other is called *ring inversion*. Recently, the ring inversion of small molecules has been studied by NMR. These measurements provide a value of the free energy of activation $\Delta F^{\ddagger}$. A wealth of data are available for cyclohexane and deuterated cyclohexanes. These data have been discussed by Anet and Bourn[4] and Binsch.[5] The best value that can be obtained from these data for the activation energy, $E_a^{\ddagger}$, of ring inversion is 11.4 kcal/mol. This is the barrier for the transition of one chair conformer of cyclohexane to the other.

In a polymer molecule the motion of a cyclohexyl ring is more restricted than in a liquid. In poly(cyclohexyl methacrylate) (PCHMA), the cyclohexyl group C_6H_{11} is connected to the main chain via a—COO—group. In PCHMA the motion of the —COO—group is very much restricted; if at all possible, it begins at higher temperatures than the ring inversion. Therefore, one end of the cyclo-hexyl ring is fixed. The resulting motion of the six-membered ring is sketched in Figure 2. As the ring is inverted, an axial bond becomes equatorial and vice versa.

For the following it is important to note that, when the ring is inverted, a bond linking a substituent to the ring in the 4-position retains its direction in space, shifting parallel to itself, whereas a bond linking a substituent to the ring in the 2-position is turned through an angle of 109°.

THE CYCLOHEXYL LOSS PEAK AND ITS MOLECULAR ORIGIN[6]

Figure 3 compares the shear storage modulus and the damping of poly-(cyclo-hexyl methacrylate) (PCHMA) with those of poly(phenyl methacrylate) (PPhMA), both at 1 Hz. The two polymers have about the same softening range (near 400 K), but they differ strikingly in the region of 190 K, where PCHMA shows a sharp damping peak that is not shown by PPhMA. In this temperature region poly-(cyclopentyl methacrylate) does not show a loss peak either.[7] This is a clear indication that the peak is due to the cyclohexyl ring. Since the peak is shown by nearly all polymers containing a cyclohexyl ring, we shall call it the *cyclohexyl maximum* or cyclohexyl loss peak.

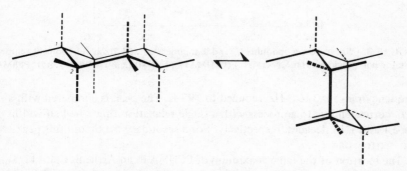

FIGURE 2   The chair—chair transition of the cyclohexyl group with carbon atom 1 in a fixed position.

Figure 4 gives the shear modulus and the losses of PCHMA as a function of temperature for a series of frequencies, from creep to ultrasonic wave propagation. The picture is dominated by the cyclohexyl maximum. At very low temperatures there appears a so-called δ-maximum, whereas the loss curve for 10^{-4} Hz near 220 K shows a faint remnant of the β-maximum of polymethacrylates.

In Figure 5 the loss modulus $G'' = G' \tan \delta$ is plotted as a function of $1/T$ for five frequencies; its half-width decreases with increasing frequency whereas its height remains practically constant; only at 0.8 MHz is it higher, because of the vicinity of the glass transition.

Figure 6 gives the results of torsion pendulum measurements of G'' in the

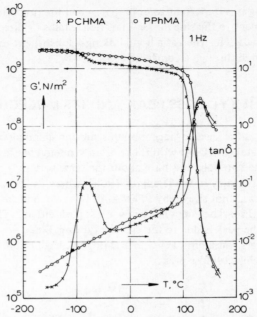

FIGURE 3   Shear storage modulus G' and damping tan δ at 1 Hz as functions of temperature for poly(cyclohexyl methacrylate) (PCHMA) and poly(phenyl methacrylate) (PPhMA).

frequency range 0.3 to 3 Hz, reduced to 193 K. The peak is compared with a
peak corresponding to a process with a single relaxation time; the half-widths
are 2.1 and 1.14 decades, respectively. For a secondary maximum this peak is a
very narrow one.

The location of the tan δ maximum of PCHMA in an Arrhenius plot ($1/T_{max}$
vs log ν) is illustrated in Figure 7. The maxima are obtained from tan δ vs T
curves at constant frequency. It is surprising that over the very broad frequency
range of 10 decades the points fall on a straight line. Using the equation
$\nu = \nu_0 \exp(-E_a^{\ddagger}/RT)$, we find an apparent activation energy $E_a^{\ddagger}$ of 11.3 kcal/mol,
and a log ν_0 of 13.0, which is a reasonable value for a molecular vibration frequency. $E_a^{\ddagger}$ agrees very well with the $E_a^{\ddagger}$ of 11.4 kcal/mol found from NMR
measurements for the chair–chair transition of cyclohexane.

Interestingly, the data of poly(cyclohexyl acrylate) (PCHA), a polymer whose
T_g is about 80 K lower than that of PCHMA, fall on the same line. Even the points
which Karpovich[8] and Waterman[9] obtained from ultrasonic measurements on
cyclohexanol lie very close to this line. This near-coincidence indicates that the
location of the cyclohexyl maximum is remarkably insensitive to the molecular
environment.

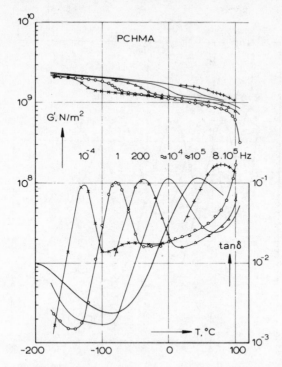

FIGURE 4   Shear modulus G' and damping tan δ of PCHMA as functions of temperature at six frequencies.

This insensitivity is also confirmed by Figure 8, in which a comparison is given of the storage and loss modulus in shear of three compositions: PCHMA plasticized by dimethyl phthalate (DMP) and by diphenyl phthalate (DPP), and poly(methyl methacrylate) (PMMA) plasticized by dicyclohexyl phthalate (DCHP). First it is seen that the location of the cyclohexyl loss maximum of PCHMA is not influenced by the addition of a large amount of plasticizer; moreover, a polymer with a plasticizer containing a cyclohexyl group shows the loss maximum at the same temperature. Clearly it does not matter whether the cyclohexyl group belongs to the macromolecule or to the plasticizer. The height of the loss peak does not depend on it either.

This fact is illustrated by Figure 9, which shows the height of the loss peak, corrected for background losses, as a function of cyclohexyl concentration. The filled circles are for polymer compositions containing the cyclohexyl ring in the plasticizer molecule. It is seen that they fall on the same line that is found for compositions containing the cyclohexyl ring in the polymer molecule. I think

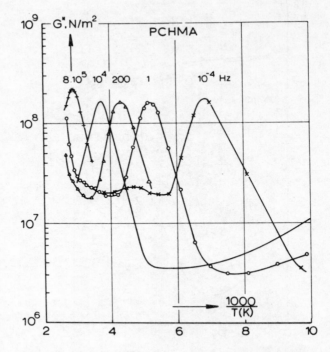

FIGURE 5   Loss modulus in shear G'' as a function of $1/T$ for PCHMA at five frequencies.

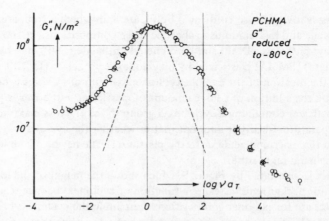

FIGURE 6   Reduction of G'' for PCHMA to 80°C. Dashed line: maximum for a single relaxation time.

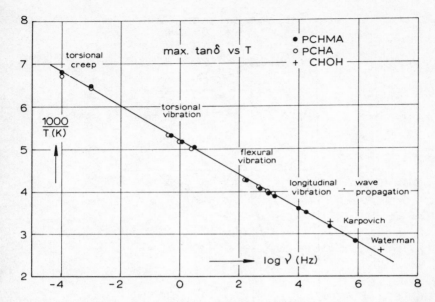

FIGURE 7   Arrhenius plot of the loss maxima of poly(cyclohexyl methacrylate) (PCHMA), poly(cyclohexyl acrylate) (PCHA) and cyclohexanol (CHOH).

this fact is important for the understanding of the interaction between the mechanical stress and molecular motions in the glassy state.

We now return to the arguments that the cyclohexyl loss peak is caused by the chair—chair transition. In Figure 10 we see that poly(4-t-butylcyclohexyl methacrylate) does not show a cyclohexyl peak. (The dimunitive bump at −110°C might be a very weak remainder of the cyclohexyl peak). We have seen that for t-butyl the axial position is very unlikely. It follows that, when one of the chair conformers is not present and therefore the chair—chair transition becomes impossible, the cyclohexyl maximum disappears. This also means that the cyclohexyl maximum is not due to a chair—boat transition, since such a transition is not made impossible by a t-butyl substituent.

It is known that the barrier to the ring inversion of cyclohexane is strongly reduced by introduction of a ketogroup.[10−13] From Figure 11 it is seen that for a plasticizer containing a keto substituted cyclohexyl ring, the maximum has shifted from 190 K to about 110 K; the activation energy has decreased from 11.3 to 6.3 kcal/mol.[14] Thus, when the barrier to the ring inversion is decreased, the activation energy of the motion corresponding to the loss peak shifts accordingly.

Figure 11 also shows that enlargement of the moving ring by a dioxyethylene

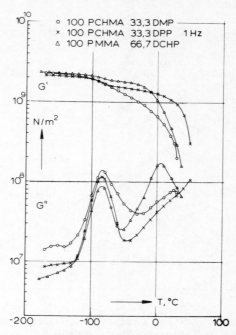

FIGURE 8   Storage (G') and loss (G'') modulus at 1 Hz of PCHMA plasticized with dimethyl phthalate (CMP) and with diphenyl phthalate (DPP) and of PMMA plasticized with dicyclohexyl phthalate (DCHP). The amounts given are parts by weight.

group does not affect the temperature location of the maximum; this is not shifted to a higher temperature by the increased steric hindrance.

Figure 12 gives an interesting example of a composition which contains a cyclohexyl group unhindered in its flipping motion, but nevertheless showing no cyclohexyl maximum. The unsubstituted dioxaspirodecane, dissolved in PMMA, shows no maximum at 190 K, 1 Hz, but it does show a different one near 140 K, 1 Hz. The explanation probably is that this small molecule rotates or oscillates about its longitudinal axis, thereby giving rise to the low-temperature loss peak. Since the rotation frequency is higher than that of the flipping motion of the six-membered ring, the mechanical stress field "sees" no difference between the two cyclohexyl conformers. This explanation is confirmed by the shape of the other damping curve in Figure 12, which shows the reappearance of the cyclohexyl maximum when rotation of the small molecule is blocked by introduction of phenyl groups, which are bulky and rigid.

PMMA containing cyclohexane in solid solution also shows no cyclohexyl maximum. Evidently, for a cyclohexyl loss peak to appear, the small molecules must to some degree be anchored to the polymer matrix.

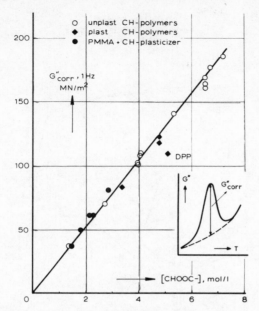

FIGURE 9   Corrected height, G''_{corr}, of the cyclohexyl maximum, as a function of cyclo-hexyl concentration for polymers containing the cyclohexyl-carbonyl group (CHOOC—) in the polymer molecule or in a plasticizer molecule. Insert: determination of G''_{corr}.

A last argument for the molecular mechanism of the cyclohexyl loss peak is derived from a comparison of mechanical and dielectric losses, which are given in Figure 13 for two polymers, one substituted with chlorine in the 4-position and the other in the 2-position. The heights of the mechanical maxima are nearly equal; the maxima lie at a somewhat higher temperature than that of PCHMA. However, on comparing the dielectric losses, we see that the polymer substituted in the 2-position has a cyclohexyl loss maximum about a factor of 3 higher than that substituted in the 4-position. These relative heights are just what we would expect from a chair—chair transition, when we recall from Figure 2 that on ring inversion the bond to the 2-substituent considerably changes its direction in space, whereas the bond to the 4-substituent only shifts parallel to itself. This constitutes another argument against the chair—boat transition.

CONCLUSIONS

We now summarize our evidence that the cyclohexyl loss peak is due to the chair—chair transition:

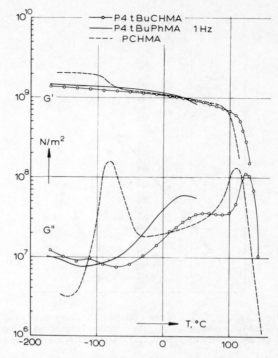

FIGURE 10   Storage (G') and loss (G'') modulus at 1 Hz for poly(4-t-butylcyclohexyl methacrylate) (P4tBuCHMA) and for poly(4-t-butylphenyl methacrylate) (P4tBuPhMA). The curves for PCHMA are given for comparison.

1) The peak is typical of cyclohexyl; it is not observed with phenyl or cyclopentyl.

2) It is not shown by polymers in which a chair—chair transition is known to be impossible.

3) It gives the correct energy barrier.

4) A lowering of the barrier causes a corresponding shift of the peak.

5) The relative heights of dielectric and mechanical loss peaks of 2- and 4-chloro derivatives are in qualitative agreement with the detailed mechanism.

Further evidence can be obtained from the relation between relaxation strength and conformational energy.[6,15]

In my opinion the evidence presented is sufficiently strong to warrant the

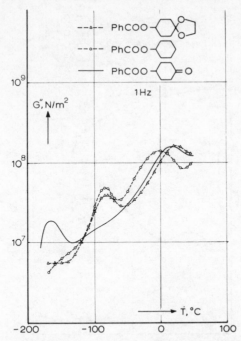

FIGURE 11   Loss modulus G'' at 1 Hz as a function of temperature for PMMA plasticized with cyclohexyl benzoate, 4-oxocyclohexyl benzoate and (4,4'-dioxyethylenecyclohexyl)-benzoate. 33.3 parts by weight of plasticizer per 100 parts of PMMA.

conclusion that it is the chair—chair transition of the cyclohexyl ring which give rise to the mechanical maximum near 190 K at 1 Hz.

References

1. For a more detailed discussion, see, e.g., J. D. Roberts and M. C. Caserio, *Basic Principles of Organic Chemistry* (W. A. Benjamin Inc., New York—Amsterdam, 1965) p. 103.
2. H. Sachse, *Ber.* **23**, 1363 (1890).
3. E. L. Eliel, N. L. Allinger, S. J. Angyal and G. A. Morrison, *Conformational Analysis* (Interscience, New York, 1965) p. 36.
4. F. A. L. Anet and A. J. R. Bourn, *J. Amer. Chem. Soc.* **89**, 760 (1967).
5. G. Binsch, *Top. Stereochem.* **3**, 97 (1968).
6. J. Heijboer, Doctoral Thesis, Leiden, 1972, C. L. Communication 435, Chapter 5.
7. J. Heijboer, *J. Polymer Sci. C* **16**, 3413 (1968).
8. J. Karpovich, *J. Chem. Phys.* **22**, 1767 (1954).
9. H. A. Waterman, personal communication, 1965.
10. R. Bucourt and D. Hainaut, *Bull. Soc. Chim. France* 1967, 4562

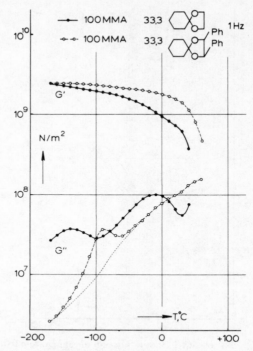

FIGURE 12 G' and G'' of PMMA compositions containing 1,4-dioxaspiro[4.5]decane and its 2,3-diphenyl derivative. Parts by weight.

11. F. R. Jensen and B. H. Beck, *J. Amer. Chem. Soc.* **90**, 1066 (1968).
12. F. A. L. Anet, G. N. Churny and J. Krane, *J. Amer. Chem. Soc.* **95**, 4423 (1973).
13. M. Bernard, L. Canuel and M. St-Jacques, *J. Amer. Chem. Soc.* **96**, 2929 (1974).
14. J. Heijboer, Secondary loss peaks in glassy amorphous polymers, this volume pp. 75−102.
15. J. Heijboer, Doctoral Thesis, Leiden, 1972, C.L. Communication 435, Chapter 7.

DISCUSSION

M. Litt (*Case Western Reserve Univ.*): Were your substituted cyclohexyl rings in the *cis* or the *trans* form? This would of course be important in the interpretation of your data.

J. Heijboer: As far as this paper is concerned, your remark applies only to the examples given in Figures 10 and 13. The configuration of these compounds had not been determined, but this circumstance does not affect the reasoning presented. Data on the difference between *cis*- and *trans*-isomers, which is indeed very important, are given in Reference 1.

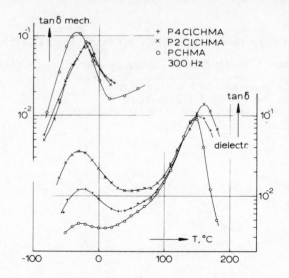

FIGURE 13   Mechanical losses in flexure and dielectric losses, both at 300 Hz, for PMMA, poly(2-chlorocyclohexyl methacrylate) (P2C1HMA) and poly(4-chlorocyclohexyl methacrylate) (P4C1CHMA).

Y. S. Papir (*Richmond College*): Would the chair–chair transition still be operable if you put the cyclohexyl groups in the main (backbone) chain? I am especially interested in the case where the cyclohexyl groups are all para in the backbone chain.

J. Heijboer: Yes, sometimes it is. Baccaredda et al.[2] observed pronounced cyclohexyl loss peaks in poly(1,4-cyclohexylene ether) and in a copolymer of 1,4-cyclohexylene ether. In the crystalline homopolymer the loss peak lies at a temperature about 20 K higher than in PCHMA; in the amorphous copolymer it lies at the same temperature as in PCHMA.

In poly(1,2-cyclohexeneoxide) we found (unpublished observations) a low but distinct cyclohexyl loss peak (maximum tan δ = 0.014 at 188 K).

However, in other polymers containing 1,4-cyclohexylene units in the main chain, no clear evidence has been found that the chair–chair transition contributes to the losses.[3]

I seriously doubt that the loss peaks of methyl-substituted poly(1,4-cyclohexylene ethers) described by Chapoy and Kops[4] are due to chair–chair transitions of the six-membered rings.

References

1. J. Heijboer, Doctoral Thesis, Leiden, 1972, C.L. Communication **435**, Chapters 5 and 7.
2. M. Baccaredda, P. L. Magagnini and P. Giusti, *J. Polymer Sci. A−2* **9**, 1341 (1971).
3. (a) A. Hiltner and E. Baer, *J. Macromol. Sci. Phys. B* **6**, 545 (1972).
   (b) W. Reddish, *Pure Appl. Chem.* **5**, 723 (1962).
   (c) J. Heijboer, *Brit. Polymer J.* **1**, 3 (1969), Fig. 20.
4. L. L. Chapoy and J. Kops, *J. Polym. Sci. B* **11**, 515 (1973).

Free Volume and its Relationship to Zero Shear Melt Viscosity: A New Correlation

M. H. LITT

Department of Macromolecular Science, Case Western Reserve University, Cleveland, Ohio 44106, U.S.A.

Over the past twenty-five to thirty years, a large number of semi-empirical equations have been developed relating zero shear melt viscosity with temperature.[1-13] In the main, they can be grouped into two types. One class emphasizes the activation energy as the main barrier to diffusive motion of a translational unit from one energy minimum to the next.[1,2] The others consider the free volume as the most important factor in the unit diffusion. This can be a generalized free volume,[3-5] or localized into holes of various sizes. In such a case, the main barrier is then entropic, the probability that a hole is found adjacent to a properly oriented unit, which can then move.[6,9] While both classes of theories have had their successes, the free volume approach, popularized by Doolittle, and Williams, Landel and Ferry (WLF) has had the most application.[3-5] Simple equations which reproduced the temperature/viscosity curves reasonably well were developed and have been used extensively for the past twenty years[3-5,14,15] These all postulate that all the free volume participates in molecular relaxations.

Most of the present theories using free volume are derived from Doolittle through WLF, or else try to rationalize equations similar to WLF using more sophisticated approaches.[6,11,12] Except for Doolittle,[3,4,13] and Macedo and Litowitz[7] who measured their "free volumes", the other investigators use some variation of a linear extrapolation, $f = a + bT$, where f is the free volume and a and b are constants. Quite often it is written as $f = f_g + (\alpha_l - \alpha_g)(T - T_g)$ where f_g is the free volume frozen in at the glass temperature, α_l and α_g are respectively the cubical coefficients of expansion of the liquid and glass and T_g is the glass temperature.

The validity of the free volume expressions given above, which assume linear dependence with temperature, can be seriously questioned. There is much evidence that points to a $T^{3/2}$ dependency of the excess volume for most liquids of interest. The deviation from first order dependence was mentioned by Gee,[16] and an expression was derived by Bueche which gave a $T^{3/2}$ dependence.[8] Simha and Weil[17] showed that their theoretical volume/temperature master curves had almost a $T^{3/2}$ dependency. Close examination of the temperature dependence of the excess volume of polymeric and molecular liquids from the literature shows a $T^{3/2}$ dependency. This will be demonstrated later.

The experimental validation of a non-linear dependence of free volume on temperature means that the rationale for most of the viscosity expressions based on the Doolittle approach has vanished. It is the purpose of this paper, using an approach developed by Bueche,[8] to develop a modified viscosity equation which fits the data both for experimental free volume and viscosity with a minimum number of adjustable parameters.

COEFFICIENT OF EXPANSION OF A LIQUID

The following discussion is adapted from Bueche.[8] I have changed his definition of the number of molecules but used his formulations. Since his initial paper he has changed his mind.[18]

Consider the following model: a "liquid" at absolute zero has no free volume. It will have n molecules in a unit volume, say 1 cm,[3] each occupying a portion of the total volume, $1/n$. For any other volume V_0, the number of molecules, N will be nV_0. At higher temperatures, when the liquid has expanded, these N molecules will occupy a larger volume V, where $V = V_0(1 + f)$. (For a polymer, substitute n "oscillating segments".) Each of these molecules (or segments) will occupy a volume $v = V/N$ at a temperature, T. This volume per molecule (or segment) will consist of the Van der Waals volume, V_0/N, plus *free space*, $V_0 f/N \, (= f/n)$, since the molecules, in vibrating anharmonically about their equilibrium positions, can exclude other molecules from taking up positions too close to them. The *free space* can be considered a small hole within the time scale of the oscillation. Under the random thermal motions of the molecules, the free space may aggregate due to the pulling apart of several molecules to generate a larger hole.

An average amount of free space, $V_f \, (= f/n)$, may be assigned to each of the n molecules. As an initial approach, we assume that the holes are approximately spherical with a radius, b. The major contributor to the energy of hole formation (in the absence of external pressure) is the surface tension of the polymer. For a hole of radius b to form, energy $4\pi b^2 \epsilon$ must be expended, where ϵ is the surface tension of the liquid. There are two modifications when the surface tension ϵ is compared to the bulk surface tension. The postulate is that the material around the hole is dense, in Van der Waals contact, and randomly oriented. At the surface, there may be specific orientations. It is also true that at separations of several Angstroms, there is still considerable energetic interaction across the hole. The first modification raises the surface tension compared with the bulk, while the second lowers it.

The argument presented here postulates that the internal pressure, the attracting forces between the molecules, can be considered mainly in terms of the energy necessary to create surfaces when separating the molecules. This is a standard approach in fracture theory. The inclusion of a volume term in the energy of hole formation, as must be done when pressure is applied, is not necessary at one atmosphere since it is so much smaller than the surface energy term.

Such holes would be random in size and the distribution of sizes should follow the Boltzmann distribution law. The chance that a molecule (or segment) will have a hole of radius, b, associated with it is then proportional to exp $(-4 \pi b^2 \epsilon / kT)$. The average volume of a hole associated with a single molecule (or segment) in the liquid is therefore:

$$V_f = \frac{\int_0^\infty (4/3) \pi b^3 \exp(-4\pi b^2 \epsilon / kT) \, db}{\int_0^\infty \exp(-4\pi b^2 \epsilon / kT) \, db}. \tag{1}$$

Integrating, we find

$$V_f = (1/6\pi)(kT/\epsilon)^{1.5}. \tag{2}$$

The total free space associated with a unit volume of molecules (or segments) is therefore:

$$f = nV_f = (n/6\pi)(kT/\epsilon)^{1.5}. \tag{3}$$

Equation (3), developed by Bueche,[8] predicts that the change in liquid volume has a $T^{3/2}$ dependency, which has been found experimentally for liquids (see later). This is therefore a good relationship to use for consideration of zero shear melt viscosity and other parameters.

Before progressing to melt viscosity, we want to modify Eq. (3) to get it in a form that is closer to an equation of state. This can be done by introducing a concept, used by Cohen and Turnbull,[6] and by Bueche.[9] One can postulate that there is a minimum hole size of radius b^*, which is just large enough for the molecule (or segment) to move from one low energy position to another one, occupying the hole. (From this point onward, the present paper diverges from Bueche.)

The energy necessary to form a hole of the critical size is $4\pi b^{*2} \epsilon$. Introducing

b^* into Eq. (3) to generate the critical surface energy as a term, we find:

$$f = [(4/3)\pi b^{*3}n] \, [kT/4\pi b^{*2}\epsilon]^{3/2} \pi^{-1/2}. \tag{4}$$

The term, $4\pi b^{*2}\epsilon/k$ is a characteristic temperature, which we will call T^*, which is proportional to the energy necessary to generate the critical hole size. It is, in fact, the reducing temperature for an equation of state. (RT^*, in fact, is the energy necessary to generate one mole of such holes.)

The first term of the equation, $(4/3)\pi b^{*3}n$ is the product of the volume of a critical hole and the number of molecules (or segments) that are packed into a unit volume at 0 K. It is reasonable to expect that a molecule (or segment) can occupy a hole only if it is about the same size, or larger. Therefore, the critical volume for a jump, times n, will just about equal the unit volume, that hypothetically occupie by the "liquid" at 0 K. Thus, for any volume, V, $n \approx V/(4/3)\,\pi b^{*3}$ per V and $(4/3)\,\pi b^{*3}n \approx 1$. For generality, we write $(4/3)\pi b^{*3}n = A$, where A is close to one and can be considered a *packing factor*. A need not exactly equal 1, as hole and molecular shapes and volumes are not necessarily identical.

We can then write:

$$f = A\pi^{-1/2}(T/T^*)^{3/2} \tag{5}$$

and

$$V = V_0(1+f) = V_0(1 + A\pi^{-1/2}(T/T^*)^{3/2}). \tag{6}$$

V_0 is the hypothetical volume of the liquid at 0 K.

With only volume/temperature information, A and T^* cannot be determined separately. However, the prediction that volume should vary as $T^{3/2}$ is definite and it was tested on several polymer liquids and one molecular liquid where data were available over a wide temperature range. Tables are given below for polystyrene,[19] poly(o-methyl styrene),[20] polydimethyl siloxane,[21] vinyl acetate[22] and tris-α-napthyl benzene.[23] Similar results have been found for polystyrene measured by Simha and Quach.[20] V_0 for polystyrene found by us, 0.857 agrees well with that found by Haward, Breuer and Rehage,[24] 0.85, from pressure data.

While only selected values are given to shorten the tables, the whole range is shown. The standard deviation, S.D., is based on all points given by the authors. As can be seen, it is within the last digit of the measurement for four of the five liquids except for polyvinyl acetate. There it is 7×10^{-5} for volumes measured to five significant figures.

However, our calculations for data on isotactic polymethyl methacrylate[25] give $V = 0.6579 + 5.20 \times 10^{-4}\,T$, and α_l would have a negative dependence on temperature. If the backbone of the polymer is hindered, the basic assumptions of the theory seem to break down. The segment size probably changes with temperature.

TABLE I

Specific volume of polystyrene (*selected values from Ref. 19).

$T\,^{\circ}C$	V_{obs}/cm^3g^{-1}	V_{calc}/cm^3g^{-1}	$\Delta \times 10^{-4}$
96.1	0.978	0.9789	9
100.6	0.981	0.9812	1
115.4	0.989	0.9886	−4
136.7	1.000	0.9996	−4
145.2	1.004	1.0040	0
162.3	1.013	1.0131	1
178.7	1.023	1.0220	−10
202.8	1.036	1.0354	−7
229.0	1.050	1.0503	3
248.9	1.061	1.0618	8

S.D. = 5.4×10^{-4}

$V = 0.8572\,[1 + 2.0015 \times 10^{-5}\,T^{3/2}]$

TABLE II

Specific volume of poly o-methyl styrene[20]

$T\,^{\circ}C$	V_{exp}/cm^3g^{-1}	V_{calc}/cm^3g^{-1}	$\Delta \times 10^4$
139.4	1.0117	1.0118	+1
149.4	1.0171	1.0171	0
159.2	1.0224	1.0223	−1
167.8	1.0272	1.0272	0
179.2	1.0333	1.0333	0
187.6	1.0379	1.0379	0
197.7	1.0435	1.0436	1

S.D. = 0.5×10^{-4}

$V = 0.8669\,[1 + 1.996 \times 10^{-5}\,T^{3/2}]$

TABLE III

Specific volume of polydimethyl siloxane (*selected values from Ref. 21).

$T\,^{\circ}C$	V_{obs}/cm^3g^{-1}	V_{calc}/cm^3g^{-1}	$\Delta \times 10^{-4}$
20	1.0265	1.02674	+2.4
40	1.0453	1.04530	0
60	1.0646	1.06445	−1.5
80	1.0844	1.08419	−2.1
100	1.1046	1.10449	−1.1
120	1.1254	1.12535	−0.5
140	1.1467	1.14674	+0.4
160	1.1686	1.16866	+0.6
180	1.1910	1.19109	+0.9
200	1.2140	1.21402	+0.2

S.D. = 1.2×10^{-4}

$V = 0.84844\,[1 + 4.1860 \times 10^{-5}\,T^{3/2}]$

TABLE IV

Specific volume of polyvinyl acetate (*selected values
from Ref. 22).

$T\,^\circ C$	$V_{obs}/cm^3 g^{-1}$	$V_{calc}/cm^3 g^{-1}$	$\Delta \times 10^{+5}$
40	0.84870	0.84878	+8.8
50	0.85485	0.85482	−3.4
60	0.86104	0.86094	−10.3
70	0.86723	0.86715	−7.9
80	0.87343	0.87346	+2.6
90	0.87986	0.87985	−8.5
100	0.88622	0.88634	+11.5

S.D. = 7×10^{-5}
$V = 0.723925\ [1 + 3.1118 \times 10^{-5}\ T^{3/2}]$

TABLE V

Specific volume of tris-α-naphthyl benzene (*selected
values from Ref. 23).

$T\,^\circ C$	$V_{obs}/cm^3 g^{-1}$	$V_{calc}/cm^3 g^{-1}$	$\Delta \times 10^{+4}$
69.2	0.8668	0.8670	+2
89.2	0.8761	0.8761	−
109.2	0.8854	0.88547	+0.7
124.3	0.8929	0.8927	−2
144.3	0.9026	0.9025	−1
160.0	0.9104	0.91035	−0.5
184.8	0.9230	0.92305	+0.5
199.8	0.9308	0.93085	+0.3
238.5	0.9516	0.95165	+0.3
277.0	0.9732	0.97325	+0.5
310.6	0.9929	0.99265	−2.4

S.D. = 1.1×10^{-4}
$V = 0.7648\ [1 + 2.116 \times 10^{-5}\ T^{3/2}]$

It is seen that a case can be made for saying that the volume dependence of
liquids of polymers at temperatures up to twice T_g is proportional to $T^{3/2}$, if
the backbone is not too hindered. Of course, this should be validated on many
more polymers and molecular liquids before it can be considered a general
feature of liquids. However, the data presented show such a close fit to the
calculated dependence, that Eq. (6) should prove to be very useful. Simha and
Wilson[26] have compared their theories with this one and found that this
approach gives a better fit to the volumes in most cases.

The data for the five liquids are summarized in Table VI below. If we assume,
for the time being, that A, the packing factor, equals 1, we can calculate T^*
and compare it with T_g for the particular material. T_g/T^* has the average value
of 0.38 ± 0.05. Again, if this is confirmed by more data, it means that we can

TABLE VI

Summary of relationships between volume, its temperature dependence and T_g for several liquids.

Material	V_0	Absolute volume expansion coefficient[a]	T^* K	T_g K	T_g/T^*
Polydimethyl siloxane	0.84844	4.1860×10^{-5}	566	150	0.27
Polyvinyl acetate	0.72393	3.1118×10^{-5}	690	301	0.436
Polystyrene	0.8572	2.0015×10^{-5}	926	368	0.397
Poly o-methyl styrene	0.8669	1.996×10^{-5}	939	404	0.430
Tris-α-naphthyl benzene	0.7648	2.1116×10^{-5}	894	335	0.375

[a]Calculated from eq. 6, $V = V_0 [(1 + A/\pi^{-1/2} (T/T^*)^{3/2}]$

predict from the T_g the coefficient of expansion of most liquids to about 15% accuracy.

When Doolittle's data on paraffins are tested[3] for volume/temperature dependence, the higher boiling materials have a region at the lower end of the table where $\Delta V \propto T^{3/2}$. As the critical temperature, or a high temperature is approached the volume expands more rapidly than the $T^{3/2}$ power. The reason for this, based on the present model, is that after a given amount of free space is present , there are no longer true holes but an interpenetrating network of voids, and the microscopic surface tension, ϵ, must drop above that temperature. The largest hydrocarbon of the table, $C_{64}H_{130}$, follows the $T^{3/2}$ dependence from 100 to 250°C. Above 250°C, when the free space, $f \approx 0.4$, the expansion increases. V_0 calculated according to this theory is 1.028, compared to Doolittle's value of 1.0112.

APPLICATION TO ZERO SHEAR MELT VISCOSITY

When Doolittle tested his equation, he used normal paraffins more than 200°C above their T_g. At these temperatures practically all of the free space should be in volume packets large enough to allow easy molecular or segmental jumps. However, near the T_g only a small fraction of the free space could be aggregated into such large holes. Since we are considering only equilibrium properties, segmental motion does not disturb the hole size distribution. In such a case, the following postulate should apply:

While the Doolittle approach still holds true, only that fraction of holes with radii $\geq b^*$ can contribute to segmental motion and therefore to the local relaxation processes. Thus the zero shear melt viscosity should depend, not on total free space, but only on that fraction found in holes of radii $\geq b^*$.

Such a fractional free space can be called the effective free space, f_{ef}. Doolittle's equation can be redefined as:

$$\ln (\eta) = B/f_{ef} + C \approx 1/f_{ef} + C \tag{7}$$

f_{ef} can be calculated easily, as we have already defined the average free space in Eqs. (1) and (3). It is given in Eq. (8). B is expected to be equal to 1.

$$f_{ef} = f \; \frac{\displaystyle\int_{b^*}^{\infty} (4/3)\,\pi b^3 \exp\,(-4\pi b^2 \epsilon/kT)\,\mathrm{d}b}{\displaystyle\int_{0}^{\infty} (4/3)\,\pi b^3 \exp\,(-4\pi b^2 \epsilon/kT)\,\mathrm{d}b}. \tag{8}$$

Equation (8) can be integrated in closed form, Eq. (9).

$$f_{ef} = (n/6\pi)(kT/\epsilon)^{1.5}(1 + 4\pi b^{*2}\epsilon/kT)\exp\,(-4\pi b^{*2}\epsilon/kT). \tag{9}$$

Rewriting Eq. (9) in terms of A and T^*, we obtain Eq. (10)

$$f_{ef} = A\pi^{-1/2}(T/T^*)^{3/2}(1 + T^*/T)\exp\,(-T^*/T) \tag{10}$$

The transformed Doolittle equation is therefore

$$\ln(\eta) = 1/f_{ef} + C = (\pi^{1/2}/A)(T^*/T)^{3/2}(1 + T^*/T)^{-1}\exp\,(T^*/T) + C. \tag{11}$$

Equation (11) should be a master curve for all liquids which follow the volume/temperature relationships from the T_g to the temperature where the basic assumptions break down. It has one material parameter in the form of a reducing temperature, T^*, and a packing parameter A. These together should govern both melt viscosity and the coefficient of expansion. In addition, if the melt viscosity is measured in poises, C should be close to zero if the modification of the Doolittle equation proposed here is valid. When the shift factor is used for the calculations, C will depend on the normalization temperature.

TESTING OF THE MASTER CURVE

Since Eqs. (5) and (11) are in terms of normalized temperatures, the comparison with experimental data cannot be carried out as easily as with the WLF approach. If a coefficient of expansion at one temperature is known, a relationship between A and T^* can be calculated from Eq. (6).

$$\alpha_1 T = \frac{1}{V}\frac{\mathrm{d}V}{\mathrm{d}t} = \frac{3/2(A/\pi^{1/2})(T/T^*)^{3/2}}{1 + (A/\pi^{1/2})(T/T^*)^{3/2}}. \tag{12}$$

Here T is the center of the temperature range where α_l was measured. From Eq. (12), one can write.

$$A = \frac{\alpha_l T}{1.5 - \alpha_l T}\frac{\pi^{1/2}}{T^{3/2}} \cdot T^{*3/2}. \tag{13}$$

A can now be replaced in Eq. (11) with a function of T^*. For any given T^*, log η can be calculated as a function of T and compared with the experimental values. When a plot of $(\log \eta)_{calc}$ versus $(\log \eta)_{exp}$ has a slope of 1.0, a value of T^* and therefore A has been determined which fits both the coefficient of expansion and viscosity (or shift factor) data. If the full volume/temperature data are available, the coefficient of the $T^{3/2}$ term from Eq. (5) or (6) can be used to define the relationship between A and T^* which is substituted into Eq. (11). For best results, the specific volume and viscosity data should be determined on the same batch of polymer.

If the volume/temperature curve for a specific material does not have a $T^{3/2}$ dependence, the equations may no longer be valid. However, in at least one case where this is true they work well, as will be shown.

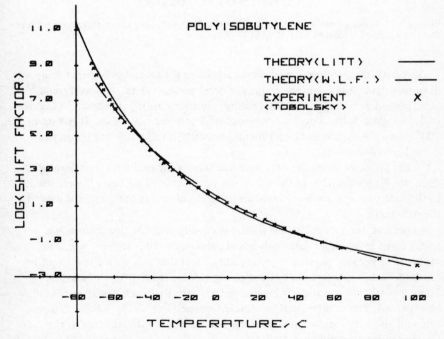

FIGURE 1   Comparison of literature data[27] for polyisobutene versus present theory (–) and WLF values (- -), Ref. 15.

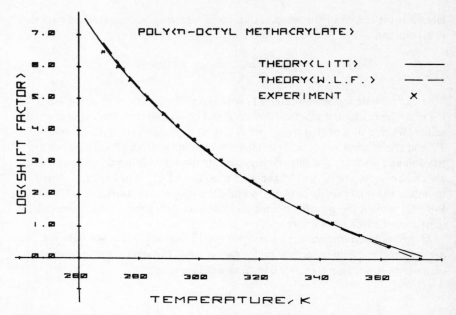

FIGURE 2   Comparison of literature data[28] for poly (*n*-octyl methacrylate) versus present theory (-) and WLF values (- -), Ref. (15).

Viscosity/temperature curves were calculated for several polymers using literature data. Polyisobutene,[27] poly(*n*-octyl methacrylate),[28] polystyrene,[29, 30] poly(methyl acrylate)[31] and poly(methyl methacrylate)[32] are shown in Figures 1 to 4. In Figure 5, the data for tris-α-naphthyl benzene[23] is shown. It fits up to 200° above the T_g, but the experimental viscosity levels while the theoretical curve keeps decreasing.

In the polymer cases, the data here has been compared with that developed from the WLF equation. In the main, the values quoted by Ferry[15] were used. In the case of polystyrene, new values were calculated as the quoted values fit the data badly.

In general, the two equations fit about equally well. As this approach is trying to fit much more data with less constants, the fit is slightly worse in some cases, notably poly(methyl acrylate). I feel that this may come about because the data for α_l and shift factor or log η come from different sources. There is sufficient variation from source to source that such deviations could be expected. The problem can be attacked properly by doing volume/temperature and viscosity measurements on the same sample. With those data, the theory can be properly evaluated.

The data for two inorganic glasses[33,34] are shown in Figures 6 and 7. Here, with a three dimensional melt, the fit is perfect over an 800°C temperature range.

Values for A and T^* calculated from the combined viscosity and expansion coefficients are given in Table VII. The average value of T_g/T^* from the combined $\log \eta$ and α calculations is also about 0.38, as was true for coefficients of expansion alone. We then find an average value for $\alpha_1 T_g$ of about 0.17. One can calculate A based on this value, and this is also given in Table VII. $\alpha_1 T_g$ as a constant for most liquids was proposed in a preliminary version of this paper,[35] and was considered by Simha and Boyer,[36] who decided at that time that $\Delta \alpha T_g$ was the correct approach to use.

However, $\alpha_1 T_g$ is more constant than $\Delta \alpha T_g$. All the values for the poly(alkyl methacrylates) quoted by Simha and Boyer, which vary widely using $\Delta \alpha T_g$, fit very well $\alpha_1 T_g = 0.17 \pm .013$.

DISCUSSION AND CONCLUSIONS

The theory presented here has two points in its favor. First, it reconciles volume with viscosity over a wide temperature range for many liquids. It does this with

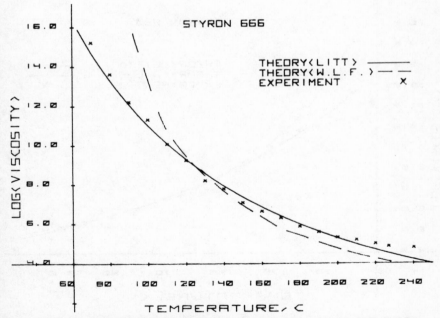

FIGURE 3a   Comparison of literate data[29] for polystyrene (Styron 666) versus present theory (-) and WLF values (- -), Ref. 15.

TABLE VII
Master curve parameters for selected liquids

Material	T_g K	$10^4 \alpha_1$	$\alpha_1 T_g$	A calc.	from $\alpha_1 T_g/0.17$	T_g/T^*	T^* K	Ref.
PIB	200	(6.9)	(0.138)	(0.63)	0.80	0.418	479	27
P(n-oct) meth.	253	6.67	0.169	0.95	0.98	0.383	660	28
Styron 666	362	4.94	0.179	1.03	1.04	0.380	954	29
Styron 690	362	4.94	0.170	1.07	1.04	0.369	982	30
PMA	282	5.6	0.158	0.89	0.92	0.371	761	31
PMMA	375	5.32	0.200	(2.05)	1.16	(0.260)	1443	32
B_2O_3	513			(1.0)		0.384	1335	33
Na/Ca/SiO$_4$ Glass, #10	810			(1.0)		0.369	2194	34
Tris α-naphthyl benzene	342	5.18	0.177	1.24	1.03	0.332	1030	23

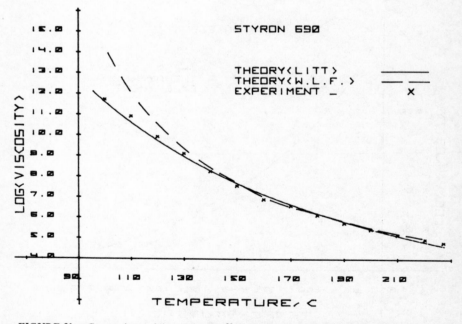

FIGURE 3b   Comparison of literature data[30] for polystyrene (Styron 690) versus present theory (-) and recalculated WLF values (- -), $C_1 = 13.96$, $C_2 = 62.8$, $T_0 = 102$ and $\log \eta_0 = 3.27$.

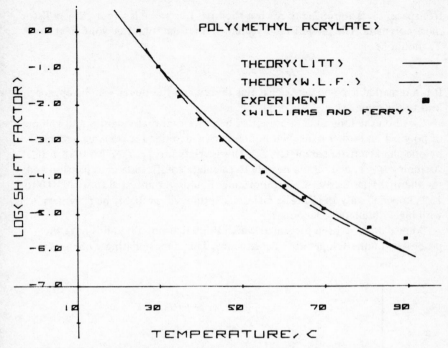

FIGURE 4a   Comparison of literature data[31] for poly(methyl acrylate) versus present theory (-) and WLF values (- -), Ref. 15.

a minimum of adjustable parameters. There are two characteristic terms, A and T^*, needed for both and the specific terms V_0 and log η_0 as the normalizing parameters in volume and viscosity.

In addition it removes the paradox implicit in the WLF approach. Here, the implication is that the viscosity will reach infinity at some value about 50°C below T_g, if one could ever achieve equilibrium. This discontinuity appeared because of the definition of free volume used in the WLF approach, $f \approx f_g + \Delta\alpha(T - T_g)$. The theory presented here avoids the paradox, as the effective free space reaches zero only at 0 K.

This approach also has the advantage of reconciling two opposing definitions of free volume. The Doolittle/WLF approach finds a free volume of about 2.5% at the T_g. Most other approaches such as the Simha–Boyer[36] based on expansion coefficients, or Litt–Tobolsky[37] based on crystalline/amorphous volumes, and others find free volumes of 5 to 15%. In particular, Simha and Boyer found $\Delta\alpha T_g$ = 0.113. The work outlined in this paper distinguishes between total free volume (free space here) and effective free space. Thus, at the T_g, the effective free space, f_{ef}, is about 2.5 to 3% but the total free space is about 12% at T_g for most polymers. The present theory predicts that the total free volume at the T_g should be:

$$f_g \approx 2/3 \ \alpha_1 T_g = 0.113.$$

It is a remarkable coincidence that this theory and the Boyer–Simha approach lead to the same value.

Based on the limited data presented here, we find that viscosity and volume of polymer and other molecular liquids can be described reasonably accurately by one characteristic parameter, T^*; This is related to T_g, $T_g/T^* = 0.38 \pm .02$. Assuming $A = 1$, one should be able to calculate coefficients of expansion and the thermal shift factor of most molecular liquids yet untested to within 10 to 15%, knowing only the T_g. The major exception will probably be polymers with highly hindered backbones.

Some data have been presented which show that not all liquids have the proposed volume/temperature dependence. This is most glaringly obvious in

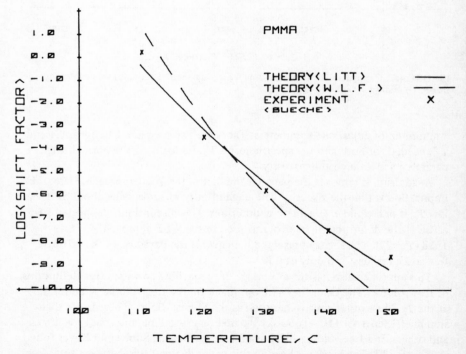

FIGURE 4b   Comparison of literature values for poly(methyl methacrylate)[32] with present theory (-) and WLF values (- -), Ref. 15.

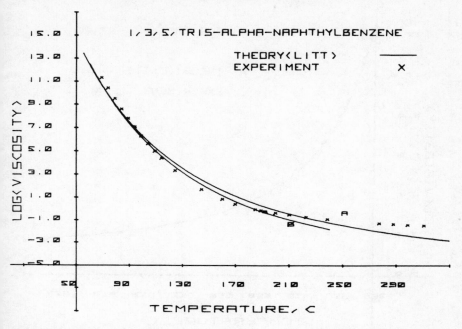

FIGURE 5   Comparison of literature values for 1, 3, 5-tris α-naphthyl benzene[23] vs. present theory. A—best fit up to 300°C; B—best fit up to 240°C.

the inorganic glasses, where the coefficients of expansion almost reach zero at the higher temperatures.[7, 33, 34]. The viscosity fit here is certain empirical, but the agreement is excellent.

Much work must be done to test the theory properly. Volumes and viscosities should be measured on the same samples to avoid the problems found when making a wide range of measurements. Trapped solvent, monomer, additives, etc. can change both viscosity and expansion behavior. Even if samples are rigorously purified, enough variation can remain to make both measurements on the same polymer highly desirable.

Acknowledgements

I thank Dr. R. Simha for stimulating conversations and encouragement in this work. I would also like to thank Paul Tong for writing the programs and performing the calculations in this work.

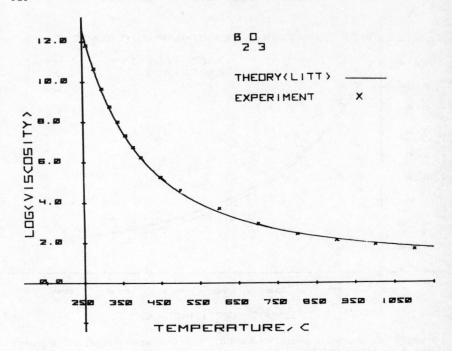

FIGURE 6   Comparison of literature values for boric oxide[33] with master curve for present theory.

References

1. T. Ree and H. Eyring, *J. Appl. Phys.* **26**, 793 (1955).
2. (a) N. Hirai and H. Eyring, *J. Appl. Phys.* **29**, 810 (1958);
   (b) N. Hirai and H. Eyring, *J. Polymer Sci.* **37**, 51 (1959).
3. A. K. Doolittle, *J. Appl. Phys.* **22**, 1471 (1951).
4. A. K. Doolittle, *J. Appl. Phys.* **23**, 236 (1952).
5. M. Williams, R. F. Landel and J. D. Ferry, *J. Amer. Chem. Soc.* **77**, 3701 (1955).
6. M. H. Cohen and D. Turnbull, *J. Chem. Phys.* **31**, 1164 (1959).
7. P. B. Macedo and T. A. Litovitz, *J. Chem. Phys.* **42**, 245 (1955).
8. F. Bueche, *J. Chem. Phys.* **21**, 1850 (1953).
9. F. Bueche, *Physical Properties of Polymers* (Interscience Publishers, New York, 1962), pp. 85–110.
10. J. H. MaGill and Hin-Mo Li, *Polymer Letters* **11**, 667 (1973).
11. S. M. Breitling and J. H. MaGill, *J. Appl. Phys.* **45**, 4167 (1975).
12. I. C. Sanchez, *J. Appl. Phys.* **45**, 4204 (1974).
13. A. K. Doolittle and D. B. Doolittle, *J. Appl. Phys.* **26**, 901 (1957).
14. A. V. Tobolsky, *Properties and Structure of Polymers* (J. Wiley, New York, 1960).
15. J. D. Ferry, *Viscoelastic Properties of Polymers*, 2nd Ed. (J. Wiley, New York, (1970).
16. G. Gee, *Polymer [London]* **7**, 177 (1966).

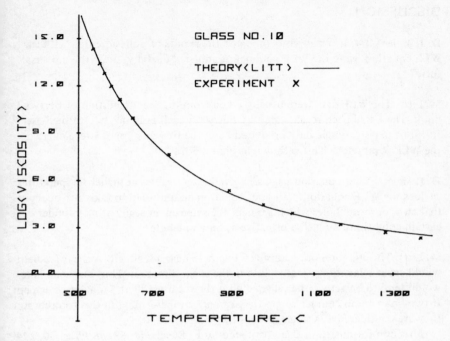

FIGURE 7   Comparison of literature values for Glass 10[34] with master curve for present theory.

17. R. Simha and C. E. Wiel, *J. Macromol. Sci. Phys. B* **4**, 215 (1970).
18. F. Bueche, *J. Chem. Phys.* **36**, 2940 (1962).
19. K. H. Hellwege, W. Knappe and P. Lehmann, *Kolloid-Z. Z. Polym.* **183**, 110 (1962).
20. A. Quach and R. Simha, *J. Appl. Phys.* **42**, 4592 (1971).
21. H. Shih and P. Flory, *Macromolecules* **5**, 758 (1972).
22. J. McKinney and R. Simha, *Macromolecules* **7**, 894 (1974).
23. D. J. Plazec and S. H. MaGill, *J. Chem. Phys.* **45**, 3038 (1966).
24. R. N. Haward. H. Breuer and G. Rehage, *Polymer Letters* **4**, 375 (1966).
25. P. S. Wilson and R. Simha, *Macromolecules* **6**, 902 (1973).
26. R. Simha and P. S. Wilson, *Macromolecules* **6**, 908 (1973).
27. A. V. Tobolsky and E. Catsiff, *J. Polymer Sci.* **19**, 111 (1956).
28. W. Dannhauser, W. C. Childe and J. D. Ferry, *J. Colloid Sci.* **13**, 103 (1958).
29. H. J. Karam and J. C. Bellenger. *Trans. Soc. Rheol.* **8**, 61 (1964).
30. H. J. Karam, K. S. Hyun and J. C. Bellenger, *Trans. Soc. Rheol.* **13**, 209 (1969).
31. M. L. Williams and J. D. Ferry, *J. Colloid Sci.* **10**, 474 (1955).
32. F. Bueche, *J. Appl. Phys.* **26**, 738 (1955).
33. G. S. Parks and M. E. Spaght, *Physics* **6**, 69 (1935).
34. H. A. Robinson and C. A. Peterson, *J. Amer. Ceram. Soc.* **27**, 129 (1944).
35. M. Litt, *ACS Polymer Preprints* **14**, 109 (1973).
36. R. Simha and R. F. Boyer, *J. Chem. Phys.* **37**, 1003 (1962).
37. M. H. Litt and A. V. Tobolsky, *J. Macromol. Sci. Phys. B* **1**, 433 (1967).

DISCUSSION

D. J. Meier (*MMI*): When you compared the results of your equation with the WLF equation, were the WLF parameters adjusted or did you use the universal form?

M. Litt: The WLF data were usually taken from the second edition of Ferry's book. These had been recalculated by Ferry for each polymer to give the best fit. For the polystyrene data presented here, the fit was so poor I recalculated the WLF parameters. This is shown in Figure 3b.

D. J. Meier: Your equation begins to give inexact values at higher temperatures as does the WLF equation. Can your equation be modified to take into account that two, or more, holes which are near the same chain segment may, under certain conditions, be just as effective as only one hole?

M. Litt: The equation only considers one hole near a segment. A shear gradient would cause relaxation to take place in the shear direction. Thus only one hole would tend to be active, that aligned with the shear gradient for a given segment. If two holes are in contact where they can act cooperatively, in this formulation they are considered as a single.

A second formulation, also postulated by F. Bueche (*J. Chem. Phys.* **36**, 2940 (1962)), is that the number of holes greater than the critical size rather than the volume is most important. After all, if a hole is 10% greater in volume than the segment, only one segment can relax into it at any moment. Since the energy barrier is related to hole formation, rather than to segmental motion, the probability of segmental motion should not change drastically with hole size.

I have worked out a preliminary formulation using this approach. (unfortunately, it has no analytical solution). I get agreement with the data and the approach presented in this paper up to $100-150°$ above T_g with, however, one more adjustable parameter. This is the average number of segments which are positioned to relax into a given hole; it seems to about 1.4. The two approaches are equal initially but the number of holes of the appropriate size increases more slowly at higher temperatures than the free space. Thus, a better fit to the viscosity at higher temperatures should be found. However, I have not yet carried out detailed calculations over the whole temperature range.

Nomenclature for Multiple Transitions and Relaxations in Polymers

RAYMOND F. BOYER[†]

The Dow Chemical Company, Midland, MI 48640, U.S.A.

The diversity of nomenclature used to designate the numerous loss peaks in the anelastic and dielectric spectra of polymers is well recognized[1] and has certainly been the cause of considerable confusion.

Anticipating that various speakers at the symposium would be using slides based on different nomenclature systems, a large wall chart was prepared for display at the front of the lecture hall during the entire Symposium. Also, 8–1/2 x 11" copies were available for all attendees. A slightly modified version is seen in Figure 1.

Several concepts were employed.

a) Different systems of equivalent nomenclature were portrayed, trying to select the ones most frequently seen.

b) A reduced temperature scale was employed, using $T_g(K)$ as the reference temperature.

c) The major loss peaks common to many amorphous and semi-crystalline polymers were shown at appropriate locations along the T/T_g scale, according to selection principles which we had recently summarized,[2] i.e.,

$$T_\beta \equiv T < T_g \cong 0.75\ T_g\ (100\ \text{Hz})$$

$$T_{1,1} \cong 1.20\ T_g$$

$$T\alpha_c \cong 0.85\ T_M\ (100\ \text{Hz})$$

[†]Currently representing Boyer and Boyer, 415 West Main Street, Midland; also a Research Affiliate at Midland Macromolecular Inst.

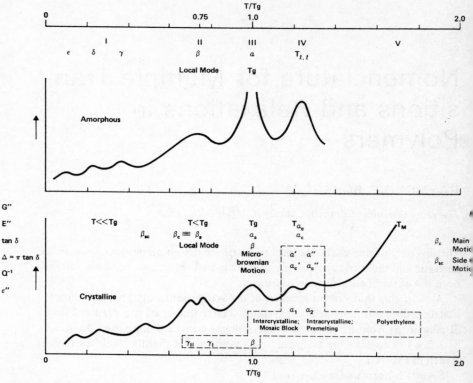

FIGURE 1   Generalized anelastic or dielectric spectra of amorphous and semi-crystal polymers plotted against a reduced temperature scale. The various mechanical loss quantities, E'', G'', tan δ, Δ = log *decrement* and Q^{-1} = tan δ are those most commonly used. While E'' and G'' are related numerically to tan δ, they have a different physical significance. Dielectric loss ϵ'', is not always readily related to mechanical loss. The subscript, a, signifies that the process occurs in the amorphous phase. The subscript, c, is sometimes used to denote a crystalline phase relaxation, or, as indicated here, a main chain motion. The designations inside the dashed line box have been use mostly for polyethylene. The loss curves as drawn are for mechanical tan δ.

d) A portion of the spectrum could always be pointed to in the case of real confusion about meaning.

In selecting the several nomenclatures shown, heavy reliance was placed on McCrum *et al.*[1] and the numerous recent papers of Takayanagi and of Illers. Also shown are two devices we have used recently:[2]

a) A set of Roman numerals to indicate the four key regions in an amorphous polymer and the five key regions in a semi-crystalline polymer.

b) The qualitative locations with respect to T_g of the rather common amorphous phase loss peak around 0.75 T_g which we designate as $T < T_g$; and the very low temperature $T \ll T_g$ loss peaks arising from side chain motion or crystal defects.

It is not assumed that Figure 1 is the ultimate solution to an old problem or that it would gain widespread acceptance. Eventually when the molecular basis of all loss peaks is known, some attempt can be made at a rational, universal system of nomenclature.

Since the large wall chart and handout were part of the background for the symposium, inclusion of Figure 1 seemed justified.

Finally, a word about the $T_{l,l}$ loss peak which is frequently considered as controversial. Joint efforts on polystyrene with Gillham and his students, both published[3,4] and unpublished[5] have shown the existence of $T_{l,l}$ as a dynamic peak and have elucidated its salient characteristics such as mol. wt. dependence repeatability,[3] and polyblending[3,4] as well as response to plasticizer.[5]

Also, a current literature survey indicates that $T_{l,l}$ like phenomena occur in other polymers than PS. See last section of Ref. 3, especially Table IX.

A $T_{l,l}$ by torsional braid analysis on one fraction of PMMA and one fraction of polyvinylacetate has been found, in each case at about 1.2 T_g.[6] Finally, $T_{l,l}$ has been found dielectrically in fractions of Poly(ethylmethacrylate) and n-butylmethacrylate.[7]

Hence, while the molecular origins of such $T > T_g$ relaxations are not understood, their generality to many, if not all, polymers seems assured.[8]

Note: Extra copies of Figure 1 are available from the author by writing to Midland Macromolecular Institute, 1910 West St. Andrews Drive, Midland, MI48640.

References

1. N. G. McCrum, B. E. Read, and G. Williams, *Anelastic and Dielectric Effects in Polymeric Solids* (Wiley, New York, 1967).
2. R. F. Boyer, *J. Polymer Sci. [Symposia]* C 50, 189 (1975).
3. S. Stadnicki, J. K. Gillham, and R. F. Boyer, *J. Appl. Polymer Sci.* 20, 1245 (1976). See also this volume p. 359 (Gillham) and p. 354 (Boyer).
4. C. Glandt, S. Toh, J. K. Gillham, and R. F. Boyer, *J. Appl. Polymer Sci.* 20, 1277 (1976).
5. Experimental work by J. Benci and J. K. Gillham, to be published.
6. Experimental work by J. K. Gillham and R. F. Boyer, May 11—June 4, 1975 at Princeton University. See this volume.
7. S. Strella and R. Zand, *J. Polymer Sci.* 25, 97 and 105 (1957). We are indebted to Dr. Dale Meier, MMI, for calling these references to our attention.
8. R. F. Boyer, *J. Polymer Sci. [Symposia]* C 50, 198 (1975).

Molecular Motion in Polystyrene

RAYMOND F. BOYER[†] and S. G. TURLEY

The Dow Chemical Company, Midland, MI 48640, U.S.A.

1 INTRODUCTION

Atactic and isotactic polystyrene, derivatives and copolymers, have been studied extensively over the past 25 years by most of the physical chemical tools which can give information about internal molecular motion, i.e.,

dynamic mechanical spectroscopy,
nuclear magnetic resonance,
dielectric loss,
attentuation of ultrasonic waves,
thermal expansion, and
neutron scattering.

One attractive feature of the polystyrene system is the ease with which polar or non-polar groups can be substituted in the ring or in the side chain, thus providing, on the one hand, controlled steric hindrance to motion of the side chain phenyl group or the backbone chain atoms; and, on the other hand, a handle which is more responsive to external force fields, especially dielectric fields, for polar substituents such as halogens.

Our purpose is to recount the main history of attempts to elucidate molecular motion by the chemical structure variation — inferential approach in contrast to modern sophisticated tools such as C^{13} NMR or neutron scattering which measure such motion more directly. Aside from the historical value, we wish to prepare a base for approaching the $T_{1,1}$ relaxation phenomena to be discussed in the following paper.[1]

We have reviewed the transition and relaxation behaviour of polystyrene on several earlier occasions.[2−4] The present review is both an up-dating and an

---

[†]Now at Midland Macromolecular Institute, 1910 West St. Andrews Drive, Midland, MI 48640, U.S.A.

expansion, in some respects, of these earlier studies. A review of the anelastic and dielectric data on polystyrene by McCrum et al.[5] covers much of the early literature and is not, for the most part, repeated here.

2　HISTORICAL DEVELOPMENT OF THE RELAXATION SPECTRUM OF POLYSTYRENE AND ITS DERIVATIVES

Figure 1 is a very condensed history of the development of the relaxation spectrum of polystyrene. A schematic spectrum is shown starting with the $T_{l,1}$ or liquid-like transition above T_g and then the β, γ and δ glassy state transitions below T_g. The different peak heights vary depending on the thermal history and other details about the polymer history such as orientation and exact composition, including diluents. There is still controversy about the relative strengths of both the γ and the δ relaxation processes. Factors affecting the strength of the β process will be considered later in some detail. The $T_{l,1}$ process is not generally seen by conventional torsion pendulum experiments.

The scheme used in Figure 1 is to underline the name(s) of each author with a horizontal line which covers the temperature range used in that particular investigation starting at the lower and going to the upper temperature limit. The types of polymers and other experimental conditions are given after each name in the right-hand column according to the coding at the bottom of the figure.

The classical paper by Schmieder and Wolf[6] in 1953 contained data on a large number of polymer systems including polystyrene. They clearly show the α, β, and γ loss peaks. Subsequent to this, Illers and Jenckel[7,8] and Illers[8,9] alone extended the work of Schmieder and Wolf by including derivatives of polystyrene, crosslinked polystyrene and the effect of diluent or solvents. In 1962 Sinnott[10] examined the cryogenic region and discovered the δ loss peak which has since been extensively confirmed by many other investigators.

Turley[11-13] investigated a large number of polystyrene derivatives and copolymers and also one example of a crosslinked polystyrene. Some of his results will be shown later. He also observed rubber modified polystyrenes (both mechanical blends and copolymers).[11-15] He confirmed and extended the original observations of Buchdahl and Nielsen[16,17] that blends of polystyrene with an incompatible elastomer gave two glassy loss peaks, one for PS, the other for the elastomer. Turley observed the $T_{l,1}$ in polystyrene.[2] His actual curve appears here as Figure 2. Some of Turley's studies in regard to the β relaxation will be detailed later.

Crissman, Sauer and Woodward[18,19] studied a number of styrene derivatives and at least one copolymer by dynamic mechanical means over a rather extensive temperature range. They in particular were able to point out the effect of ring substitution and side change substitution on the temperatures and

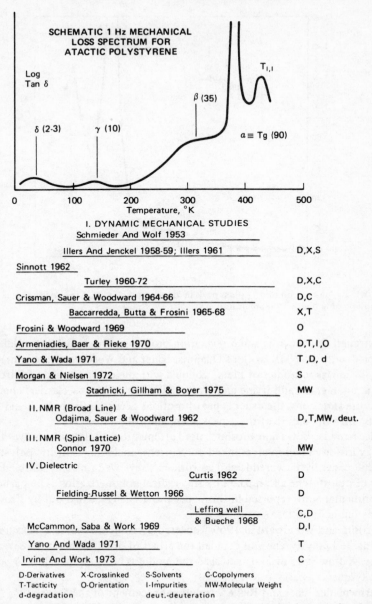

FIGURE 1   Historical survey of some key studies which aided in depicting and under-standing the relaxation spectrum of atactic and isotactic polystyrene and some of its co-polymers. The horizontal bar below the names of the investigators signifies the temperature range covered. The symbols at the right (explained at the bottom) illustrate the key variables studied. The numbers in parenthesis beside the greek letter loss peak designations are the apparent activation enthalpies in kcal/mol for the four main glassy state loss peaks.

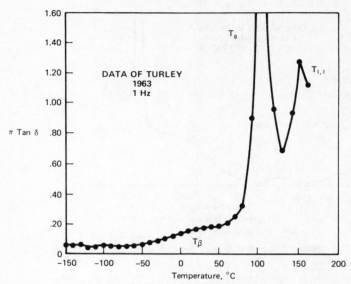

FIGURE 2   Indication of a $T_{l,l}$ loss peak in atactic polystyrene. Creep of the specimen prevented going higher in temperature.

the strengths of the β, γ, and δ relaxation processes. These anelastic studies, coupled with the NMR work of Odajima, Sauer and Woodward (see later); plus the anelastic studies of Illers[9] coupled with dielectric studied of Curtis, McCammon *et al.*, and Irvine and Work, helped to define, by classical chemical structure variations, the exact types of motions occurring at the α, β, and δ loss peaks, with the nature of the γ peak then, and still, unresolved.

Baccaredda, Butta and Frosini[21] used attenuation of ultrasonic waves to study energy loss in radiation crosslinked polystyrenes including isotactic polystyrene. A 1968 paper by Baccaredda and coworkers[21] reviewed ring and side chain derivatives of polystyrene all prepared by free radical polymerization. The particular experimental work reported in this paper was actually carried out by Frosini and Magagnini in 1966.[2]

Frosini and Woodward in 1969[23] covered dynamic mechanical measurements on oriented polystyrene and detailed the effect of orientation on the several loss peaks. A draw ratio of 3–1 increased the modulus but suppressed the δ peak of polystyrene.

Armeniades, Baer and Rieke[24] found enhancement of the δ peak by α, β, β trifluorination, in each case compared with atactic PS. They also showed that mineral oil at a level of 2% introduces a relatively strong new loss peak at 110 K. These authors show no evidence for a γ peak.

Yano and Wada[25] present a comprehensive anelastic and dielectric study on thermal, anionic, isotactic and thermally degraded polystyrenes. They report two new loss peaks below T_g: a γ peak found only by dielectric measurements and ascribed to impurities such as oxygen atoms in the chain; a peak at 25 K at 10 kHz whose origin is not clear. We shall discuss their findings later.

Morgan and Nielsen[26] made a systematic study of 15 organic liquids, usually at 10–15% concentration, on the low temperature loss peaks in atactic PS. Some diluents suppress the δ peak, others intensify it. Some diluents cause a sharp loss peak at or above where the pure diluent might be expected to show a T_g. They suggest that the γ loss peak may arise from the presence of styrene monomer. First, they note that addition of 14.5% styrene causes a loss peak in the 105–120 K region. Second, they cite the illusive nature of the γ peak which is seen by some workers but not by others. However, Illers[27] has shown that the 1 Hz torsion pendulum γ peak does not change significantly in temperature or intensity for as-polymerized thermal PS and the same polymer after precipitation from solution. Moreover, Yano and Wada,[25] not cited by Morgan and Nielsen, show well-developed γ peaks for thermal, anionic, and isotactic polystyrenes. Their thermal PS was precipitated from benzene solution with methanol.

The work of Stadnicki et al., as reported elsewhere in this symposium,[1] concerns the T_g and $T_{1,1}$ 1 Hz torsional braid data on anionic and thermal polystyrene from molecular weight 600 to 2×10^6. Stadnicki[28] has also covered the β and γ regions for these same anionic and fractionated thermal PS's. A large hysteresis loop appears for the cooling to liquid N_2 followed by heating regime, especially at lower molecular weights. Since the β peak is much more intense on the heating cycle; and since the hysteresis loop is repeatable, it was concluded that crazing was occurring on cooling below T_g but healing on subsequent heating. This view was strengthened by the fact that higher molecular weight specimens showed much smaller hysteresis loops.

A molecular weight dependent loss process was found below the normal γ region.[28] Since its intensity increased as molecular weight decreased, it was ascribed to the butyl end group, possibly complicated by the presence of crazes.

A NMR studies

The classical study by Odajima, Sauer, and Woodward[29] on broad line NMR narrowing as a function of temperature covered 9 derivatives, including deuteration. In addition to the line narrowing associated with T_g, a partial narrowing set in at and above 77 K, associated with limited phenyl or substituted phenyl group motion. Such motion was inhibited to varying extent by the nature of the substituents. Three classes of behaviour were found as indicated in Table I. Methyl group rotation was observed but is not pertinent here. These results suggest strongly that a hindered motion of the phenyl ring about the axis connecting it to the chain is being observed between 77 K and T_g.

TABLE I

Effect of substituents on the NMR line narrowing above 77 K for polystyrene

1) Non-hindering of phenyl group motion
   Atactic PS
   Annealed and guenched isotactic PS's
   Poly(β-deuterostyrene)
   Poly(α, β, β-trideuterostyrene)
   Poly(p-chlorostyrene)
   Poly(p-methylstyrene)

2) Partial hindering
   Poly(o-methylstyrene)
   Poly(m-methylstyrene)

3) Complete hindering
   Poly(2,4-dimethylstyrene)
   Poly(2,6-dimethyl-4-tert. butylstyrene)

Connor[30] measured spin lattice relaxation times on a series of anionic PS's with molecular weights from 900 to 1.8×10^6. He found the expected dependence of T_g on molecular weight. He also observed a low temperature process whose intensity increased with decreasing molecular weight, thus suggesting that the signal arose from motion of the butyl end group. Crist[31] showed that this effect arose from rotation of the CH_3 group at the end of the n-butyl chain end. This is probably the origin of the low temperature loss peak found by Stadnicki.[29]

Connor[30] also found a non-exponential decay of relaxation above T_g which suggested, after elimination of other less plausible mechanisms, a manifestation of the $T_{1,1}$ relaxation.

Schaeffer's room temperature study of phenyl group motion as reported elsewhere in this symposium[32] showed a vast improvement in sophistication of NMR techniques suitable for such studies.

B   Dielectric studies

An excellent introduction to this subject appears in McCrum et al.[5] Curtis[33] studied the poly(3 chlorostyrenes) and poly(4 chlorostyrenes) through the β and T_g regions. He concluded, from the presence of a β peak in the 3-chloro but not in the 4-chloro derivatives that the β relaxation involved relaxation of the chlorophenyl ring about the bond connecting the ring to the chain. Curtis further concluded that there was an additional relaxation process (a low frequency one) above T_g for both polymers. This is presumably the $T_{1,1}$ relaxation.

Fielding-Russell and Wetton[34] appear to confirm the conclusions of Curtis for poly(3-chlorostyrene) as regards the β region. They also cover poly(3-methyl) and poly(4-ethoxystyrenes) and include anelastic data.

Leffingwell and Bueche[35] followed the dielectric behavior of three styrene 2-chlorostyrene copolymers containing 91, 25, and 50 vols 2-chlorostyrene per 100 vols of styrene. They obtain a well resolved β peak which is consistent with conclusions reached by Curtis for the 3-chloro derivative.

Yano and Wada[25] find thermal and anionic polystyrenes to be dielectrically active at 110 Hz in the β region, thus seemingly contradicting the conclusions of Curtis regarding poly(4-chlorostyrene). Yano and Wada conclude that the β process in PS involves local mode motion, i.e., motion about the chain backbone. They cite the absence of a dielectric β peak in isotactic PS as further proof of their position since a perfect 3_1 helix should not be expected to exhibit a resultant dipole.

Such marked differences in conclusions about the molecular origin of the β process — local mode vs. phenyl ring rotation — is typical of the present state of the field. The following two papers should clarify the nature of phenyl ring motion, especially for the δ peak.

McCammon, Saba, and Work[36] prepared extremely pure specimens of PS and its three monochlorinated isomers. These were measured electrically from 0.1 to 20 kHz between 4 and 300 K, with special emphasis on the δ region. All four polymers were dielectrically active with slightly different values for T_δ and for apparent enthalpy of activation. Since PS and poly(4-chlorostyrene) lack a resultant dipole for pure phenyl group rotation or oscillation, these authors conclude that a combined phenyl group oscillation with wagging of the backbone chain is involved.

Irvine and Work[37] prepared a series of random copolymers from 4-chloro-styrene and 4-methylstyrene and measured dielectric properties from 1.6 to 300 K, again from 0.1 to 200 kHz. The strength of the δ relaxation was found to vary linearly with composition with a constant ΔH of 2.7 ± 0.7 kcal mol^{-1} and a constant shape for the relaxation curves. They conclude that the phenyl groups relax independently of each other in pure poly(p-chlorostyrene) as well as in the copolymers.

Anticipating the next section, one might conclude that if the δ process at ~50 K involves a combination of hindered group oscillation of the phenyl group coupled with wagging of the backbone; and if, as indicated by NMR, such hindered motion increases continuously above 77 K, then it would appear that the β relaxation, with its much higher activation energy, involves a more complex process than hindered rotation or oscillation of the phenyl groups.

3 MOLECULAR ORIGINS OF THE PS LOSS PEAKS

A General considerations

A PS unit in the polymer chain is depicted in Figure 3 as exhibiting three types of hindered motion: rotation about the chain and side chain bonds indicated by open arrows, and a precessional motion shown by the circle which involves some wagging of the backbone. The phenyl group is constrained to be in a plane containing the α hydrogen and the connecting carbon atom. Tonelli[38] has offered calculations concerning phenyl group rotation possibilities arising from nonbonded interactions. When all backbone bonds are in one of the *trans* or *gauche* states, the phenyl group is restricted to $\pm20-30°$ angular motion about its normal position. Complete phenyl group rotation is possible only for sterically unfavorable backbone conformations which should be present in equilibrium amounts of below 1%. The barrier to partial rotation is calculated as $9-10$ kcal for both PS and PαMS. Poly(o-methylstyrene) does not have any conformation which permits complete backbone rotation. m- and p-methyl substitution does not pose any new problems over those present in PS. Tonelli believes that complete phenyl group rotation can be ruled out for the δ and γ relaxations in PS and its derivatives. Reich and Eisenberg[39] had earlier made less extensive calculations which arrived at a barrier of 8.0 kcal in the most favorable conformation. Since the measured ΔH for the γ process is ~10 kcal, they assigned phenyl group rotation as the origin of the γ peak.

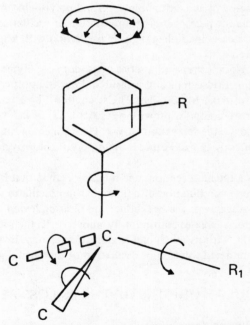

FIGURE 3   Most likely motions of a monomer unit in a PS chain. The open arrows show hindered oscillation about the bonds in question. The closed circle at top with arrows, suggests precessional motion of phenyl group with wagging of backbone. The ϕ ring tends to be in the plane defined by ϕ—C̵—H.

Hindrance to such phenyl group rotation is clearly intrachain in origin. Heijboer[40] is of the opinion that most $T < T_g$ relaxations respond to intrachain barriers. This likelihood is seen clearly in the Illers 1 Hz dynamic loss data on poly(p-halogenated styrenes).[9] The temperature and intensity of the β peak scarcely changes on going from H, F, Cl, Br to I whereas T_g, generally considered to involve interchain barriers (as well as intra-) increases from 106 to 156°C more or less in proportion to the Van der Waals radius of the halogen group. However, the intensity of the γ peak increases in the series H, F, Cl and Br, while tending to increase slightly in temperature.

The remaining general question concerns the number of mer groups moving at each relaxation. Dielectric data indicate[37] that at the δ peak single mer units are involved, with no apparent coupling between adjacent mers. With this as a base, we use the increase in activation energies shown in Figure 1 as a measure of the increasing size and/or increasing complexity of the motion at high lying relaxations.

Heijboer has shown[40] especially for secondary relaxations, that ΔH increases linearly with the temperature of the relaxation. In the case of polystyrene, ΔH increases exponentially for the sequence δ, γ, β and α.[41] The meaning of ΔH is not clear, beyond being the slope of the Arrhenius plot in its linear portion. The reason is precisely that the molecular weight of the moving group is not known. However, $\Delta H(T_g) > \Delta H(T_\beta)$ as a general rule for many polymers. One may properly conclude that the size of the moving moiety in a given polymer increases as the temperature of the relaxation increases, at least when one is dealing with backbone motions in C—C backbone polymers.

We have used[42,43] a precursor argument to explain why

$$T_\beta \ (100 \ \text{Hz}) \sim 0.75 \ T_g \ (100 \ \text{Hz})$$

for C—C backbone polymers. The presumption here is that motion at the β relaxation is similar to but more local in nature than at T_g. We also argue that T_g and $T\alpha_c$ in semi-crystalline polymers both anticipate the melting process at T_M.[42,43] It seems reasonable to suppose that the δ relaxation in PS anticipates both β and T_g processes. We will not speculate in similar fashion about the γ process since facts concerning it are so conflicting.

B The δ process

We consider this the simplest and best understood relaxation in PS consisting, as stated earlier, of hindered partial rotation of a phenyl group along with wagging of the phenyl group and the associated chain backbone.[36,37] This can be decomposed into its two components under suitable conditions by considering dielectric data on the polychlorostyrenes.[36] The combined motion occurs at ~45–50 K (10^3 Hz) in PS and P(pClS)[36] and P(pCH$_3$S)[37] with a ΔH in the range of 3 kcal.

However, it has dropped to 18 K for the $P(3C1S)^{36}$ derivative with a $\Delta H \sim 0.5$ kcal. This presumably represents the simple phenyl group oscillation without backbone wagging. Poly(2-chlorostyrene) is intermediate at 33 K and ΔH = -1.3 kcal presumably because any phenyl oscillation is inhibited and the wagging is the active loss mode.

Mechanical loss on $P\alpha MS^{19}$ shows the δ region as consisting of a double peak: one at ~ 18 K which is presumably phenyl motion and the other at 140 K which arises from a wagging motion highly hindered by the intrachain crowding of the α methyl group.[19]

Several groups, namely Salinger and his colleagues, and Dillinger with his colleagues, have reported a specific heat anomaly below 4.6 K which is attributed to motion of the phenyl side group. They have also measured sound wave velocities. Choy, Hunt, and Salinger[44] propose, as a model, pendant phenyl groups moving independently near voids. They calculate a barrier to rotation of 0.3 to 0.5 kcal mol^{-1} which agrees quite well with the dielectric data for poly(3-chlorostyrene)[36] that we ascribe to simple phenyl group oscillation. Madding, Fehl and Dillinger.[45] postulate a torsional coupling of the pendant phenyl groups along the chain. This is contrary to the conclusion of Irvine and Work.[37] In an earlier paper Zoller, Fehl and Dillinger[46] estimated that one phenyl group in 40 might be involved. This is not too inconsistent with Tonelli's[38] estimate of $<1\%$ of phenyl groups free to rotate.

Since Cp and sound velocity results can be fitted by a number of models, whereas the dielectric data[36,37] are best interpreted by a single simple model which is not completely inconsistent with specific heat and sound velocity data, we accept the dielectric interpretation of combined phenyl group oscillation and wagging for PS.

C   The γ relaxation

Since the very existence of the γ relaxation is in question,[47] it not being seen by some investigators and being ascribed to the presence of styrene monomer by others,[27] one cannot expect to reach many firm conclusions about T_γ at this time.

It was first proposed by Illers and Jenckel[7, 8] that the γ process arose from a moiety, $-(CH_2-CH_2)-$, resulting from termination by coupling of two growing PS chains, or by an occasional head-to-head insertion of monomer. PE had a very strong relaxation, also called a γ relaxation, at the same temperature.

This concept of Illers and Jenckel can be re-examined from several points of view.

1) It is now known that the benzyl radical is the active chain end so a $-(CH_2-CH_2)-$ could not form by termination, and is unlikely by a head-to-head

insertion. It is still possible by thermal initiation according to a new mechanism by Pryor et al.[48] who suggested that 0.01 to 1% of all chains are initiated by the following di-radical formed from styrene monomer.

$$2 \text{ Ar}-\text{CH} = \text{CH}_2 \longrightarrow \text{Ar}-\dot{\text{C}}\text{H}-\text{CH}_2-\text{CH}_2-\dot{\text{C}}\text{H}-\text{Ar}.$$

2) Illers[49] has recently found that the γ relaxation in polyethylene involves three methylene units in branched PE and a greater number in linear. One can write the next step in the Pryor et al. mechanism as:

$$
\begin{array}{cc}
\text{Ar}-\dot{\text{C}}-\text{H} & \text{Ar}-\dot{\text{C}}\text{H} \\
| & | \\
\text{CH}_2 & \text{CH}_2 \\
| & | \\
\text{Ar}-\text{CH}-\text{CH}_2-\text{CH}_2 & -\text{CH}-\text{Ar}
\end{array}
$$

which gives the required number of methylene sequences, albeit in quite low concentrations.

3) Unfortunately for this and related concepts, the γ peak has been found in polystyrenes initiated with peroxides, butyl lithium, and Ziegler–Natta catalysts. It apparently occurred to three groups of workers[22,26] and ourselves, at about the same time to make this kind of check. Our own work was not published but confirmed the published literature already discussed.[22,26] Actually, Illers and Jenckel[8] had earlier noted that such a mechanism couldn't hold in isotactis PS.

The work of Connor[31] who found an end group effect in the γ region suggested to us the possibility that the γ process might arise from an end group normally occurring in PS. The molecular dependence of PS is given by a Fox and Flory type equation

$$T_g = T_g^{\infty} - K_g \bar{M}_n^{-1}$$

with $T_g^{\infty} = 373$ K and K_g in the range of (8 to 17) x 10^4. One can, therefore, calculate the value of $\bar{M}n$ corresponding to a T_g (chain end) of $T_\gamma = 150$ K. The values of $\bar{M}n$ are 359 and 762 respectively for the extreme values of K_g. These values are within a small factor (3–7) of what might be expected as an end-group in thermal initiation.[48] Pezzin et al.[50] estimate the volume of the end group in PS as contributing about 20% more free volume than a mid-chain monomer unit.

Inspection of Figure 8 in the Yano and Wada paper[25] showed that the γ peak height and the area under the γ peak (both after subtracting an assumed background curve) decreased as molecular weight increased from (1.2 to 3.8 to 4.6 x 10^5. The respective initiators are thermal, anionic, Ziegler–Natta. This was one of several factors which led to a collaborative program with Gillham on low molecular weight PS's where an end group effect would be greatly enhanced.[1] As already stated, the results of this study are inconclusive because of crazing and the butyl end group dominance at low molecular weights. The thermal PS's have yet to be measured.

Nozaki *et al.*[51] report a definite end group effect by dielectric loss at 10^3 Hz for telomerized PS's with specific end groups other than a proton. The authors did not state molecular weights but since the telogens possessed active chlorines, the molecular weights are certainly low. In four examples studied, the pairs of chain ends are

1) benzyl and Cl

2) trichloromethyl and H or

3) dichloromethyl and chlorine

4) trichloromethyl and Cl

Number 1 was most active in the γ region, Number 4 least active. The presence of chlorine on the polymer was determined by activation analysis. A copolymer of styrene with 3% p-chlorostyrene was completely inactive in the γ region. Since these telomers were precipitated from benzene solutions with methanol several times, the absence of monomer and telogen can be assumed.

The fact that the γ peak intensity increases[9] in the series para H, F, Cl and Br, while T_γ increases from 133 to 137 to 141 to 156, is indicative of a mass effect on an end group. The p-I derivative is out of line. All polymers were prepared with benzoylperoxide so that both ends should be catalyst fragments, C_6H_5 or C_6H_5COO. The effect of substituents on the monomer would have to involve the penultimate group. This does not rule out a residual monomer effect which might respond in similar fashion to substituent groups.

Whereas Tonelli[38] stated that neither the δ or γ relaxation could likely result from free phenyl group rotation because of the low probability of a conformation permitting it, Reich and Eisenberg[39] turned this argument around: the activation energy of the γ process, i.e., $\sim 7-10$ kcal, is correct for such phenyl group rotation, the very weak intensity of the γ process being explained by the low probability of favorable conformations. Illers[9] showed the γ peak intensity for quenched isotactic PS to be substantially less than for atactic PS, consistent with conformation probability considerations.

However, an activation enthalpy argument, *per se*, cannot provide a convincing basis for assigning a relaxation mechanism. As Heijboer has shown[40]

$$\Delta H/T \cong constant$$

especially for $T < T_g$ relaxations. This relationship states that any process occurring at ~ 150 K will have an activation enthalpy of about 8 kcal/mol[40] regardless of molecular mechanism.

Given that T_δ represents the torsional wagging of a mid-chain monomer unit, a similar motion at a chain end should occur at $T < T_\delta$, perhaps the ϵ relaxation found by Yano and Wada.[25] The combined motion of a foreign group with torsional wagging of the penultimate monomer unit (or units) might constitute a γ process with the low intensity determined by the low number of end groups.

There is no doubt that styrene[27] and various low molecular weight constituents will cause mechanical[24,26] and dielectric[36a,b] loss in polystyrene and its derivatives in the γ region. We doubt that this is the sole source of the γ relaxation for the following reasons:

1) No change in intensity on precipitation.[27]

2) Present in precipitated thermal polystyrene, anionic PS and isotactic PS.[25]

3) Present in precipitated telomers.[51]

4) The NMR experiments showing an end group effect.[30]

5) A very specific end group electric loss with Telomers.[51]

While an end group origin for the γ loss is certainly not proven, the idea should be pursued. Since type and number of end groups can vary, this could be a cause of conflicting literature data.

D   The glass transition

The glass temperature of polystyrene is probably as well understood in terms of molecular motion as that for any polymer, possibly excepting PVC and polypropylene whose internal motions are simpler.

A pictorial summary of the intra- and interchain barriers which influence T_g is given in Figure 4.

With the exception of the oxide(s), hydrocarbon derivatives have been chosen to minimize polarity differences. Several general statements can be made based on a vast body of data but readily illustrated by the examples in Figure 4:

1) Any substituent(s) which hinder oscillation of the phenyl group about the C—ϕ bond, i.e, ortho methyl or hydrogenation raises T_g.

2) Any factor which reduces intrachain barriers to rotation about the C—ϕ bond or C—C chain bonds, i.e., ether oxygen in the chain or the methylene bridge C—CH_2—ϕ, lowers T_g.

3) Substituents which increase barriers to rotation about the C—C chain bonds, i.e., α-CH_3, p-tert. butyl, increases T_g.

4) p-CH_3— has little effect but this may represent competition between reduced polarity and increased size of side group. However, introducing polarity with no change in size as with p—F— has essentially no effect. p-Cl raises T_g about 18°C.

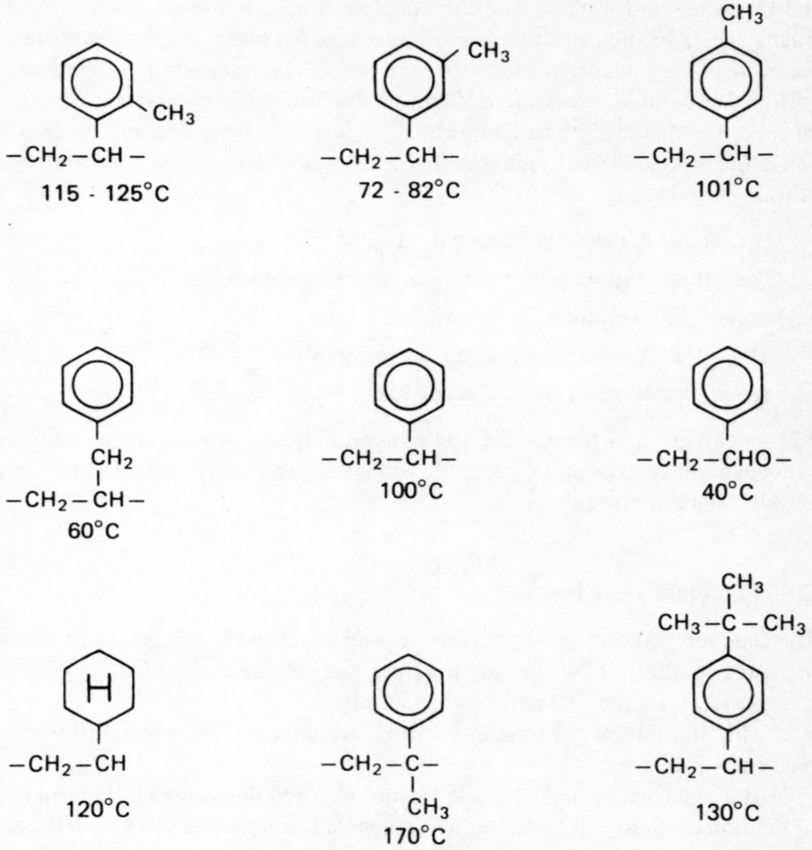

FIGURE 4   Effect of chemical structure variations on the T_g of polystyrene derivatives. Substituents which restrict oscillation about the C—ϕ bond, such as σ—CH_3—, or rotation about the backbone C—C bonds, such as α-CH_3—, raise T_g. Substituents which promote phenyl group oscillation, such as —CH_2— or rotation about the backbone, such as —O—, lower T_g. T_g values are not available for a phenyl ether or phenylthioether side group.

5) The role of m-CH_3— in reducing T_g is not clear beyond a reduced density compared with PS[19] and the obvious one of reduced polarity. The latter can be ruled out since m-Cl has a similar effect on T_g. We have speculated elsewhere[52] that since isotactic poly(m-methylstyrene) forms a 4_1 helix in contrast to a 3_1 helix for isotactic PS, the increased size of the helix is possibly responsible for increased freedom of phenyl group rotation. We assumed a tendency for 4_1 helicity in atactic poly(m-methylstyrene). We have since noted that Privalko[53] has made the generalization that the most stable form of monomeric units in isolated chains is identical with the most stable crystalline conformations.

It therefore seems clear that the T_g process involves some complex combination of phenyl group oscillation about the C—ϕ bond and oscillation or rotation of the ϕ group around the chain axis. NMR data[30] indicate that ϕ group rotation, presumably in all planes – is not complete until above T_g. Tonelli[38] estimated a ΔH of $>$100 kcal mol^{-1} for phenyl group rotation in the normal backbone conformation for PS. ΔH for T_g is 90–100 kcal/mol but this may be a coincidence. Also, as stated earlier, ΔH increases with T_g monotonically and hence a ΔH value is not evidence *per se* for a mechanism.

Moreover, to the extent that our analysis of the m-CH$_3$— or m-Cl— effect is correct, any conclusions reached about either the β or the T_g relaxation mechanisms in a meta-substituted poly(styrene) is not necessarily indicative of the situation in PS.

E   The β relaxation

1) General   Discussion of T_β was deferred until after T_g because many of the structural and experimental variables affect both β and T_g in the same way. While much is known about T_β, its origin seems less certain than in the case of T_g. On a previous occasion[54] we tried to summarize a number of these factors, using in part unpublished data by Turley.[11] Some of this data is presented here for the first time in Figures 2 and 4 through 10.

Figure 5 shows the comparison between a quenched and an annealed thermal polystyrene of low volatile content. This effect has been studied in considerable detail by Goldbach and Rehage.[55] One notes a double loss peak in the quenched specimen, consistent with DSC observations by Wunderlich.[56] Evidence for this double peak also appears in Figure 7.

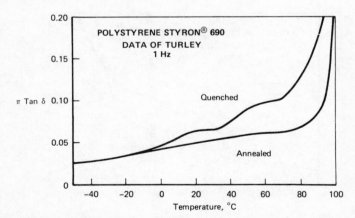

FIGURE 5   Comparison of quenching and annealing on the β region loss of an atactic polystyrene (thermal initiation followed by devolatilization).

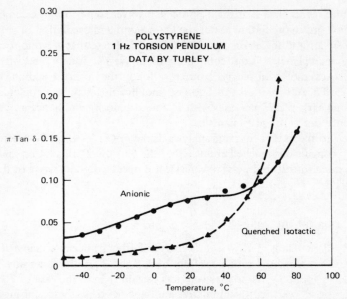

FIGURE 6   The β loss region for slow-cooled anionic PS and a quenched isotactic PS.

Figure 6 compares anionic — slow cooled — with quenched isotactic. Even quenching done to reduce crystallinity fails to develop a β peak. This is consistent with PMMA data of Gillham et al.:[57] syndiotactic PMMA has a strong β peak, isotactic PMMA has none.

Figure 7 compares general purpose thermal PS with a lightly crosslinked PS.

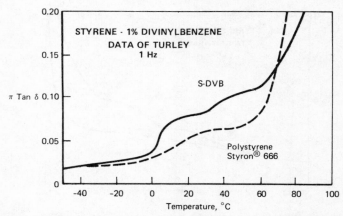

FIGURE 7   The β loss region for a copolymer of styrene with 1% divinylbenzene compared with a commercial atactic PS. Both specimens cooled slowly.

Illers and Jenckel[7] found that crosslinking with 1, 3 and 9% divinylbenzene increases the β peak height progressively compared to the polystyrene blank. By comparing Figures 5 and 7 at any given temperature in the β region, one sees that 1% crosslinking is more effective than quenching in increasing the β peak.

We suggest that under some conditions crosslinking may increase free volume, contrary to first expectations. We located a 2 cm diameter cast rod of 5–5.5% DVB prepared under our supervision 30 years ago by a very slow curing schedule. It had a measured density of 1.01 g cm^{-3} compared with the normal 1.05 for PS. It seems likely that the sample did not shrink in diameter once the gel point was reached. Internal shrinkage with creation of free volume resulted as the remaining monomer polymerized. This rod lacks the sharp ring of PS and has about 50% higher tensile strength than PS.

A further but less dramatic indication of this same effect is afforded in the data of Millar[58] on S-DVB 18–50 mesh beads prepared in suspension. The density goes through a mild minimum at 2 to 4% DVB before starting to climb sharply. This could result from the competition between the shrinkage effect already discussed and the normal tendency of crosslinkers to increase density.

Much earlier than this Ueberreiter and Otto–Laupenmuhlen[59] cast a series of S-DVB rods with 5, 10, and 15% DVB. The densities (g cm^{-3} at 20°C) were respectively 1.047, 1.021 and 1.005. The final thermal treatment for these cast rods was a very slow cooling from 200°C to room temperature.

This increase in free volume and the enhancement of the β loss peak may well be an artifact of specimen preparation. Mason[60] has shown that the post-crosslinking of prepolymerized Hevea results in a net decrease in free volume.

Such rarefied low density form of PS may well constitute a new polymeric state to consider for PVT and various mechanical properties investigations.

Anionic PS appears to give a stronger β peak for similar thermal treatment than does thermally catalyzed PS (compare Figures 6 and 7). One possible reason is that dimers, trimers, and tetramers which are formed in the thermal kinetics but not in anionic[61] can act as antiplasticizers to suppress the β peak intensity. Illers and Jenckel[7] found that plasticizers did tend to suppress the β peak intensity, and lower its temperature.

Concerning the number of consecutive monomers participating in the "local mode" β relaxation, Figure 8 indicates the β peak to be enhanced in an essentially alternating copolymer of styrene and acrylonitrile. While the VCN groups space out phenyl groups, there is still some non-bonded interaction, presumably strong, between ϕ and $-C\equiv N$ side groups. A copolymer of styrene with 15 wt. % (about 30 mol %) butadiene leads to a complete absence of a β peak.[62] Such units in the chain may greatly reduce the intrachain barriers which the β relaxation process involves. Precision studies on carefully selected and prepared copolymers might help to resolve this matter.

In a different type of approach, Rabold[63] has used two nitroxide free radical

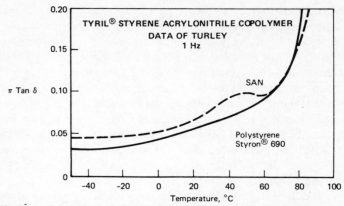

FIGURE 8   Comparison of the β loss region for atactic PS and an equimolar copolymer of styrene with acrylonitrile (SAN). Both materials were thermally initiated, devolatilized, and slow-cooled after molding.

probes of mol wt 154 and 254, respectively, dissolved at 10 ppm in atactic PS. An experimental determination is made with ESR equipment of the temperature, T_w, at which these probes are tumbling freely at an effective frequency of $f \sim 10^7$ Hz. T_w for the higher mol wt probe is 130°C and hence well above T_g. as it should be in view of the higher frequency. However, T_w for the smaller probe is 45°C, indicating that it is responding rather freely below T_g at a frequency of 10^7 Hz in the T_β region. T_β will not attain this frequency until about 150°C (see Figure 2, p. 279 of Ref. 5). However, the probe is not encumbered with intrachain barriers to rotation as is the β motion. Hence there is available near T_β a combined free volume plus swinging volume[64] capable of permitting free motion of the smaller but not of the larger probe.

We have estimated an activation volume for the β process in PS at 173 cm^3 mol^{-1}, contrasted with 836 cm^3 mol^{-1} for T_g.[65] These volumes correspond to spherical diameters of $\sim$7 and $\sim$11.5 Å, respectively.

While these several approaches can give no more than a first approximation, it appears that one monomer unit, or possibly

$$-CH_2—CH\phi—CH_2-$$

might be involved in the β motion.

Figure 9 shows dynamic loss curves for the three isomers of poly(chlorostyrene). These effects are similar to what was discussed about CH$_3$— substituents in connection with Figure 4 and T_g. Ortho clearly suppresses T_β, meta enhances it. Both are consistent with oscillation about ϕ—C at T_β.

However, Figure 10 shows that α-CH$_3$— diminishes the intensity and raises the temperature of T_β. It must be assumed that backbone motion, however limited, is also involved in the β process.

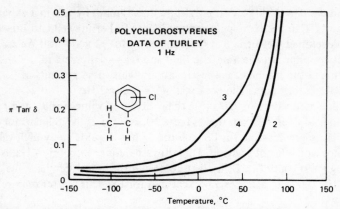

FIGURE 9   The β loss region for the three isomers of atactic poly(monochlorostyrenes). These specimens are considered to have approximately equal volatile contents and thermal histories.

2) *Discussion of nature of β loss in PS*   A combination of data such as in Figures 5–10, along with cited literature results, now permits a more detailed discussion of the molecular significance of the β loss motion.

The absence of a dynamic loss β peak in poly(*o*-chlorostyrene), Figure 9, should not be construed to imply the non-existence of such a peak. Quach and

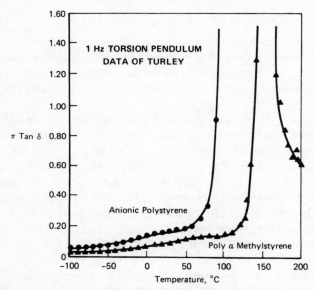

FIGURE 10   The β loss region for poly(α-methylstyrene) prepared anionically, with anionic PS shown as a reference material.

Simha[66] found the β peak at 10°C in poly(o-methylstyrene) by thermal expansion, in contrast to −18°C in PS. More pertinent to Figure 9, Leffingwell and Bueche[35] found a well developed dielectric β at frequencies up to ~2 x 10^4 Hz. We assume that the direct coupling between electric field and Cl—ϕ dipole and the absence of such coupling in the mechanical experiment explains these two sets of seemingly contradictory data. The peak was at 72°C at ~3 Hz.

We have calculated values of ΔH using their tabulated values for T_g and f and estimating values of f for T_β from their Figure 10. ΔH for T_g is 98 kcal/mol and hence only slightly greater than for PS. ΔH for T_β has the extremely high value of 172 kcal/mol which is not realistic. Their lowest frequency point, 4 Hz, is off the line for $\log f = f(1000/T_\beta)$. Using that point and the Quach-Simha value, 283°K at $f \sim 10^{-4}$ leads to a $\Delta H \sim 33.5$ kcal/mol for PS. This matter requires clarification.

Graham[67] has been able to resolve a mixed dielectric $T_g - \beta$ process in poly-(n-butylmethacrylate) into its two components by application of hydrostatic pressure. Since dT_β/dP is less than dT_g/dP for poly(o-methylstyrene) (see Table VI of Ref. 66), hydrostatic pressure should help to resolve T_β and T_g in dynamic mechanical experiments. Tabor and coworkers[68] are making dynamic measurements under pressure but have not yet published such an effect on ortho substituted polystyrenes.

3) The ratio T_β/T_g in (K/K)   Whereas PS itself falls generally on the line

$$T_\beta(K, 100 \text{ Hz}) \sim 0.75 \, T_g \, (K, 100 \text{ Hz})$$

derivatives pose some problems.

According to Figure 10, the 1 Hz β peak temperature for PαMS is about 75°C, T_g is 150°C. This T_g is low and probably results from monomer generated on slow heating. A more likely 1 Hz value is 180°C. Using the latter value gives a ratio of 0.77.

However, o-CH$_3$—, o-Cl— and t-butyl— derivatives lack a mechanical β peak, presumably because it lies under the T_g peak even at 1 Hz. It is not known to us whether tension or combined tension—torsion would bring out these mechanical β peaks. We have already noted that an electric field reveals a β peak even at high frequencies[36] in the 2-chlorostyrene copolymers. The dielectric T_β/T_g ratio is 0.88 at 100 Hz but not too frequency sensitive. The ratio for PS itself is also high, which can be explained in part by the high ΔH of its β peak.

At the moment one must state that the T_β/T_g ratio, and the related precursor argument, are not well delineated for the polystyrene family.

4) Effect of tension, pressure and densification   It has been apparent for some time that a dynamic tensile stress enhances the β process. For example, Takayanagi[67] was able to observe the β relaxation at 110 Hz in the Rheovibron (tension) whereas the torsion pendulum experiments of Adam first cited by Illers and Jenckel[7,9] indicated that it had already merged with the T_g peak at 40 Hz.

Sternstein and coworkers[68] have now made a systematic study of the combined effects of tension and torsion. Tension does enhance the β peak through an increase in free volume on dilatation.

Dale and Rogers[69] densified PS by applying pressure to PS above T_g and cooling under pressure. A densification of about 1% thus achieved leads to a suppression of the β peak intensity. One should study also the results of Yourtee and Cooper.[70]

5) The quasi-equilibrium value of T_β   Quach and Simha[66] have reported a P−V−T study on both atactic PS and atactic poly(o-methylstyrene). One can consider such volumetric measurements as quasi-equilibrium with an effective frequency of ∼10^{-4} Hz. They have measured T_β and its pressure dependence. Huang and Koenig[73] have determined infrared absorbance as a function of temperature for PS and find evidence for a transition at 235 K. It appears that this is related to thermal expansion. There are many other quasi-equilibrium methods which give discordant results.[72] Table II is a limited comparison of several sets

TABLE II

Quasi-equilibrium values of T_β

Technique	Physical units	Atactic PS	Atactic P(2 CH$_3$S)	Ref.
1) P−V−T				
T_g	K	374	404	66
T_β	K	255	283	
dT_β/dP	K/(100 bar)	65[a]	44[a]	
dT_g/dP	K/(100 bar)	74.2[a]	73.0[a]	
2) Infrared	K	235	−	73
3) Electrical conduction[b]	K	270	−	75
	K	256		

[a]Values for their low pressure glass.
[b]Thermally stimulated electrical discharge (TSD) gives a sharp current spike at 251 K. See, for example, *Rikagaku-Kenkyusho Hokoku* 46, No. 3, 47 (1970); or Toyoseiki brochure "Electret Thermal Analyzer" Atlas Electric Devices Co., 4114 Ravenswood Ave., Chicago, Illinois 60613.

of data by very low frequency, and hence quasi-equilibrium methods.

Since dT_β/dP is substantially less for σ CH$_3$ than for PS, one must conclude that T_β is less sensitive to changes in free volume than T_β for PS. The real barrier to motion is intra chain and is only slightly dependent on free volume.

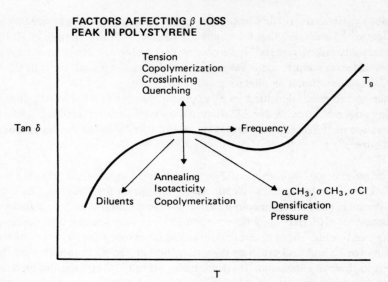

FIGURE 11   Summary of some key structural, compositional, thermal and external
parameters which influence the location and intensity of the dynamic mechanical β peak.
The schematic curve shown is visualized for a low frequency (0.1 to 1.0 Hz). Most of the
factors shown involve either free volume or intrachain barriers to rotation.

6) Summary of factors influencing β relaxation   Figure 11 summarizes the
experimental and structural variables which influence both the temperature and
intensity of the β peak. Aside from chemical modifications such as ring or side
chain substituents, cross-linking and copolymerization, the other factors seem to
relate clearly to free volume. Even crosslinking may result in an increase in free
volume as suggested earlier. One subtle aspect of free volume has been noted by
Crissman *et al.*[19] The densities of the ring substituted PS's are less than that for
PS, suggesting an increase in free volume. The results discussed must represent a
further interplay between steric factors and free volume. In the case of ortho, the
steric effect dominates for T_g and T_β whereas, they note, free volume may
dominate at T_δ .

The frequency effect follows a simple Arrhenius plot over about 5 decades
(see Figure 10.35 of Ref. 5), indicative of a simple relaxation process. Assigning
an effective frequency of 10^{-4} Hz to the Quach—Simha value of $-20°C$,[63] the
point lies reasonably close to the line, adding 2 more decades of linearity.

The chemical structure variations, including copolymerization, seem to relate
to changes in intrachain barriers to rotation, although new systematic studies
would be needed to quantify this aspect.

7) The $T_{1,1}$ process   The $T_{1,1}$ process, and its salient characteristics are adequately covered elsewhere.[1] Whereas the δ, β, and T_g processes appear to involve ever longer portions of the PS chain, the $T_{1,1}$ process seems to involve the entire chain. We deduce this from the fact that $T_{1,1}$ changes character[1] when the molecular weight exceeds that for chain entanglement, M_c. For example, a plot of $T_{1,1}$ against T_g with mol wt as the common parameter, (using data in Table III of Ref. 1) is linear with positive slope until M_c is reached, after which the slope increases exponentially, $T_{1,1}$ continues to rise whereas T_g is leveling off. And yet polyblend experiments show that $T_{1,1}$ varies linearly with M_n^{-1} but curvilinearly with M_w^{-1}.[76]

In some exploratory torsional braid experiments[77] blends of one part anionic PS of mol wt 2×10^6 (which does not exhibit a $T_{1,1}$ loss in the temperature range employed, i.e., up to 250°C) with four to five parts of anionic polystyrenes with any of the following molecular weights: 10,000, 20,000, 37,000 and 110,000 were made. Since the T_g's of the two components coincided, one observed a single T_g with a $T_{1,1}$ peak characteristic of the lower molecular weight species, usually 5–10°C lower in temperature. In brief, a guest species of mol wt of 20,000 behaves in an unentangled fashion even though dispersed in an entangled host. A guest species with mol wt $> M_c$ behaves as if entangled. Aging experiments in 10% solution rule out the likelihood of improper mixing.

Stockmayer speculated[78] that $T_{1,1}$ might represent the principle Rouse mode of vibration, involving the entire chain. Starkweather suggested[78] that T_g and T_β might then correspond to higher Rouse modes. The dependence of $T_{1,1}$ on M_n would appear to rule out the principle Rouse mode or any of its overtones.

A value of ΔH for the $T_{1,1}$ process is not known to us.

4 CONCLUDING REMARKS

We have attempted an overview of representative literature data intended to elucidate the nature of molecular motion occurring in the solid state of PS, its derivatives and selected copolymers. With the exception of the δ process, very little of a detailed nature can be said. It seems that powerful new techniques such as ^{13}C as discussed by Schaeffer,[32] will be needed before such motion can be elucidated in detail.

Various arguments are cited to suggest that the size of the moving group and/ or the complexity of its motion, increases along the temperature scale from δ to β to T_g to $T_{1,1}$.

Essentially four new concepts have been suggested tentatively in this paper, all requiring further study:

1) The γ process might arise from end groups whose number varies with mole-

cular weight and branching; and whose type depends on catalysts and chain transfer process (Section 3C).

2) The enhancement of intensity of the β process by divinylbenzene crosslinking may be a direct result of increased free volume caused by inability of the network polymer to shrink to a normal density following cooling (Section 3E, Part 1).

3) The apparent enhancement of the β process in anionic PS compared to thermal PS may well result from the antiplasticizing effect of monomers, dimers, and trimers present in the latter but not the former (Section 3E, Part 1).

4) Attention is directed to a new thermal initiation mechanism[48] which would introduce small amounts of phenyl substituted methylene sequences into the polystyrene chain (see Section 3C).

Acknowledgements

R. F. Boyer expresses appreciation for a two-year collaboration with Prof. J. K. Gillham of Princeton University and several of his graduate students on a detailed Torsional Braid Analysis of the ~1 Hz mechanical relaxation spectrum of PS as a function of molecular weight; and for the privilege of spending the period of May 11 to June 4, 1975 in Prof. Gillham's Laboratory.

S. G. Turley acknowledges the assistance of R. A. Tribfelner in obtaining the torsion pendulum data.

References

1. S. J. Stadnicki, J. K. Gillham, and R. F. Boyer, this volume, page 359.
2. R. F. Boyer, *J. Polymer Sci. C* 14, 267 (1966).
3. R. F. Boyer, Preprints, Organic Coatings and Plastics Div., *Amer. Chem. Soc.* 30 1 (1970).
4. R. F. Boyer, *Encyclopedia of Polymer Science and Technology, Vol. 13*. edited by N. Bikales (Wiley-Interscience, New York, 1970) pp. 277–289.
5. N. G. McCrum, B. E. Read, and G. Williams, *Anelastic and Dielectric Effects in Polymeric Solids* (Wiley, New York, 1967) especially pp. 409–420.
6. K. Schmieder and K. Wolf, *Kolloid-Z. Z. Polym.* 134, 139 (1953).
7. K.-H. Illers and E. Jenckel, *Rheol. Acta* 1, 322 (1958).
8. K.-H. Illers and E. Jenckel, *J. Polymer Sci.* 41, 528 (1959).
9. K.-H. Illers, *Z. Elektrochem.* 65, 679 (1961).
10. K. M. Sinnott, *Trans. Soc. Plastics Engrs* 2, 65 (1962).
11. S. G. Turley, The Dow Chemical Company, unpublished observations using a 1 Hz torsion pendulum. Some of his data appear later in Figures 5 to 10.
12. S. G. Turley, *J. Polymer Sci. C* 1, 101 (1963). There was subsequent participation in the IUPAC Working Party: see T. T. Jones, *J. Polymer Sci. C* 16, 3845 (1968).
13. S. G. Turley and H. Keskkula, *J. Polymer Sci. C* 14, 69 (1966).
14. H. Keskkula, S. G. Turley, and R. F. Boyer, *J. Appl. Polymer Sci.* 15, 351 (1971).
15. H. Keskkula, *Encyclopedia of Polymer Science and Technology*, Vol. 13, edited by N. Bikales (Wiley–Interscience, New York, 1970) pp. 395–425, especially Figures 3 and 4.

16. R. Buchdahl and L. E. Nielsen, *J. Polymer Sci.* 15, 1 (1955). This refers to earlier work.
17. L. E. Nielsen, *Mechanical Properties of Polymers* (Reinhold, New York, 1962, 2nd Edition, 1975).
18. J. M. Crissman, J. A. Sauer, and A. E. Woodward, *J. Polymer Sci. A* 2, 5075 (1964).
19. J. M. Crissman, A. E. Woodward, and J. A. Sauer, *J. Polymer Sci. A* 3, 2693 (1965).
20. M. Baccaredda, E. Butta, and V. Frosini, *Polymer Letters* 3, 189 (1965).
21. M. Baccaredda, E. Butta, V. Frosini, and S. dePetris, *Mat. Sci. Engng.* 3, 151 (1968/69).
22. V. Frosini and P. L. Magagnini, *Europ. Polymer J.* 2, 129 (1966).
23. V. Frozini and A. E. Woodward, *J. Polymer Sci. A−2* 7, 525 (1969).
24. C. D. Armediades, E. Baer, and J. K. Rieke, *J. Appl. Polymer Sci.* 1$\mathcal{S}$, 2635 (1970).
25. O. Yano and Y. Wada, *J. Polymer Sci. A−2* 9, 669 (1971).
26. R. J. Morgan and L. E. Nielsen, *J. Polymer Sci. A−2* 10, 1575 (1972).
27. K.-H. Illers, BASF, Ludwigshafen, private communication to the authors.
28. S. J. Stadnicky, Ph.D. Thesis, Princeton University, 1974.
29. A. Odijima, J. A. Sauer, and A. E. Woodward, *J. Polymer Sci.* 57, 107 (1962).
30. T. M. Connor, *J. Polymer Sci. A−2* 8, 191 (1970).
31. B. Crist, Jr., *J. Polymer Sci. A−2* 9, 1719 (1971).
32. J. Schaefer, this volume, pp. 103. J. Schaefer, E. O. Stejskal, and R. Buchdahl, *Macromolecules.* 8, 291 (1975).
33. A. J. Curtis, *SPE Trans.* 2, 82 (1962).
34. G. S. Fielding-Russell and R. E. Wetton, *Molecular Relaxation Processes* (Chemical Soc., Burlington House, London).
35. J. Leffingwell and F. Bueche, *J. Appl. Phys.* 39, 5910 (1968).
36a. R. D. McCammon, R. G. Saba, and R. N. Work, *J. Polymer Sci. A−2* 7, 1721 (1969).
36b. See also Ph.D. thesis of R. G. Saba, Penn. State University, 1967, for further details.
37. J. D. Irvine and R. N. Work, *J. Polymer Sci. Polymer Phys. Ed.* 11, 175 (1973).
38. A. E. Tonelli, *Macromolecules* 6, 682 (1973).
39. S. Reich and A. Eisenberg, *J. Polymer Sci. A−2* 10, 1397 (1972).
40. J. Heijboer, This volume, pp. 75; see also TNO Central Laboratory Communication 508, Delft, Holland.
41. R. F. Boyer, Encyclopedia of Polymer Science and Technology, Vol. 13, edited by N. Bikales (Wiley−Interscience, New York, 1970) Fig. 3 on p. 280.
42. R. F. Boyer, Parts I and II, *Plastics and polymers* 41, 15, 71 (1973). Parts I−IV published as 7th Annual Swinburne Award Lecture, available from The Plastics Inst., 11 Hobart Place, London or as a reprint from the author.
43. R. F. Boyer, *J. Polymer Sci. C* 50, 189 (1975).
44. C. L. Choy, R. G. Hunt, and G. L. Salinger, *J. Chem. Phys.* 52, 3629 (1970).
45. R. P. Madding, D. L. Fehl, and J. R. Dillinger, *J. Polymer Sci., Polymer Phys.* 11, 1435 (1973).
46. P. Zoller, D. L. Fehl, and J. R. Dillinger, *J. Polymer Sci., Polymer Phys.* 11, 1441 (1973).
47. R. J. Morgan and L. E. Nielsen, *J. Polymer Sci. A−2* 10, 1584 (1972).
48. W. A. Pryor, J. H. Coco, W. H. Daly and K. N. Houk, *J. Amer. Chem. Soc.* 96, 5591 (1974).
49. K.-H. Illers, *Kolloid-Z. Z. Polym.* 252, 1 (1974).
50. G. Pezzin, F. Zilio-Grandi, and P. Sanmartin, *Europ. Polymer J.* 6, 1053 (1970).
51. M. Nozaki, K. Kamisako, and T. Takase, *Repts Progr. Polymer Phys. Japan* 14, 453 (1971).
52. R. F. Boyer, *Encyclopedia of Polymer Science and Technology,* Vol. 13, edited by N. Bikales (Wiley−Interscience, New York, 1970) p. 287.
53. V. P. Privalko, *Macromolecules* 6, 111 (1973).
54. R. F. Boyer, *Encyclopedia of Polymer Science and Technology,* Vol. 13, edited by N. Bikales (Wiley−Interscience, New York, 1970) pp. 233−234.
55. G. Goldbach and G. Rehage, *Kolloid-Z. Z. Polym.* 216−217 56 (1967).

56. B. Wunderlich and D. M. Bodily, *J. Polymer Sci.* C 6, 137 (1964).
57. J. K. Gillham, S. J. Stadnicki, and Y. Hazony, *ACS Polymer Preprints* 15, 562 (1974); Y. Hazony, S. J. Stadnicki, and J. K. Gillham, *ACS Polymer Preprints* 15, 549 (1974); S. J. Stadnicki, J. K. Gillham, and Y. Hazony, *ACS Polymer Preprints* 15, 556 (1974).
58. J. R. Millar, *J. Chem. Soc. [London]* 1311 (1960).
59. K. Ueberreiter and E. Otto-Laupenmühlen, *Kolloid-Z. Z. Polym.* 133, 26 (1953).
60. P. Mason, *Polymer* 5, 625 (1964).
61. R. F. Boyer, *Encyclopedia of Polymer Science and Technology,* Vol. 13, edited by N. Bikales (Wiley–Interscience, New York, 1970) pp. 163–165.
62. L. E. Nielsen, R. Buchdahl, and G. C. Claver, *Ind. Eng. Chem.* 43, 345 (1951).
63. G. Rabold, *J. Polymer Sci. A–1* 7, 1203 (1969).
64. G. Kanig, *Kolloid-Z. Z. Polym.* 190, 1 (1963), see especially Figure 5.
65. R. F. Boyer and P. L. Kumler, manuscript in preparation. See page 225 of this volume.
66. A. Quach and R. Simha, *J. Appl. Phys.* 42, 4592 (1971).
67. G. Williams and D. A. Edwards, *Trans. Faraday Soc.* 62, 1329 (1966).
68. E. Jones Parry and D. Tabor, *Polymer [London]* 14, 617, 623, 628 (1973).
69. Data cited by S. Matsuoka and Y. Ishida, *J. Polymer Sci.* C 14, 297 (1966).
70. S. Sternstein and T. C. Ho, *J. Appl. Phys.* 43, 4370 (1972). This particular paper concerns PMMA; J. J. Pulerno, Jr., The Toughening Mechanism in Rubber Modified Glassy Plastics Ph.D. Dissertation, Rensselaer Polytechnic Institute, Troy, New York, March 1970.
71. W. C. Dale and C. E. Rogers, *J. Appl. Polymer Sci.* 16, 21 (1972).
72. J. B. Yourtee and S. L. Cooper, *J. Appl. Polymer Sci.* 18, 897 (1974).
73. Y. S. Huang and J. L. Koenig, *J. Appl. Polymer Sci.* 15, 1237 (1971).
74. R. F. Boyer, *Encyclopedia of Polymer Science and Technology,* Vol. 13, edited by N. Bikales (Wiley–Interscience, New York, 1970) Table 3, p. 284.
75. V. Adamec, *J. Polymer Sci.* B 6, 687 (1968).
76. C. Glandt, H. Toh, J. K. Gillham, and R. F. Boyer, *AIS Polymer Preprints* 16, 126 (1975).
77. J. K. Gillham and R. F. Boyer, unpublished results.
78. Prof. W. H. Stockmayer, Dartmouth University, Dr. H. W. Starkweather, DuPont Experimental Station and R. F. Boyer in an informal discussion after Prof. Stockmayer's lecture on application of The Rouse Model to Block Copolymers.

The T_{ll} ($>T_g$) Transition of Atactic Polystyrene†

S. J. STADNICKI and J. K. GILLHAM

Polymer Materials Program, Department of Chemical Engineering, Princeton University, Princeton, New Jersey 08540, U.S.A.

and

R. F. BOYER‡

Dow Chemical Company, Midland, Michigan 48640, U.S.A.

Torsional braid analysis (TBA) ($\sim$0.3 Hz) and differential thermal analysis (DTA) data are presented for the temperature region 0–200°C for two series of atactic polystyrenes with narrow molecular weight distributions: (a) anionic series, $\bar{M}_n$ = 600 – 2 x 10^6, $\bar{M}_w/\bar{M}_n \simeq 1.1$; (b) fractionated thermal series, $\bar{M}_n$ = 2,000 – 1.1 x 10^5, $\bar{M}_w/\bar{M}_n < 1.25$. Preliminary results on bimodal blends are also reported. Heating and cooling cycles were employed with TBA; only the heating mode was used with DTA. In addition to a dynamic mechanical loss peak at T_g, a higher temperature loss peak was also found. Designated the T_{ll} or liquid–liquid transition (relaxation), its temperature is 1.1 to 1.2 T_g (°K) for polymers with molecular weight below the critical molecular weight (M_c) for chain entanglements. Above $M_c \simeq 35,000$, it rises steeply, being $\simeq$200°C for $\bar{M}_n$ = 110,000. The common dependence of T_g and T_{ll} on $\bar{M}_n^{-1}$ below M_c suggests a common molecular origin. The two facts, (a) that $T_{ll} > T_g$ and (b) that T_{ll} reflects chain entanglements, further suggests that T_{ll} involves a longer chain segment length and possibly the entire molecule. Comparison of T_{ll} versus log M plots with T versus log M isoviscous state plots based on zero-shear melt viscosity data from the literature implies that T_{ll} as measured by the TBA technique corresponds to an isoviscous state of 10^4 –10^5 poises. The employment of narrow molecular weight polymers is presumably responsible for both the linear variation of the T_{ll} transition with $\bar{M}_n^{-1}$ (which suggests a free volume

---

†The full text has been published in *J. Appl. Polymer Sci.* **20**, 1245–1275 (1976).
‡Present address: Midland Macromolecular Institute, Midland, Michigan 48640.

basis for the relaxation) and the form of the variation of the T_{ll} transition with log M (which suggests an isoviscous basis for the relaxation). The sharpness of the T_{ll} loss peak by TBA decreases with increasing molecular weight and dispersity. The DTA endothermic event corresponding to T_{ll} is clearly related to the occurrence of flow since the fused films which result from heating granules to 200°C and cooling to R.T. do not reveal a T_{ll} on reheating. If a fused film is crushed, a T_{ll} event is observed on heating. For bimodal blends with $\bar{M}_n < M_c$ for both components, the T_{ll} transition was averaged; with one component less than and one greater than M_c, the T_{ll} transitions of the components appeared to occur independently at temperatures corresponding to those of the isolated components. In accordance with Ueberreiter and Orthmann, T_g appears to separate a glassy state from a fixed liquid state, whereas T_{ll} separates the fixed liquid from a true liquid state. Possible molecular interpretations for the T_{ll} process are discussed. Systematic bodies of data from the literature which indicate the presence of the T_{ll} process in other polymers are summarized.

DISCUSSION

J. F. Jansson (*MIT*): We have seen a transition in poly(vinyl chloride) above T_g. Would you care to compare or speculate on the similarity to T_{ll} in poly(styrene)?

J. K. Gillham: L. A. Utracki has reported (see L. A. Utracki, *J. Macro. Sci., Phys.,* **B-10**, 477 (1974)) on the effect of plasticizer on the T_{ll} relaxation in poly(vinyl chloride). The summary of Table VIII demonstrates the generality of T_{ll}.

E. Baer (*Case Western Reserve Univ.*): Could T_{ll} be a manifestation of ordering in the "liquid" phase? Could structures such as the nodules of Greg Yeh explain this transition?

J. K. Gillham: The techniques (DTA, TBA) used in the present contribution are not able to discriminate between changes which might occur throughout a continuum and the collapse of discrete domains. Careful morphological experiments should be conducted on systematic series through their T_{ll} transitions to answer this vital question.

Polystyrene: Mechanical loss versus temperature.

Top: logarithmic decrement (Δ), exponential damping constant (α) and out-of-phase shear modulus (G'') *versus* temperature

Middle: G'' (expanded) *versus* temperature

Bottom: $d(\omega^2)/dT$ *versus* temperature

These plots were obtained from the raw digital print-out of the paper tape[1] by transfer to an interacting computing facility which utilizes a graphic extension of APL on a Tetronix 4013 display terminal. The derivative curve was obtained from a piece-wise polynomial fit of five consecutive data points centered at the point of evaluation.[2]

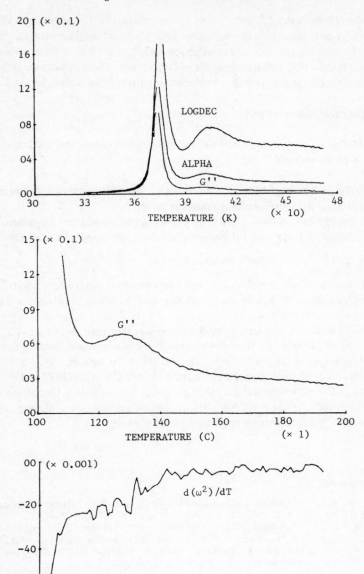

R. Simha (*Case Western Reserve Univ.*): Concerning the T_{ll} process and thermal expansivity α, it is noteworthy that in poly(cyclohexyl methacrylate) a change in the *slope* of the α-T line is seen at about 443 K or $1.17\, T_g$, similar to the Höcker–Blake–Flory observation for poly(styrene). There is such an indication also for the cyclopentyl polymer in the same temperature region, i.e. at $1.27\, T_g$.

J. K. Gillham: Needs no comment by authors.

N. G. McCrum (*Oxford Univ.*): Why was logarithmic decrement used as the measure of mechanical loss?

J. K. Gillham: The logarithmic decrement (Δ) function was employed because the hard-wired data analyzer[1] computes it directly.

G'' is really the preferred function. For example, transition assignments obtained from dG'/dT equal those obtained from the temperature of the maxima of G''.[2] G'' can be calculated from $G'' \approx \dfrac{\Delta}{\pi}\, G'$ $\left(\text{or } G'' \approx \dfrac{\Delta}{\pi}\, \dfrac{1}{P^2} \sim \omega^2\right)$.

Under certain circumstances Δ can be a more sensitive parameter than G'' (the out-of-phase shear modulus) and α (the exponential decay constant for a damped wave).

A comparison of the various modes for presenting mechanical loss [Δ, α, G'' and $d(\omega^2)/dT\,(\sim dG'/dT)$] in the T_{ll} region is demonstrated for data from these laboratories[3] on a polystyrene sample [$\bar{M}_n = 10{,}300$, $\bar{M}_w/\bar{M}_n = 1.12$]. The TBA specimen was prepared by preheating to 200°C and cooling (under nitrogen). The experimental data shown are for the increasing temperature mode under nitrogen. $\Delta T/\Delta t = 1°C/\text{min}$. (Note: numerical constants relating the G'', α, Δ, and $d(\omega^2)/dT$, have been neglected in the presentation.)

References

1. C. L. M. Bell, J. K. Gillham and J. A. Benci, *Amer. Chem. Soc., Polymer Preprints* **15**, 542 (1974). Also: J. K. Gillham, A.I.Ch.E. Journal, **20**(6), 1066 (1974).
2. Y. Hazony, S. J. Stadnicki and J. K. Gillham, *J. Appl. Polymer Sci.* **21**, 401 (1977).
3. J. K. Gillham, R. F. Boyer and Y. Hazony, unpublished results. We are grateful to Professor M. Szwarc, State University of New York, Syracuse, New York for providing the characterized sample of "monodisperse" polystyrene.

Effect of Dispersity on the T_{ll} ($>T_g$) Transition in Polystyrene†

C. A. GLANDT, H. K. TOH‡ and J. K. GILLHAM

Polymer Materials Program, Department of Chemical Engineering, Princeton University, Princeton, New Jersey 08540, U.S.A.

and

R. F. BOYER §

Dow Chemical Company, Midland, Michigan 48640, U.S.A.

The presence of a high temperature ($> T_g$) relaxation in amorphous polystyrene has been investigated further. In the previous work, the techniques of differential thermal analysis (DTA) and torsional braid analysis (TBA) were employed to study polystyrene as a function of "monodisperse" molecular weight. The occurrence of the T_{ll} transition appeared to be associated with the attainment of a critical viscosity level which also corresponded with a free volume level. An entanglement network developed at a critical value of molecular weight, M_C, giving a break in the T_{ll}-versus-M plots. The present work deals with the influence of dispersity on the T_{ll} transition, below and above M_C. A series of binary blends of "monodisperse" anionically polymerized polystyrenes with systematic changes in $\bar{M}_n$ and heterogeneity index ($\bar{M}_w/\bar{M}_n$) was tested by TBA. The results show that when both components have molecular weights below M_C, single and average values of T_g and T_{ll} are observed which are linearly related to $\bar{M}_n^{-1}$, as predicted by free volume arguments. Although a single T_g is observed when one component has a molecular weight above and the other has a molecular weight below M_C, the components appear to undergo the T_{ll} relaxation independently. The results indicate that both the glass transition and the T_{ll} transition are basically governed by the same type of molecular motion but at different length ranges.

† The full text has been published in *J. Appl. Polymer Sci.* **20**, 1277–1288 (1976), and ibid **20**, 2009 (1976).
‡On leave from the Malaysian University of Science, Pinang, Malaysia.
§ Present address: Midland Macromolecular Institute, Midland, Michigan 48640.

Structure–Property Relationships in Polymers and the Preparation of Ultra-High Modulus Linear Polyethylenes

I. M. WARD

Department of Physics, University of Leeds, Leeds

INTRODUCTION

The subject of structure–property relationships in polymers is an extremely wide one, and it is therefore possible to select only a few topics for detailed discussion in this paper. In particular some recent studies of oriented polymers will be described. In spite of the greater complexity of the behaviour of oriented polymers compared with isotropic polymers, I would submit that this is the part of the subject where our understanding of structure–property relationships has advanced most definitively in recent years. Two aspects have been of especial interest:

1) a general understanding of the mechanical anisotropy at low strains,

2) the mechanism of plastic deformation.

These two aspects have come together very closely in our own recent research on the preparation and properties of ultra high modulus fibres and films. It is therefore intended to give a brief historical survey of the situation and follow this by a more detailed account of the high modulus oriented polyethylenes.

Before the advent of the ultra-high modulus oriented polymers it appeared that there were two classes of oriented polymers where it was not too difficult to achieve a reasonable understanding of the situation. There are amorphous polymers and crystalline polymers with a clear lamellar texture. As this has already been discussed in detail in previous publications[1,2] only a brief recapitulation will be made here.

AMORPHOUS POLYMERS AND POLYMERS OF LOW CRYSTALLINITY

In amorphous polymers, and polymers which can be oriented without much increase in crystallinity, (i.e., care is taken not to crystallise the polymer greatly during the orientation processes) overall molecular orientation is the predominant factor influencing the mechanical anisotropy. The properties of the oriented polymer can be regarded as arising from a partially oriented aggregate of aniso-tropic units of structure.[3] This model enabled the mechanical anisotropy to be directly linked with methods for determining molecular orientation such as broad line nuclear magnetic resonance[4] and laser-Raman spectroscopy.[5] An example of this is shown in Figure 1, where the measured stiffness constants of oriented samples of polymethylmethacrylate (PMMA) are shown to be close to half way between the Reuss and Voigt bands, i.e. the bounds obtained by

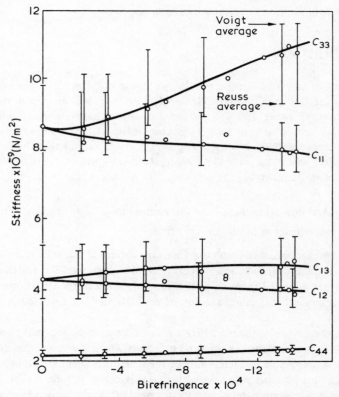

FIGURE 1   Stiffness constants of PMMA as a function of birefringence in a series of uniaxially oriented samples.

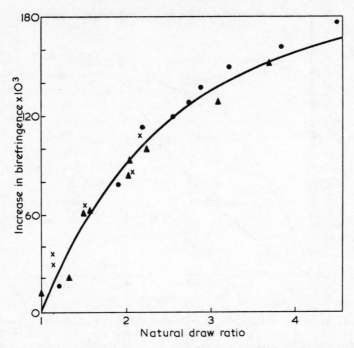

FIGURE 2   Variation in birefringence with natural draw ratio for PET fibres. x, o, Δ, experimental data for drawing at −20, +24 and 62°C respectively; ——, theoretical curve based on pseudo-affine deformation scheme.

averaging compliances and stiffnesses on the aggregate model. The molecular orientation was determined by broad line NMR.[6] The aggregate model has been applied with some success to mechanical anisotropy in polyvinyl chloride by Hennig,[7] to polyethylene terephthalate (PET) by the present author[1, 8, 9] and to PMMA, PVC, PET and polyacrylonitrile by Kausch.[10, 11]

It has also become clear that for these materials the development of molecular orientation during drawing can be understood in terms of continuum models. Uniaxial cold drawing can be described by a deformation scheme in which the unique axes of the anisotropic units (considered to be transversely isotropic) rotate towards the draw direction in the same way as lines joining pairs of points in the macroscopic body, which deforms uniaxially at constant volume. This scheme has been called pseudo-affine, to distinguish it from the affine deformation scheme of rubber elasticity. Figure 2 shows the variation in birefringence with natural draw ratio for PET fibres, and the good agreement with the prediction of the pseudo-affine deformation scheme.[9] Hot drawing, on the other hand, follows the affine deformation scheme up to comparatively high

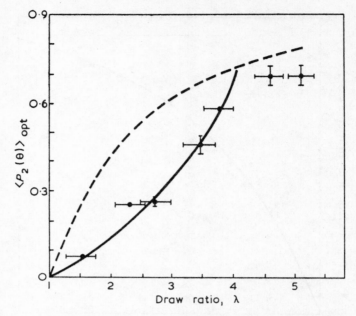

FIGURE 3 $\langle P_2(\theta)\rangle_{opt}$ as a function of draw ratio, showing fit to rubberlike (number of random links, N = 4.8) network (—) at low draw ratios and turn over to pseudo-affine deformation scheme (- - -) at high draw ratios.

draw ratios, as shown by Figure 3, which is taken from some recent drawing studies of PET.[12] In this case the drawing was carried out at 80°C.

Irrespective of whether the mechanism of reorientation follows the affine or the pseudo affine deformation scheme, the draw ratio, and hence the degree of orientation achievable, appears to be explicable in terms of stretching a network formed as the polymer cools from the melt, to an ultimate degree of extension. This ultimate degree of extension is quite small (~400%) and has been shown to be irrespective of the division of extension between the spinning stage and the drawing stage (or drawing stages, if more than one is involved).[9, 13] This means that the natural draw ratio in cold drawing can be markedly reduced by the introduction of stretching at the spinning stage. Because the development of molecular orientation is so much greater for a given degree of stretching in cold drawing than it is in rubber-like stretching, spinning orientation markedly reduces the final orientation achieved.

Even when such polymers as PET are drawn on textile machinery at high temperatures, the natural draw ratio cannot be much exceeded. This means that, even in spite of the changes in chain conformation and orientation which occur (and these have been quantitatively monitored by infra-red spectroscopy)[12] the molecular alignment is still severely restricted compared with that achievable for highly crystalline polymers, as will be discussed in detail below.

ORIENTED CRYSTALLINE POLYMERS WITH A CLEAR LAMELLAR TEXTURE

Crystalline polymers can be transformed into oriented structures with a clear lamellar texture in a number of equivalent ways, e.g. drawing and annealing; drawing, rolling and annealing; crystallisation under strain. These structures can be considered as composite solids, as has been recognised recently in the isotropic case by other workers, notably Halpin and Kardos.[14]

For oriented polymers, recognition of the analogy with composite materials, dates to the earlier work of Takayanagi.[15] Takayanagi found that the modulus along the draw direction for cold drawn and annealed linear polyethylene became less than that perpendicular to the draw direction at high temperatures. He proposed that these results could be explained in terms of what is now a well-known model. In its simplest form the model considers that the amorphous material, whose stiffness falls markedly at high temperatures, is in series with the crystalline regions for the situation where stress is applied along the draw direction and in parallel for the perpendicular situation.

In detailed studies of mechanical anisotropy in oriented branched polyethylenes[16,17] this composite model was developed by including the recognition that in physical terms it is the shear strains which are dominating the behaviour at high temperatures. Specifically, the term interlamellar shear was introduced to emphasise that the dominant physical mechanism is shear of the material between the lamellae. The orientation of the lamellae is therefore critical in determining the patterns of anisotropy. This could be seen most dramatically in the comparison of the loss anisotropy in a series of specially oriented sheets.[17] Figure 4 shows the dynamic mechanical loss spectra for thin strips cut from these sheets in the indicated directions. It can be seen that the magnitude of the interlamellar shear process P is greatest where the resolved shear stress parallel to the lamellar planes is a maximum. Dielectric relaxation measurements were then undertaken on sheets similar to those used for the dynamic measurements,[18] but using branched polyethylene polymer which had previously been subjected to some oxidation so that a few carbonyl groups were introduced into the polyethylene chain. The interlamellar shear relaxation showed no dielectric anisotropy (Figure 5) which confirms that the mechanical anisotropy arises entirely from the arrangement of the lamellae.

This work was then followed by the development of simple theoretical models for predicting the loss anisotropy from the distribution of lamellar orientations[19] and by an attempt to predict the moduli for the parallel lamellae sheet taking into

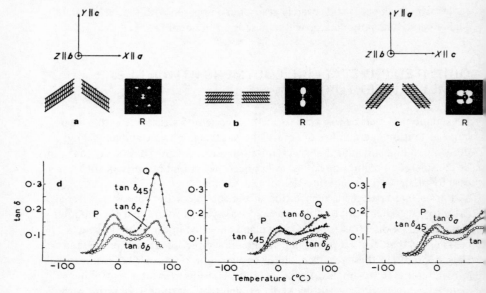

FIGURE 4   Schematic structure diagrams and mechanical loss spectra. (a) and (d): for
b–c sheet; (b) and (e): for parallel lamellae sheet; (c) and (f): for a–b sheet. P, interlamellar
shear process; Q, c-shear process; R, low angle x-ray diagram, beam along Z.

account *pure* shear as well as *simple* shear.[20] In addition the same ideas were
applied with some success to linear polyethylene,[21] polypropylene[22] and
oriented annealed PET.[23] The loss anisotropy model has also been applied to
linear polyethylenes by Buckley and McCrum.[24]

We have also studied the mechanical anisotropy of oriented linear polyethylenes produced by crystallisation under strain.[25,26]. In a typical case cold
drawn LPE was radiation cross-linked and crystallised by slow cooling from the
melt at permanent strains of ~1000%. Following the work of Stein and Judge[27]
and Keller and Machin,[28] it was anticipated that a "shish-kebab" structure
would be produced, with a core of extended chain crystallites surrounded by a
parallel lamellae texture. It might have been expected that such materials would
be appreciably stiffer than the parallel lamellae textures in the same polymer
produced by hot drawing. Comparison of the results on the strain crystallised
LPE with those for hot drawn LPE is made in Figure 6. It can be seen that the
two materials do not differ markedly which suggests that there is not a very
significant proportion of extended chain crystallites in the strain crystallised
polymer, since mechanical behaviour would be expected to be a sensitive test of
such a change in structure. It was concluded[25,26] that the strain crystallised
materials were also explicable in terms of a composite model.

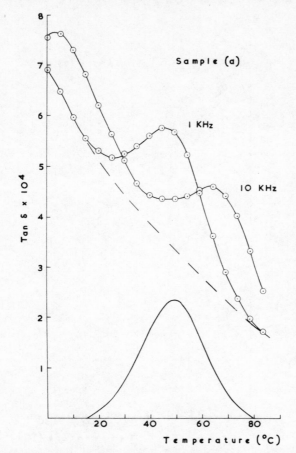

FIGURE 5   The dielectric loss tangent of oxidized, branched polyethylene. Sample (a) Isotropic, annealed at 105°C.

From the viewpoint of obtaining high modulus it can be seen that such composite structures are very limited, since their overall stiffness is so greatly reduced by the fraction of low stiffness non-crystalline material which is "in series" with the draw direction.

ULTRA-HIGH MODULUS FIBRES AND FILMS

The comparatively low values for the Young's modulus of oriented polymer films and fibres, compared with those of inorganic materials and theoretical estimates for perfectly oriented polymers, have considerably limited their

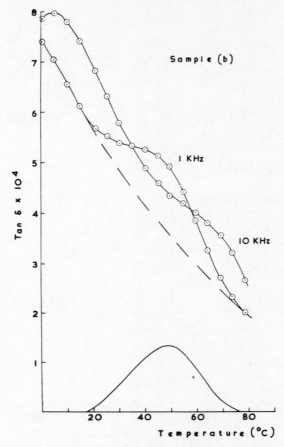

FIGURE 5 Sample (b): c axis normal to plane of disc, prepared at 104.5°C.

technological applications. The discussion has indicated how this arises for amorphous polymers and a polymer such as PET which is drawn from the amorphous isotropic state, due to the inability of the drawing process to produce anything approaching aligned extended molecular chains. It also appears that although high crystallite orientation can be produced in the oriented, annealed crystalline polymers, the Young's modulus is always limited by the composite nature of the polymer, because there is low stiffness material between the crystalline lamellae.

This leaves the area of cold drawn oriented polymers, where the structural studies (notably by Peterlin),[29] have suggested that the initial texture is broken up during drawing, so that chain folds can be pulled out both during and

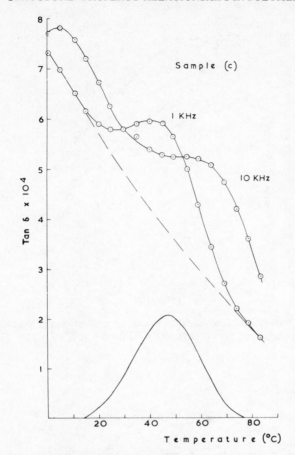

FIGURE 5 Sample (c): c axis in plane of disc, prepared at 105°C.

following initial plastic deformation processes in the crystalline regions and the reorientation of the original lamellar texture.

During the last few years, there has been an extensive programme of research at Leeds University concerned with understanding the drawing of crystalline polymers with especial regard to the structure and properties of the oriented polymers produced. The research has highlighted the importance of three inter-related factors which determine the nature and extent of the drawing process. These are

i) The lengths of the molecular chains, and, perhaps for want of a better description, we describe the effects in terms of the number average and weight average molecular weights, $\bar{M}_n$ and $\bar{M}_w$.

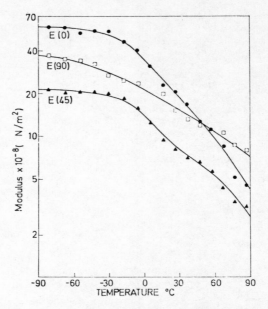

FIGURE 6(a)   Variation of extensional 10 s modulus with temperature. Strain-crystallised LPE (radiation cross-linked). E(0), E(45), E(90) are moduli at 0, 45 and 90° to initial draw direction.

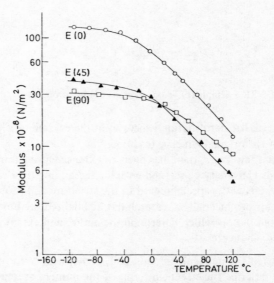

FIGURE 6(b)   LPE drawn at 117°C and annealed at 129.5°C.

ii) The crystallisation and morphology of the initial polymer before drawing. Here we have become aware of the value of the two types of information; those that describe the crystallisation (either kinetics or equilibrium) ignoring morphological factors directly but involving molecular weight and temperature, and those that consider the morphology in a descriptive sense without explicitly discussing molecular weight.

iii) The conditions of draw, i.e. such factors as applied strain rate and temperature.

When the oriented products of drawing are considered, it is clear that the relationship between structure and properties again depends on the two inter-related factors; morphology, which determines the arrangement of lamellae and the like, where we invoke analogies with the mechanics of composite materials, and the molecular situation where we are concerned with the distribution and behaviour of molecules of particular length within the morphological texture.

A comprehensive understanding of the drawing process and the structure of the drawn polymers must embrace both morphological and molecular understanding; we therefore have to combine techniques such as optical microscopy, electron microscopy and small angle x-ray scattering with molecular spectroscopy (broad line NMR, infra-red and Raman spectroscopy).

The starting point for the present studies was an examination of the cold drawing behaviour of a set of linear polyethylene (LPE) polymers,[30] which had been gathered together during a number of investigations of polymer structure using the technique of nitric acid etching followed by gel permeation chromatography.[31,32] The starting polymers were therefore very well characterised for molecular weight and molecular weight distribution, but only small quantities were available. Because of this the cold drawing was carried out on isotropic monofilaments produced by a small-scale melt spinning unit, capable of spinning ~0.5 g of polymer. Although it was attempted to spin the different polymers under comparable conditions, no particular efforts were made to control the temperature of the spinning threadline. All the monofilaments were drawn under the same conditions, extending 2 cm samples at room temperature in an Instron tensile testing machine with a cross-head speed of 1 cm min^{-1}.

The draw ratios obtained were plotted as functions of $\bar{M}_n$ and $\bar{M}_w$, as shown in Figure 7(a) and (b). It was impossible to come to any definite conclusions from data showing such a large scatter, but it can be seen that only two samples showed draw ratios substantially higher than the average value of about eight, and that both these samples had a low weight average molecular weight. It was in fact, correctly concluded that $\bar{M}_w$ is an important parameter in determining the obtainable draw ratio, but the effects were quite small, yielding maximum draw ratios of about 11–14.

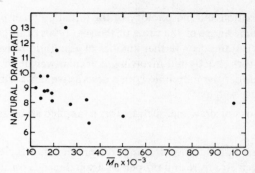

FIGURE 7(a)   Natural draw-ratio at room temperature as a function of number average molecular weight.

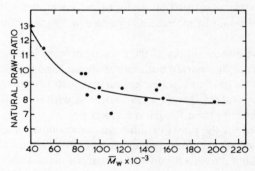

FIGURE 7(b)   Natural draw-ratio at room temperature as a function of weight average molecular weight.

When the Young's moduli of the drawn monofilaments were plotted as a function of draw ratio, a remarkably simple relationship was obtained (Figure 8). It appeared that the Young's modulus of a drawn LPE sample depended only on the draw ratio, irrespective of its initial morphology or the chemical composition in terms of molecular weight, even although these may be vital factors in determining the draw ratio and hence the value of Young's modulus achieved. Moreover, although the maximum draw ratios obtained were not high by the standards of our subsequent results, the highest value of Young's modulus observed was ~24 GN/m^2, which exceeded that obtained for commercially available synthetic fibres at that time, i.e. prior to the production of Kevlar and similar materials.

It therefore seemed very worthwhile to repeat the whole exercise, but to attempt to control and optimise both the preparation of the material prior to drawing, and the actual drawing procedure.[33] Two features were apparent at an early stage of this subsequent investigation. First, it was clear that to obtain

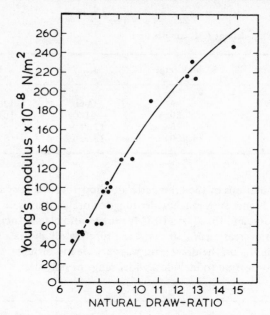

FIGURE 8   Young's modulus of the drawn monofilaments as a function of the natural draw-ratio.

most effective drawing it is convenient to draw at higher than ambient temperatures, taking care however not to approach "flow drawing" where extension occurs without corresponding molecular alignment. After some preliminary studies, the draw conditions were standardised for the purposes of the next part of the exercise, by drawing at 75°C at a draw speed of 10 cm/min on samples of initial length 2 cm for a fixed time of 60 or 90 s.

Secondly, it also now appeared that the morphology of the undrawn polymer was critical in determining the nature of the subsequent plastic deformation during drawing. There was no possibility of making samples with controlled morphology, starting with only ~1 g of polymer. For this reason it was decided to select four commercially available grades of LPE. These are listed in Table I, from which it can be seen that they form matched pairs with respect to $\bar{M}_n$ and that there is a systematic progression in $\bar{M}_w$, with the overlap situation of two samples possessing nearly identical $\bar{M}_w$ at each level of $\bar{M}_n$. It was also found most convenient to abandon melt spinning, and prepare the polymer before drawing by carefully controlled compression moulding procedures. Standard sheets were obtained by compression moulding polymer at 160°C followed either by immediate quenching into water at room temperature, or by slow cooling to a temperature T_q followed by quenching into water. Average values

TABLE I

G.p.c. molecular weight data of LPE samples used.

Sample no.	Rigidex grade	$\bar{M}_n$	$\bar{M}_w$
1	9	6060	126,600
2	50	6180	101,450
3	25	12,950	98,800
4	140/60	13,350	67,800

for several determinations of the draw ratio are shown for the four samples as a function of T_q in Figure 9. A much wider range of draw ratios is observed than in the previous work, and for $T_q = 110°C$ the draw ratio clearly increases spectacularly with decreasing $\bar{M}_w$ so that the 140/60 grade shows a value of ~36. The effect of $\bar{M}_w$ on the draw ratio was very clearly brought out by extending the investigation to include a wider range of polymers.[34] Figure 10

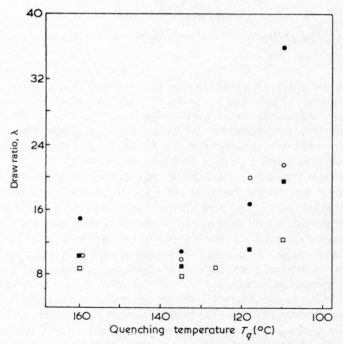

FIGURE 9   Variation of draw ratio λ with quenching temperature T_g for samples drawn at 75°C at a draw speed of 10 cm/min. □, Sample 1: ■, sample 2: ○, sample 3; ●, sample 4.

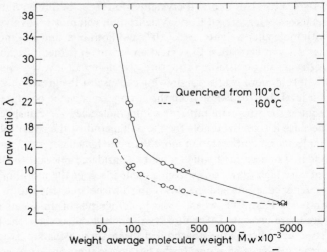

FIGURE 10   Draw ratio λ as a function of $\bar{M}_W$.

shows the variation of draw ratio with $\bar{M}_W$ for two standard compression moulding procedures, either immediately quenching into water from 160°C or slow cooling to 110°C before quenching. $\bar{M}_n$ does not produce a comparable effect, but it is possible that the reversal of the order of the draw ratios of the samples with $\bar{M}_W$ ~100,000 is due to the effect of $\bar{M}_n$, and this part will be discussed further below.

The effect of $\bar{M}_W$ is a systematic one for both compression moulding procedures, but it is most clearly discerned for the samples quenched from 110°C. It is obviously interesting to consider what reasons may be adduced for this very interesting behaviour. One line of argument can be based on the known kinetics of crystallisation.[35,36] These show that the crystallisation rate vs molecular weight curve for isothermally crystallised LPE fractions at high super-cooling, for example 121°C, goes through a maximum[36] at a molecular weight of 2×10^4. The shape of the maximum is asymmetric, so that the half life for crystallisation rises more steeply at the low molecular weight side than at the high molecular weight side. Most important the maximum becomes less clearly resolved with falling temperature and eventually cannot be detected.

The equilibrium crystallinity has also been measured as a function of molecular weight.[37] It is constant at low molecular weights, but falls off monotonically at molecular weights greater than about 10^4.

Although these results were obtained for fractionated polymers, there are indications from other work that in unfractionated polymers, fractionation can occur during crystallisation.[38] If this were so, then we can see that if the crystallisation takes place at temperatures where the rates are very sensitive to molecular weight, i.e. ~120°C, and if crystallisation stops before equilibrium crystallinities have been achieved then crystallisation will mostly involve those molecules with molecular weights *ca* 2×10^4, and both high and low molecular weight species will not crystallise to as great an extent. It is then plausible to imagine that the ensuing crystalline structure, which in fact involves chain folded lamellae, will unfold more readily on drawing because of the greater homogeneity at a molecular level. It is also clear from the strong dependence on $\bar{M}_W$ that this unfolding requires that there are rather few long molecules. We can speculate that this is because it is unfavourable for the pulling out of the chain folds to have many molecules incorporated in more than one lamella.

The second line of argument which must be considered relevant to the production of very high draw comes from a comparison of the morphology of the directly quenched sheets and the slow cooled sheets quenched from 110°C. This is illustrated in Figure 11, where optical micrographs of the sheets for Rigidex 50 grade are presented. The directly quenched sheets show the typical spherulitic structure whereas the slow cooled sheets show a coarser structure with barely recognisable spherulites. In the latter case there are large regions of uniform orientation, and the familiar banding of the typical spherulitic pattern is absent. It is known[39] that when crystallisation occurs under conditions of relatively high viscosity, the spherulitic geometry develops in a complex process of radial lamellar growth with twisting around the radial growth direction. At low viscosities, on the other hand, twisting becomes less likely and in a sense more uniform crystal growth occurs. The optical micrographs of LPE sheets in the low $\bar{M}_W$ range are consistent with these ideas, which receive further confirmation from the fact that at high values of $\bar{M}_W$, the difference between the two compression moulding procedures is not so apparent.[34]

The optical micrographs suggest that the slow cooled sheets may have a more perfect lamellar texture at a molecular level than those quenched immediately from 160°C. This is confirmed by the relative sharpness of the small angle pattern for these sheets and also by the comparatively small line width of the longitudinal acoustic mode observed in the Raman spectrum.[40, 41]

With all the above information in mind, it is not difficult to speculate in general terms that the high draw ratios are achieved for the slow cooled samples and especially for low values of $\bar{M}_W$, because the slow cooling and the low values of $\bar{M}_W$ acting in an interrelated manner through the crystallisation kinetics, give rise to a structure where the crystalline lamellae are most perfect not only with respect to their fold surfaces but also in the sense that they contain molecules of comparable length. Moreover, the structure of the polymer as a whole is such that only comparatively few long molecules are incorporated in more than one lamella. From the viewpoint of plastic deformation, this would be an ideal structure to draw to a high degree of molecular alignment because the pulling out of the chain folded molecules would not be impeded either by

RIGIDEX 50

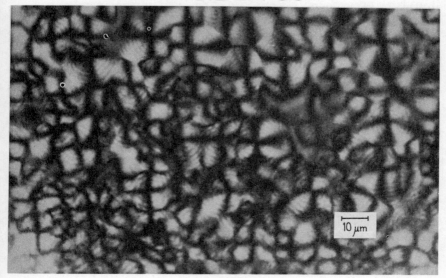

160 W

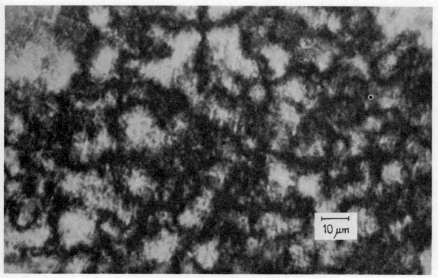

160/110/W

FIGURE 11   Photomicrographs of quenched (160 W) and slow cooled (160/110/W) sheets.

imperfections on the fold surfaces or by the intense strain hardening of too many tie molecules between the lamellae.

Although we believe that the above discussion does contain the essence of the explanation of the high draw and high modulus achieved in this work, it is necessary to redress the balance by emphasising certain points regarding the drawing process which have so far not been discussed explicitly. The molecular weight characteristics and the initial morphology do not only affect the draw ratio achieved, they also affect the whole nature of the plastic deformation process. At low values of $\bar{M}_w$, and especially for the slow cooled polymer, the deformation is very time dependent. In practical terms this means that there is no effect of drawing time at high $\bar{M}_w$, but for low and intermediate values of $\bar{M}_w$, the draw ratio may be increased by extending the drawing time. Work at present in progress suggests that with increasing drawing times the draw ratios converge to an envelope curve which defines the limiting extensibility of the initial polymer as a function of $\bar{M}_w$. In this work, the limiting extensibility is being defined in what is perhaps a special sense, in that it refers to "effective" draw with regard to producing high modulus. There is never any difficulty in merely achieving a high draw ratio as such, but, in our definition this must not be obtained under conditions of flow drawing where the modulus of the oriented polymer ceases to increase with increasing draw ratio.

This latter point turns the discussion naturally from consideration of the drawing process to consideration of the oriented polymers produced.

The most immediate result regarding the oriented polymers comes from the plot of Young's modulus as a function of draw ratio,[33] as shown in Figure 12. Values of the Young's modulus for various strains were taken from 10 s isochronal stress—strain curves, the latter being constructed from creep measurements. It can be seen that the Young's modulus is to a good first approximation dependent only on draw ratio, as found previously.[30] The different draw ratios were obtained by varying the structure of the initial polymer *and* by varying the strain-rate in drawing. Figure 12 shows results for the Rigidex 140/60 and 50 grades. As discussed above, the highest draw ratio for the 50 grade which has a higher value of $\bar{M}_w$ than the 140/60 grade, was obtained by drawing at a lower strain-rate and hence for a longer time. The oriented polymers, although much stiffer than conventional oriented LPE, still retain many of their low-strain characteristics. In particular they exhibit non-linear viscoelastic behaviour and similar temperature dependence. Figure 13 shows the Young's modulus, calculated from the 10 s isochronal stress—strain curves, as a function of draw ratio at various levels of strain for the 50 grade polymer. There is very significant non-linearity, the Young's modulus falling from ~70 to ~45 GN/m^2 as the strain level increases from 0.1 to 0.5%. Again, the major relaxations of LPE are still apparent in dynamic mechanical measurements[42] (Figure 14). These results, taken together, lead us to anticipate first that the structure of these materials is

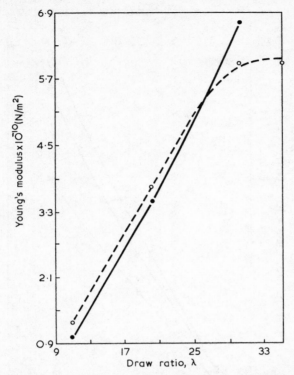

FIGURE 12   Variation of Young's modulus with draw ratio for high density polyethylenes.
•——•, $\bar{M}_n$ = 6180 and $\bar{M}_w$ = 101,450; ○--○, $\bar{M}_n$ = 13,350 and $\bar{M}_w$ = 67,800.

not different in kind from oriented LPE at much lower draw ratios, and
secondly that the interpretation of the mechanical properties in terms of the
structure may be made on somewhat similar lines to those previously proposed
for crystalline polymers, i.e. the composite model approach.

   I shall therefore summarise very briefly the present structural knowledge of
these materials and then proceed to attempt a quantitative analysis of their
mechanical properties.

   The first key information on structure came from the melting points. It was
found[33] that, although the melting points of individual high draw ratio polymers
appeared to differ slightly amongst themselves, there were in no cases melting
points above about 138°C. In spite of some difficulties in defining extended
chain crystals, it is clear that there are no extended chain crystals present in
these materials, corresponding to those which exist either in pressure-crystallised
material[43] or in stirrer crystals.[44]

   Next, we may consider the results of x-ray diffraction measurements. Wide

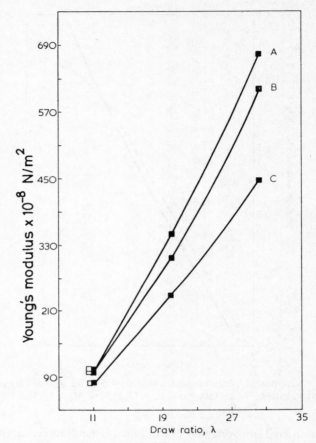

FIGURE 13   Young's modulus, calculated from the 10 s isochronal stress–strain curves, as a function of draw ratio λ, at various strains, for samples 1 (□) and 2 (■). A, $\epsilon = 0.001$; B, $\epsilon = 0.002$; C, $\epsilon = 0.005$.

angle x-ray diffraction photographs of the oriented polymers show that by the time a draw ratio of about ten is reached, the crystallite orientation is very high indeed,[45] and that higher draw therefore does not alter the structure in this respect. At a draw ratio of ten, a clear two-point small angle x-ray pattern is observed, suggesting a parallel lamellae texture. As the draw ratio is increased, the small angle pattern becomes less intense, so that longer exposure times are required for comparable intensity, but nevertheless a clear two-point pattern can always be resolved.

Characterisation of the drawn samples by density measurements and differential scanning calorimetry shows that, if we regard these techniques as

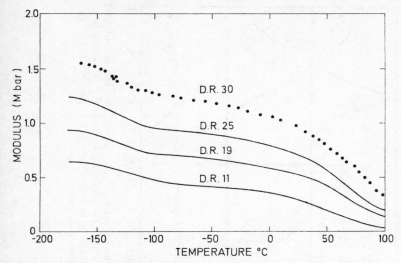

FIGURE 14(a)   Rigidex 50 modulus vs temperature (1 M_{bar} = 10^{11} N/m^2).

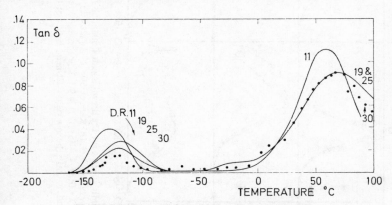

FIGURE 14(b)   Rigidex 50 Tan δ vs temperature.

providing a measure of crystallinity, the crystallinity increases with draw ratio. There is however, the intrinsic difficulty in using these particular techniques that the properties of the non-crystalline regions must be defined or assumed. We have found broad line nuclear magnetic resonance particularly valuable in providing quantitative information regarding the structure and molecular orientation in these highly oriented materials because such assumptions are not required.

As first observed by Hyndman and Origlio,[46] the broad line NMR spectra of oriented LPE shows three components. Derivative spectra for the drawn polymers produced from the four basic Rigidex grades listed in Table I are shown

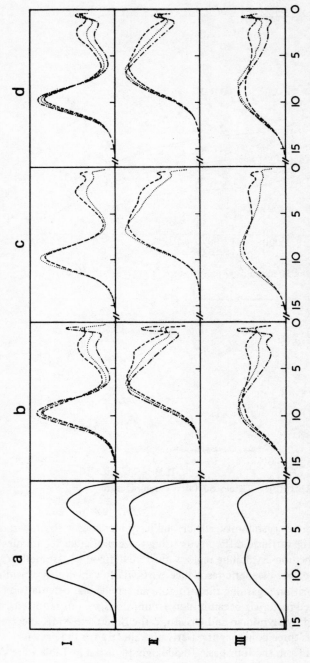

FIGURE 15   Derivative wideline NMR spectra for the drawn Rigidex materials for the 3 orientations with respect to the static magnetic field λ = 0, 45 and 90°. (a) Rigidex 9, draw ratio λ = 9; (b) Rigidex 50; (c) Rigidex 25; (d) Rigidex 140/60. - - - draw ratio λ = 11; · · · , draw ratio λ = 19, - - - - , draw ratio λ = 30. I, λ = 0: II. I, λ = 45; III, λ = 90. The abscissae are in G and the ordinates represent absorption derivative normalized to unit (integrated) intensity. Only the low field (left-hand halves of the spectra are shown.

in Figure 15, and here also three components can be discerned. The broad component is readily identified as being associated with the crystalline regions, and the narrow component can be associated with a very mobile fraction of non-crystalline material. The intermediate component is less easily identified and Hyndman and Origlio suggested that it was "strained amorphous" material. We have shown[45] that it has a second moment consistent with oriented molecules undergoing hindered rotation, and moreover that its intensity correlates with $\bar{M}_\mathrm{w}$. It is therefore suggested that it can be identified with molecules which link different lamellae, i.e. the high molecular weight fraction. It has also been found [45] that the intensity of the very mobile component correlates with $\bar{M}_\mathrm{n}$, suggesting that this component relates to the very low molecular weight fraction. This interpretation of the three components is consistent with our view of the crystallisation process, where we believe that in slow cooled sheets the low and high molecular weight material does not crystallise as much as material of intermediate molecular weight. The three components are even discernable in slow cooled *isotropic* Rigidex 50, which has one of the widest molecular weight distributions of the samples examined.[45]

The drawing process decreases the proportion of intermediate component and increases that of the broad component, and to a lesser degree increases the proportion of the narrow component. This suggests that a proportion of the high molecular weight chains now become rigid, whilst low molecular weight material (or perhaps the ends of molecules) are rejected from the crystalline regions. At

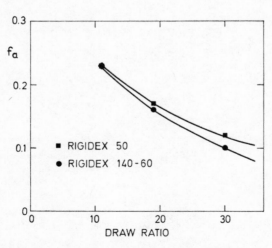

FIGURE 16 f_a as a function of draw ratio.

the same time the orientation of the intermediate component increases, approaching a limiting value at high draw ratio. The decrease in the non-crystalline fraction, obtained from the NMR data, for the Rigidex 50 and 140/60 grades, as a function of draw ratio, is shown in Figure 16.

In view of the small angle x-ray diffraction data, which suggested a parallel lamellar texture for these materials, it seemed to attempt to analyse their mechanical behaviour in terms of a single composite structure viz the series model first proposed by Takayanagi.[15] In carrying out this analysis, it was first necessary to note that the dynamic mechanical data for all samples show two plateau regions in which the modulus is reasonably independent of temperature and the loss tangent falls to zero. One of these regions, which will be called the $-160°C$ plateau region, occurs below $-160°C$, and the other, referred to as the $-60°C$ plateau region, extends from ~-20 to $\sim-100°C$. The situation will be analysed for these two plateau regions only, and the simple series model then suggests that the compliance of the oriented sheet J_T will be given by

$$J_T = f_a J_a + (1 - f_a)J_c = J_c + f_a(J_a - J_c),$$

where f_a is the fraction of non-crystalline material, and J_a, J_c are the compliances of the non-crystalline and crystalline regions respectively.

Figure 17a and b show plots of J_T as a function of f_a for the two grades of LPE polymer. The series model predicts a straight line relationship between J_T and f_a with a gradient $(J_a - J_c)$ intercepting the J_T axis at $J_T = J_c$. Although these plots are reasonable straight lines with reasonable values for the gradient $(J_a - J_c)$, they appear to be shifted from the position expected for the simple series model, i.e. they do not intercept the abscissa at $J_T = J_c$. We have therefore

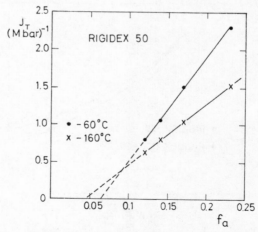

FIGURE 17(a)   Compliance as a function of amorphous fraction for Rigidex 50.

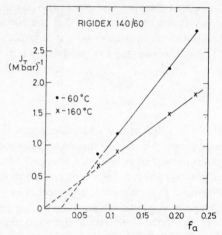

FIGURE 17(b)   Compliance as a function of amorphous fraction for Rigidex 140/60.

proposed a series—parallel model in which a fixed proportion of the non-crystalline material is in parallel (Figure 18). It has been shown that this very simple model provides a good description of the data for both grades of LPE in the two plateau regions. The values obtained for J_a at −60 and −160°C are very nearly identical for the two grades, and the values of the proportion of non-crystalline material estimated from the model at −60 and −160°C are identical for each grade of polymer. The Rigidex 50 grade, however, has a much greater proportion of non-crystalline material in parallel than the 140/60 grade. Although these results must be regarded as tentative at this stage they are

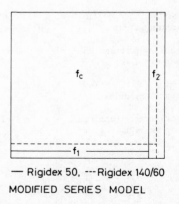

— Rigidex 50, --- Rigidex 140/60

MODIFIED SERIES MODEL

FIGURE 18   Modified series model.— Rigidex 50, --- Rigidex 140/60.

extremely encouraging in that they provide a link with our understanding of the drawing behaviour. It is very reasonable to find that the polymer with the higher value of $\bar{M}_w$ and the wider molecular weight distribution has more "ineffective material" from the viewpoint of modulus, when it is drawn.

CONCLUSIONS

This paper began with a discussion of our understanding of the relationship between the mechanical properties of conventionally oriented polymers and their structure. The development of very highly oriented linear polyethylenes has extended the field greatly, but it is reassuring to find that their low strain behaviour can be understood to a reasonable approximation in terms of the composite model which proved valuable in earlier work on crystalline polymers.

The discovery that it is comparatively straightforward to produce very high modulus films and fibres from linear polyethylene by conventional techniques does mark a new step in its own right. Quite apart from the technological applications, these materials are interesting both because they provide definitive

TABLE II
Comparison of Young's moduli and specific moduli.

Polymer	Young's modulus (Mbar)		X-ray value of Young's modulus (Mbar)	Specific modulus (Mbar x 10^{-3})
LPE (Draw ratio 30)[33, 42]	Room temperature 0.7 creep modulus		~2.5	0.71
(See also Ref. 49)	−180°C Dynamic modulus	1.6		1.7
Polyoxymethylene[48]	Room temperature 0.5 creep modulus		~0.5	0.4
(See also Ref. 50)	−150°C Dynamic modulus	0.9		0.6
Polypropylene[48]	Room temperature 0.18 creep modulus		~0.42	0.2
E glass fibres[51]	0.63		−	0.25
RAE carbon fibres[51]	4.2		−	2.1
Kevlar 49	1.32		−	0.92

structures for studies of structure/property relationships in polymers and because they throw light on the topology of the molecular chains in a bulk crystalline polymer. This is an area which is not well understood, and the time is ripe for further detailed studies.

Finally, it should be commented that these discoveries in LPE are only the starting point for the extension of this type of research both to other polymers and to other fabrication methods. Recent work at Leeds University has already been extended to the hydrostatic extrusion of LPE and polyoxymethylene,[47] and to the drawing of polypropylene and polyoxymethylene.[48] Our work takes its place alongside other similar developments, notably by Porter and his colleagues[49] and by Clark and Scott.[50]

It is of some interest to close with consideration of the present state of the art, comparing the existing Young's modulus values for these several polymers with those expected for perfectly aligned products, and with commercially available fibres (see Table II). In the first place it can be seen that in three polymers, LPE, polyoxymethylene and polypropylene, Young's moduli have already been achieved which are quite close to theoretical expectations. Secondly, in terms of specific modulus, both LPE and polyoxymethylene compare favourably with glass fibres. If we also take cost into account, these highly oriented synthetic polymers can even by considered in the same category as Kevlar 49 and carbon fibres. Indeed we can conclude with the thought that the ultra-high modulus LPE may perhaps prove to be the "poor man's carbon fibre".

Note added in proof (January 1978)

Since this paper was presented in February 1975, our understanding of the structure and properties of ultra high modulus LPE has advanced very appreciably. In particular, the interpretation of the modulus in terms of the Takayanagi series-parallel model with non-crystalline material in parallel has been superseded by a model in which the key stiffening elements are the intercrystalline bridges produced in drawing. A quantitative model has been proposed which explains the observed correlation between the $-60°C$ plateau modulus and the longitudinal crystal thickness, determined from wide angle X-ray diffraction measurements on the 002 reflection (A. G. Gibson, G. R. Davies and I. M. Ward, Polymer (in press)).

References

1. I. M. Ward, Polymer 15, 379 (1974).
2. I. M. Ward, The mechanical properties of oriented polymers: Invited paper at Polymer-physik, Leipzig, September 1974.
3. I. M. Ward, Proc. Phys. Soc. 80, 1176 (1962).
4. V. J. McBrierty and I. M. Ward, Br. J. Appl. Phys. (J. Phys. D) Ser. 2 1, 1529 (1968).
5. J. Purvis, D. I. Bower and I. M. Ward, Polymer [London] 14, 398 (1973).

6. M. Kashiwagi, M. J. Folkes and I. M. Ward, *Polymer [London]* **12**, 697 (1971).
7. J. Hennig, *J. Polymer Sci.* C 16, 2751 (1967).
8. I. M. Ward, *Text. Res. J.* **31**, 650 (1961).
9. S. W. Allison and I. M. Ward, *Br. J. Appl. Phys.* **18**, 1151 (1967).
10. H. H. Kausch-Blecken von Schmeling, *Kolloid-K. Z. Polym.* **237**, 251 (1970).
11. H. H. Kausch-Blecken von Schmeling, *Kolloid-Z. Z. Polym.* **234**, 1148 (1969).
12. A. Cunningham, I. M. Ward, H. A. Willis and V. Zichy, *Polymer [London]* **15**, 749 (1974).
13. S. W. Allison, P. R. Pinnock and I. M. Ward, *Polymer [London]* 7, 66 (1966).
14. J. C. Halpin and J. L. Kardos, *J. Appl. Phys.* **43**, 2235 (1972).
15. M. Takayanagi, K. Imada and J. Kajiyama, *J. Polymer Sci.* C 15, 263 (1966).
16. V. B. Gupta and I. M. Ward, *J. Macromol. Sci. Phys.* 2, 89 (1968).
17. Z. H. Stachurski and I. M. Ward, *J. Polymer Sci. A-2* 6, 1817 (1968).
18. G. R. Davies and I. M. Ward, *J. Polymer Sci. B* 7, 353 (1969).
19. G. R. Davies, A. J. Owen, I. M. Ward and V. B. Gupta, *J. Macromol. Sci. Phys.* 6, 215 (1972).
20. A. J. Owen and I. M. Ward, *J. Mater. Sci.* 6, 485 (1972).
21. A. J. Owen Ph.D. Thesis, Bristol University, 1972.
22. A. J. Owen and I. M. Ward, *J. Macromol. Sci. Phys.* 7, 417 (1973).
23. G. R. Davies and I. M. Ward, *J. Polymer Sci. A-2* 10, 1153 (1972).
24. C. P. Buckley and N. G. McCrum, *J. Mater. Sci.* 8, 928 (1973).
25. M. Kapuscinski, Ph.D. Thesis, Leeds University, 1974.
26. M. Kapuscinski, J. Scanlan and I. M. Ward, *J. Macromol. Sci. Phys.*, **11**, 475 (1975).
27. R. S. Stein and J. T. Judge, *J. Appl. Phys.* **32**, 2357 (1961).
28. A. Keller and M. J. Machin, *J. Macromol. Sci. Phys.* 1, 41 (1967).
29. A. Peterlin, *J. Mater. Sci.* 6, 490 (1971).
30. J. M. Andrews and I. M. Ward, *J. Mater. Sci.* 5, 411 (1970).
31. D. J. Blundell, A. Keller, I. M. Ward and I. J. Grant, *J. Polymer Sci. B* 4, 781 (1966).
32. I. M. Ward and T. Williams, *J. Macromol Sci. Phys.* 5, 693 (1971).
33. G. Capaccio and I. M. Ward, *Polymer [London]* **15**, 233 (1974).
34. G. Capaccio and I. M. Ward, *Polymer [London]* **16**, 239 (1975).
35. L. Mandelkern, J. M. G. Fatou and K. Ohno, *J. Polymer Sci.* 6, 615 (1968).
36. J. M. G. Fatou and J. M. Barrales Rienda, *J. Polymer Sci. A-2* 7, 1755 (1969).
37. J. M. G. Fatou and L. Mandelkern, *J. Phys. Chem.* **69**, 417 (1965).
38. F. R. Anderson, *J. Polymer Sci. [Symp.]* C 3, 123 (1963).
39. B. Wunderlich, *Macromolecular Physics*, Vol. 1 (Academic Press, New York, 1973) pp. 3 and 7.
40. G. Capaccio, T. A. Crompton and I. M. Ward, *J. Polymer Sci., Polym. Phys. Ed.*, **14**, 1641 (1976).
41. G. Capaccio, I. M. Ward, M. A. Wilding and G. Longman, *J. Macromol. Sci. Phys.*, in press.
42. J. B. Smith, G. R. Davies, G. Capaccio and I. M. Ward, *J. Polym. Sci., Polym. Phys. Ed.*, **13**, 2331 (1975).
43. A. J. Pennings and A. M. Kiel, *Kolloid-Z. Polymer* **205**, 160 (1965).
44. B. Wunderlich and T. Arakawa, *J. Polymer Sci. A* 2, 3697 (1964).
45. J. B. Smith, A. J. Manuel and I. M. Ward, *Polymer [London]* **16**, 57 (1975).
46. D. Hyndman and G. F. Origlio, *J. Polymer Sci.* **36**, 556 (1959).
47. A. G. Gibson, I. M. Ward, B. N. Cole and B. Parsons, *J. Mater. Sci.* 9, 1193 (1974).
48. G. Capaccio and I. M. Ward, Br. Patent Appln. No. 52644/74 and cogs.
49. N. E. Weeks and R. S. Porter, *J. Polymer Sci. A-2* 12, 4 (1974).
50. E. S. Clark and L. S. Scott, *ACS Polymer Preprints* **15**, 153 (1974).
51. W. Watt, L. N. Phillips and W. Johnson, *The Engineer* (27 May 1960).

Structure–Property Relationships in Silicone–Polyethylene Blends

JAMES R. FALENDER, SARAH E. LINDSEY
and JOHN C. SAAM

Dow Corning Corporation, Midland, Michigan 48640, U.S.A.

INTRODUCTION

Uniform stable dispersions of polydimethylsiloxane in polyethylene provide a unique opportunity for observation of the effects of morphology in polymer blends on their properties. The wide disparities between the two components in this system allow for sufficiently large and unambiguous changes in properties with composition to permit comparison with theoretical models. Particularly useful in such a study are the differences in modulus and permeability. The differences in electron absorption of the components permit observations of morphology by electron microscopy of the two phases without selective staining techniques and the attending possibilities of artifacts.

The inherent incompatibility of the component polymers is overcome by the use of heat and mechanical shear to induce formation of free radicals and grafting of the polyethylene to high molecular weight polydimethylsiloxane that contains small amounts of an unsaturated comonomer.[1] Under these conditions it is anticipated that the amount of graft copolymer formed will be sufficient to act as a "polymer-in-polymer" dispersant.[2] Crosslinking can also be anticipated within the silicone. This could lead to the formation of interpenetrating polymer networks at the interface of the two polymeric phases which might cause additional stabilization of the microdispersion.[3] Control over the extent of grafting and crosslinking may be exercised by simply changing the amount of unsaturated groups in the high molecular weight silicone. Details of the preparation and characterization of the blends are published elsewhere.[4]

EXPERIMENTAL

Samples were prepared for testing immediately after mixing by compression molding at 150°C. Mechanical properties were obtained on an Instron® testing machine operating at a crosshead rate of 12.7 cm per min and using standard dumbbell samples having a cross section of 0.317 cm by approximately 0.15 cm. A Rheovibron® operating at 3.5 Hz in a tensile mode was used for dynamic mechanical properties.

Permeability was determined at 23°C using samples of approximately 0.1 cm thickness. A mass spectrometer detected the total amount of gas passing through the sample during a 15 min interval.[5] Electron micrographs were obtained with a Hitachi HS-75 electron microscope on freeze sectioned films. Specimens were not stained. Relative crystallinity by x-ray diffraction was obtained by measuring the areas of two major diffraction peaks (corresponding to "d" spacings of approximately 4.1 and 3.6 Å) and comparing these to the total scattering area. Heats of fusion were used as a measurement of crystallinity and these were measured using a Perkin-Elmer DSC-2 differential scanning calorimeter at 10°C per min. A broad melting peak was observed and the total area was included in the heat of fusion.

DISCUSSION OF RESULTS

1 Morphology

The blends are white and homogeneous when examined visually and a distinct two-phase structure may be observed in the electron microscope (Figure 1). The dark phase is silicone and the light phase is polyethylene. A blend containing 10% silicone (top of Figure 1) shows distinct silicone particiles approximately 1 to 5 microns in diameter dispersed in a continuous polyethylene matrix. A 50% blend (middle of Figure 1) still has a continuous polyethylene phase. Finally, with 75% silicone (bottom of Figure 1) the structure has become much finer and the distinction between discrete and particulate phases is not as clear. Even at this higher level of silicone, however, polyethylene is still the continuous phase.

2 Dynamic mechanical properties

Figure 2 shows dynamic modulus data for 50% silicone–polyethylene blend and also for polyethylene. Over the temperature range of −150 to +65°C the blend is found to have a significantly lower storage modulus or stiffness than polyethylene. As an example, the blend is as flexible at −10°C as polyethylene is at room temperature. Polyethylene's loss modulus peak at about −20°C is considerably diminished by the presence of silicone while the −120°C peak is enhanced. The

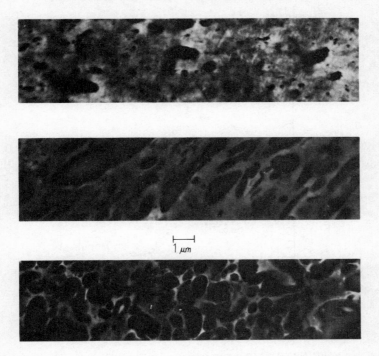

FIGURE 1   Electron micrographs of blends. Dark phase is silicone. Top: 10% silicone; middle: 50% silicone; bottom: 75% silicone.

latter observation is probably due to silicone's glass transition. Diminution of the −20°C peak could be due to dilution by the silicone or to disruption of the crystal structure as will be discussed later. The temperature of the loss peak at −20°C has not changed significantly from unmodified polyethylene, when 50% silicone is present. It is interesting that silicone's crystal transition, usually seen between −40 and −60°C, is not readily apparent.

3   Tensile properties

Tensile properties depend on composition and mixing conditions. This is illustrated in Figure 3 and shows how tensile modulus sharply decreases with small increases in silicone content, particularly when the level of silicone is less than 50%. The effect from the rubber may be predicted using the semi-empirical model described by Takayanagi.[6] Here it is assumed that the nature of deformation is partially parallel in nature and partially series. The relative portion of each depends upon the degree of mixing and for the present calculation is based on a model of

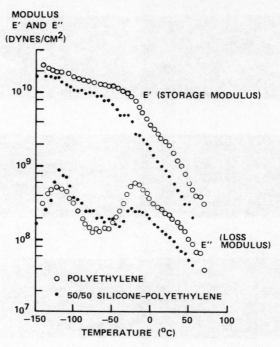

FIGURE 2   Effect of temperature on dynamic modulus.

perfectly adhering non-interacting rubber spheres dispersed in a plastic continuum.[7] The model is derived on the basis of elastic theory.

The actual modulus is consistently lower than that predicted by the model as noted in Figure 3. Somewhat more complex models have been developed, taking account of a volumetric packing parameter and a factor based on particle volumetric packing parameter and a factor based on particle shape.[8] Using these models and any reasonable combination of mixing parameters, the actual modulus continues to fall below that calculated. This can be explained by postulating a modulus in the polyethylene phase about 30% lower than that assumed. The cause for this could be inhibition of crystallization in the polyethylene or by slip between or within the crystallites promoted by the silicone.

Other evidence backs up this hypothesis. X-ray diffraction measurements show a marked decrease in crystallinity with small amounts of silicone (Figure 4). If silicone were merely diluting the crystals, then a straight line relationship would be expected between composition and crystallinity. However, the x-ray evidence suggests either a decrease in the level of crystallinity within polyethylene or else an increase in crystaline defects.

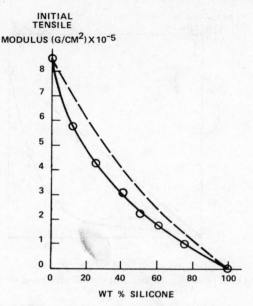

FIGURE 3   Effect of silicone content on tensile modulus. (o) observed, (···) predicted by Takayanagi model.

Measurements of the heat of fusion also show a lessening of the degree of crystallinity (Table I). The addition of 25% silicone lowers the heat of fusion from 19.7 to 11.6 cal/g. This decrease is about 21% greater in magnitude than that predicted by the lowered polyethylene content. The crystal disruption is less with greater amounts of silicone, and it is speculated that this may be due to differences in grafting or mixing. Due to broad melting peaks, it was difficult to determine exact melting points. Extrapolated temperatures for the start of melting were estimated to be 95°C for polyethylene and slightly lower for the blends.

TABLE I

Relative crystallinity calculated from heats of fusion in silicone–polyethylene blends

A percent polyethylene	Heat of fusion at T_M (cal/g)	$(B/A) \times (100)$ heat of fusion per gram polyethylene	Decrease in crystallinity calculated from heats of fusion (%)
100	19.7	19.7	–
75	11.6	15.5	21
60	10.6	17.7	10
30	9.0	18.0	9
25	4.6	18.2	8

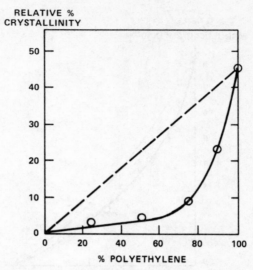

FIGURE 4   Effect of composition on degree of crystallinity based on wide angle x-ray diffraction.

TABLE II

Tensile properties

	Blend	Dow Polyethylene 130®	Silastic 55® silicone
Tensile strength (g/cm²)	91,500	154,000	91,000
Elongation (%)	560	760	600
Modulus (g/cm²)	260,000	840,000	–

Blends consist of a 50% mixture of Dow Polyethylene 130® and 4M% MeViSiO silicone mixed 20 min at 185°C/62 RPM/CAM Head/Brabender®.

Tensile strength also decreases with increasing silicone content, being roughly halved at 50% by weight of silicone. The blends act typically as rubbery thermo plastics with tensile properties shown in Table II.

4   Permeability

Permeability of the blends varies with composition as illustrated in Figure 5. Sil cone rubber is generally associated with a very high level of free volume and its unusually high permeability is attributed to this characteristic. The oxygen permeability of silicone is 730×10^{-10} cm³ (cm)/(s) (cm²) (cmHg) compared to

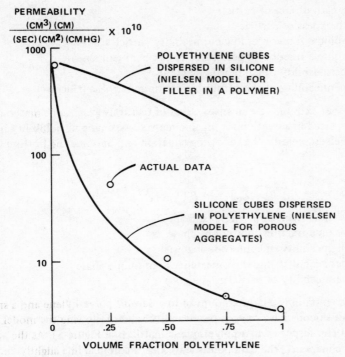

FIGURE 5   Effect of composition on oxygen permeability.

3.9×10^{-10} in polyethylene. The addition of 50% silicone makes a fairly large increase in permeability in spite of the microscope evidence that silicone is particulate in nature. The 50% blend has a permeability constant to oxygen of 11.2×10^{-10} cm³ (cm/(s)) (cm²) (cmHg).

It is interesting to compare the gas permeability of actual blends to theoretical values predicted from various types of composite structure. This should aid in understanding the structure in actual blends. Nielsen's model for permeability provides a fairly simple means of analyzing changes in structure.[10]

The first case to be considered involves small impermeable cubes (polyethylene) suspended in a silicone matrix. Since the silicone is approximately 250 × as permeable as polyethylene, it should be valid to consider the organic phase impermeable in such a system. Nielsen's equation for this case is presented below and is represented by the top of Figure 5 for polyethylene in silicone.

$$\frac{P_c}{P_s} = \frac{\phi_s}{1 + \dfrac{L}{2W}\phi_p}$$

L = length of fact of particle
W = thickness of particle
ϕ_p = volume fraction of low permeability particles
ϕ_s = volume fraction of permeable component (silicone)
P_c = permeability of composite material
P_s = permeability of the permeable continuous phase (silicone)

The other extreme case involves cubes of relatively permeable material (silicone) dispersed in a continuous organic matrix. Assuming no solubility between phases Nielson presents the following equation and one gets the bottom curve in Figure 5.

$$\frac{P_c}{P_p} = \frac{P_a \phi_s}{P_p \phi_s^{1/3} + P_s(1 - \phi_s^{1/3})} + (\phi_p)$$

ϕ_s = volume fraction particles (silicone)
ϕ_p = volume fraction continuous phase
P_c = permeability of composite material
P_p = permeability of low permeability continuous phase
P_s = permeability of particles (silicone)

Blends containing a large amount of low density polyethylene and a small amount of silicone have oxygen permeability values following the model of silicone particles dispersed in a polyethylene matrix (see Figure 5). As the amount of silicone increases, the data points leave this theoretical line slightly and go somewhat toward the inverted phase structure. However, even with 75% silicone the actual data comes closer to following the model for a continuous polyethylene phase than to the other model.

Takayanagi's models correlate properties with the degree of dispersion. His analysis leads to the conclusion that better dispersion of one phase in the other should lead to a more parallel nature of mixing. The permeability of the blends is compared to the theoretical values of parallel and series models in Table III. Using the previously discussed nomenclature, series connection of the phases will imply $1/P_c = \phi_s/P_s + \phi_p P_p$ and parallel connection will imply $P_c = \phi_s P_s + \phi_p P_p$. By comparing the actual permeabilities to the models, a percentage of parallel nature is calculated. The data in Table III shows the parallel nature to be less than 8% in all cases, and this probably means that the degree of dispersion is low.

5   Surface properties

Resistance to environmental stress cracking of polyethylene is greatly improved by the presence of silicone. A 50% polyethylene blend showed no signs of cracking after 54 h in a boiling detergent solution compared to failure after 7 h for pure polyethylene.

TABLE III

Comparison of permeability of parallel and series models

A.	Composition (volume fraction polyethylene)	0	0.25	0.50	0.75	1
B.	Permeability [$(cm^3)(cm)/(s)(cm^2)(cm\ Hg) \times 10^{10}$	730	56	11	4.8	3.9
C.	Theoretical permeability of series blend	–	15.3	7.8	5.2	–
D.	Theoretical permeability of parallel blend	–	548	367	185	–
E.	Fraction of parallel nature $\dfrac{B-C}{D-C}$	–	0.076	0.0097	0	–

Surface tension of 50% polyethylene blend was 23.5 dyn/cm. Washing the surface with hexane, however, gave an increase to 33.4 dyn/cm compared to 36.2 dyn/cm for a polyethylene control.

SUMMARY AND CONCLUSIONS

The dynamic mechanical properties of silicone polyethylene blends have been elucidated by an investigation of the molecular and morphological features. A structure is proposed consisting of microgelled and grafted particles of silicone dispersed in polyethylene. This hypothesis is based on the method of synethesis and electron micrographs. Comparison of permeability and modulus measurements to that of structural models provides added evidence for such a structure. The polyethylene crystallinity is disrupted as deduced from wide angle x-ray diffraction and by comparison of modulus data to that predicted by a mechanical model.

Features of the polyblend, as described above, are used to explain the effects of silicone on the complex modulus of polyethylene. These changes include a lowering of the storage modulus over a broad temperature range and changes in the magnitude of the loss modulus peaks.

References

1. R. J. Ceresa, *Block and Graft Copolymers* (Butterworths, Washington, 1962).
2. G. E. Molau, *J. Polymer Sci. A* 3, 4235 (1965).
3. L. H. Sperling, V. Huelck and D. A. Thomas, in *Polymer Networks: Structure and Mechanical Properties,* edited by A. J. Chompff and S. Newman (Plenum, New York, 1971).

4. J. R. Falender, S. E. Lindsey and J. C. Saam, *Polym. Eng. Sci.* **16**, 54 (1976).
5. "Modern Methods of Research and Analysis", Dow Chemical Company, Interpretative Analytical Services, Midland, Michigan 48640, p. 66.
6. M. Takayanagi, H. Harima and Y. Iwata, *Memoirs Fac. Eng. Kyushu Univ.* **23**, 1 (1963).
7. M. Takayanagi, S. Uemura and S. Minami, *J. Polymer Sci.* C 5 113 (1964).
8. L. E. Nielsen, *Rheol. Acta.* **13**, 86 (1974).
9. J. Brandrup and E. H. Immergut, eds. *Polymer Handbook* (Interscience, New York, 1966), p. V–18.
10. L. E. Nielsen, *J. Macromol. Sci.* Chem. **A1**, 929 (1967).

DISCUSSION

H. W. Starkweather (*DuPont*): Changes in the IR, DSC, and NMR spectra should give help in characterizing the two phases and the extent of phase interpenetration Have you tried any of these?

J. R. Falender: At the time of the symposium we had not characterized the samples by the mentioned techniques. Results of differential scanning calorimetry, however, are included in this publication. Briefly, it was found that the presence of grafted silicone lowered the heat of fusion of polyethylene more than that predicted by the lower polyethylene content. This provides additional evidence for our hypothesis of a disrupted crystal structure.

A. F. Burmester (*Dow Chemical USA*): Regarding the decrease in x-ray diffraction peak in terms of a change in poly(ethylene) crystallite perfection, it should be pointed out that a decrease in crystallite size to say 500 Å will cause such a decrease in peak height without the requirement for crystalline disorder.

J. R. Falender: It is true that many of our observations could be explained by either a lessening in the amount of crystallinity or a change in character of the crystal structure.

R. L. Miller (*MMI*): Did you use peak areas or peak heights in order to determine crystallinity from your x-ray data? Also, what was the density of the poly(ethylene) that you used?

J. R. Falender: Peak areas were used to determine the relative crystallinity within samples. The areas of two major diffraction peaks, corresponding to "d" spacings of approximately 4.1 and 3.6 Å, were taken as a ratio of the total scattering area.

The polyethylene portion of the samples consisted of Dow Polyethylene 130. The resin contains about 4% copolymerized vinylacetate, has a density of 0.92 and a melt index of 0.45. This material is no longer commercially manufactured.

However, we have obtained very similar results with commercial ethylene homopolymers of both low and high density.

R. F. Boyer (*Dow Chemical USA*): You showed that the γ-peak intensity increases in the polymer blend. Could this effect be caused by an increase in the surface area of the poly(ethylene) phase due to the presence of the poly(siloxane)?

J. R. Falender: Your question implies a concept of interaction at the phase boundaries. It is interesting to speculate a sharing of free volume. The rather large particles, and small interphase area, would require deep penetration to be significant. The observed changes in crystallinity would indicate that this is happening. A more likely explanation for the increase in the γ-peak intensity, however, is the glass transition of polydimethylsiloxane which falls very close to the polyethylene γ transition.

Dynamic Mechanical Characteristics of Polysulfone and Other Polyarylethers

L. M. ROBESON, A. G. FARNHAM and J. E. McGRATH†

Union Carbide Corporation, Research and Development, Bound Brook, New Jersey 08805, U.S.A.

1 INTRODUCTION

The polyarylethers comprise a family of thermoplastic polymers which have reached commercial status in the past decade. With the exception of poly (2,6-dimethyl-1,4-phenylene oxide) PPO, the commercial polymers are of the class of materials commonly referred to as poly(aryl ether sulfones). The sulfone containing polyarylethers are noted for their outstanding thermal, oxidative, and hydrolytic stability. This stability combined with rigidity and toughness over a broad temperature range allows these polymers to be utilized under demanding service conditions. Application requirements include continuous use at temperatures of 150°C or greater, resistance to highly acidic or basic environments and boiling water or low pressure steam exposure. The preparation and properties of the polyarylether sulfone have been reviewed in a recent paper.[1]

The commercial polymers include the polysulfone available from Union Carbide, Polysulfone(I), which is prepared by the step-growth polymerization of Bisphenol A and 4,4′-dichlorodiphenyl sulfone.

$$\left[O\!\!-\!\!\bigcirc\!\!-\!\!\underset{\underset{CH_3}{|}}{\overset{\overset{CH_3}{|}}{C}}\!\!-\!\!\bigcirc\!\!-\!\!O\!-\!\bigcirc\!-\!SO_2\!-\!\bigcirc \right] \tag{I}$$

$$(1)$$

The polymerization proceeds by a homogeneous mucleophilic displacement reaction as previously discussed in detail by Johnson *et al.*[2] More recently,

†Present address: *Chemistry Department, Virginia Polytechnic Institute and State University, Blacksburg, Virginia 24061.*

Bisphenol S poly ether(II), and a copolymer comprised of

$$\left[\!\!-\!\!\left\langle\bigcirc\right\rangle\!\!-\!\!\left\langle\bigcirc\right\rangle\!\!-\!\!SO_2\!\!-\!\!\right]\!\!-\qquad\text{and}\qquad\left[\!\!-\!\!\left\langle\bigcirc\right\rangle\!\!-\!\!O\!\!-\!\!\left\langle\bigcirc\right\rangle\!\!-\!\!SO_2\!\!-\!\!\right]\!\!-\right.$$

units have been introduced by ICI. Another system, Astrel 360, is available from the 3M Company. It's structure has been suggested by Rose[1] as consisting of

$$-\!\!\left\langle\bigcirc\right\rangle\!\!-SO_2,\quad\left\langle\bigcirc\right\rangle\!\!-\!\!\left\langle\bigcirc\right\rangle\!\!-SO_2,\!\!-\!\!\quad\text{and}\quad\left\langle\bigcirc\right\rangle\!\!-O\!\!-\!\!\left\langle\bigcirc\right\rangle\!\!-SO_2\text{ units.}\!\!-$$

The other commercial polyarylether PPO does not contain the sulfone moiety. It's structure

$$\left[\!\!-\!\!\left\langle\bigcirc\right\rangle\!\!\substack{-CH_3\\-CH_3}\!\!-O-\right]_n\!\!-$$

achieves a high T_g as a result of its molecular rigidity. The latter is partially due to methyl group substitution on the aromatic ring at an ortho position. This restricts the free motion of the aryl ether bond.

The purpose of this paper is to review in detail the dynamic mechanical characteristics of these materials. Emphasis is placed on the well-known −100°C (1 Hz) transition. Recent data[3] promoting a hypothesis of the structural origin of this transition in polyarylethers will be discussed. Several other unpublished dynamic mechanical results on polyarylethers will also be presented.

2   REVIEW OF DYNAMIC MECHANICAL BEHAVIOR OF POLYARYLETHERS

A prominent secondary relaxation at −100°C for polysulfone was initially noted by Heijboer.[4] The magnitude of this transition and the mechanical loss between this peak and the glass transition was found to be affected by annealing. Bisphenol A polycarbonate shows similar behavior. Heijboer also presented mechanical spectra for poly (2,6-dimethyl-1,4-phenylene oxide) which showed there was no prominent transition, although a broad transition with a maximum mechanical loss at ∿−80°C appeared in the data. Baccaredda et al.[5] observed a transition at −45°C (6100 Hz) for polysulfone which was influenced (and increased in magnitude) by water sorption. The activation energy for this relaxation was calculated to be about 12 kcal/mol. Kurz et al.[6] extended the result of Baccaredda in studies of similar polyarylethers including poly(4,4′-oxydiphenylene sulfonyl), PODS,

$$\left[\!\!-\!\!\bigcirc\!\!-\!O\!-\!\bigcirc\!\!-\!SO_2\!\!-\!\!\right]_n$$

poly(4,4′-methylene diphenylene sulfonyl), PMDPS,

$$\left[\!\!-\!\!\bigcirc\!\!-\!CH_2\!\!-\!\bigcirc\!\!-\!SO_2\!\!-\!\!\right]_n$$

and polysulfone P-1700. They observed transitions in all three polymers at $\sim-100°C$ (1 Hz) and concluded that the molecular moiety responsible for this transition was the SO_2 group as it was the logical polar moiety which would strongly adsorb water. Rather prominent transitions at $\sim0°C$ were also observed for PODPS and PMDPS. Kurz et al.[6] considered that this transition involved a more complex mechanism than simple torsional oscillations or rotation possibly involving coupling of local motion with specific undefined motions resulting from other partially mobile groups of the polymer chain.

In a study of the relaxation phenomena in aromatic polymers, Allen et al.[7] presented data on polysulfone, poly(2,6-dimethyl-1,4-phenylene oxide) PPO and Bisphenol S polyether. It was shown that the dielectric loss of polysulfone, PPO, and Bis S polyether were all influenced by the presence of water. The small relaxation corresponding to the mechanical loss transition at $-100°C$ (1 Hz) observed with all three polymers was greatly increased by the presence of water. Bis S polyether which exhibited the highest water sorption also exhibited bimodal behavior (at 2.5 wt. % H_2O sorption) in the dielectric loss-temperature data. It was observed that the peak in the dielectric loss data occurred at different temperature for the wet samples as compared to dry samples. The activation energy for the wet samples was also found to be different from the dry samples, but independent of water content. Allen et al.[7] concluded that the absorbed water molecules are bonded to the polar groups in the polymer chain and influence the β relaxation of the polymer.

Robeson and Faucher[8] found that the addition of low molecular weight diluents termed antiplasticizers greatly reduced the magnitude of the $-100°C$ transition for polysulfone as well as other polymers with prominent secondary loss transitions. This phenomena will be discussed further in a later section of this paper.

Chung and Sauers[9] investigated the low temperature mechanical relaxations in aromatic polymers including polysulfone and PPO. A larger relaxation at $-111°C$ (0.67 Hz) with a small shoulder at $\sim-67°C$ was noted for polysulfone. PPO was found to exhibit a transition with peaks assigned at $-148°C$ and $4°C$.

It was also observed that poly(p-xylylene)

$$\begin{array}{c} -\!\!\!\left[\!\!-CH_2\!-\!\!\!\left\langle\bigcirc\right\rangle\!\!-\!CH_2-\right]_{n} \end{array}$$

exhibited a large low temperature relaxation at $-114°C$ (0.54 Hz). They concluded that these relaxations were due to local reorientational motion of the
phenylene ring or combined motion of the phenylene ring and nearby chain
units. For these relaxations an activation energy of 10 kcal/mol was determined.

Thus, it is easily seen that while many investigators have studied polyarylethers, no clear agreement on the nature of the $-100°C$ transition exists. In
the course of our investigation into the structure—property relationships for
polyarylethers, many different structural variations have been examined. From
these systems, we have determined the dynamic mechanical characteristics and
have arrived at a new explanation concerning origin of the $-100°C$ transition and
the role water assumes in the resultant increase in magnitude of this transition.

The role water plays in increasing the magnitude of $-100°C$ polyarylether
transitions is an interesting characteristic. The role of antiplasticization in
decreasing the magnitude of the $-100°C$ transition, as will presently be discussed
has important ramifications concerning the relationship between this transition
and the mechanical and transport properties (i.e. diffusion) of the poly(aryl-
ethers).

3   ANTIPLASTICIZATION

The term antiplasticization has been used to describe the resultant effect of
polymer-low molecular weight diluent blends that show increased modulus and
strength with decreases in impact strength and ultimate elongation. This term
was first used in detail by Jackson and Caldwell[10−12] in studies involving polycarbonates. The addition of compounds which meet the requirements of antiplasticizers have been shown to reduce the magnitude of the secondary loss transitions in many polymers which includes polysulfone.[8,13] Robeson and
Faucher[8] illustrated that n-phenyl-2-naphthylamine resulted in the elimination
of the $-100°C$ secondary loss transiton, and further data[13] illustrated the
efficiency of Aroclor 5442 (42% chlorinated terphenyl) in the elimination of
this transition as shown in Figures 1 and 2. It was shown that the apparent
density of the antiplasticizer decreased with concentration. This indicated that
polymer free volume was also being eliminated with antiplasticizer. This interpretation is completely consistent with the loss of the molecular relaxation
process at $-100°C$. CO_2 permeability data[13] also showed a rapid decrease with
antiplasticizer addition. This further illustrates the restriction of molecular

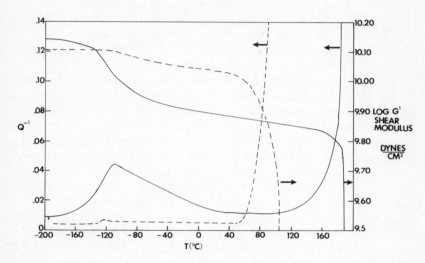

FIGURE 1   Mechanical loss curves for antiplasticized polysulfone (——— polysulfone:— — — 30% arocholor 5442 + 70% polysulfone). (Reference 3)

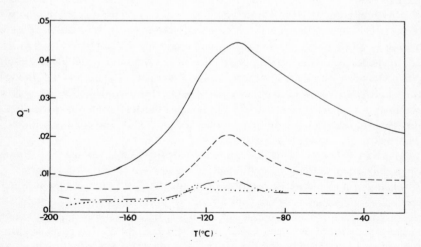

FIGURE 2   Low temperature mechanical loss transition for polysulfone–Arochlor 5442 mixtures (———0% Arochlor 5442; — — — 10% Arochlor 5442;— · — · — 20% Arochlor 5442; · · · · · 30% Arochlor 5442). (Reference 3)

motion that is observed when the $-100°C$ transition is eliminated. Water diffusion coefficient and solubility constant data on polysulfone antiplasticized with Aroclor 5442 showed that the water diffusion coefficient decreased from 4.5×10^{-8} cm^2/s. to 1.65×10^{-8} cm^2/s. with addition of 20 wt. % Aroclor 5442.[13] The solubility constant was observed to be $\sim 1/2$ that of the value of polysulfone suggesting that the potential sites of polysulfone (primarily the $+SO_2+$ group) were tied up by Aroclor interactions.

Polysulfone-poly(dimethylsiloxane) block copolymer, $+AB+_n$ of block molecular weights of 6600 and 1700, respectively, showed definite polysulfone continuity from the analysis of the modulus-temperature data.[14] It was therefore quite interesting to note that the $-100°C$ polysulfone transition was essentially eliminated. The $-125°C$ transition for poly(dimethylsiloxane) could be readily detected since the polysiloxane block is not soluble in polysulfone and is not a antiplasticizer. Therefore, it appears that the $-100°C$ relaxation is molecular weight dependant, and is possibly reflected by the well known relationship that tensile strength and impact strength exhibit with molecular weight.

Since these early publications which showed that antiplasticization eliminated the low temperature relaxation of polymers, several other investigators have studied and observed the same phenomena.[15-18] Petrie et al.[15] questioned the generalization by pointing out that Aroclor addition to polystyrene resulted in a modulus increase without the existence of a strong relaxation process. However, one may reconfirm the generality by arguing that the modulus of Aroclor extrapolated from below its T_g to the testing temperature utilized for the polystyrene–Aroclor blend is presumably higher than polystyrene. If this is indeed the case, additivity of component modulus would easily predict the expected modulus of the blend. In fact their data[15] showed no relative change in modulus difference between $-180°C$ and $25°C$ for polystyrene versus the 75/25 polystyrene/Aroclor 5460 blend. Another possibility is that the ultra low temperature relaxation of polystyrene was eliminated. We still hold to the original hypothesis that the ability to be antiplasticized with low molecular weight diluents is related to the ability to eliminate secondary loss relaxations in polymers (with appropriate corrections for the difference in modulus between the components). The cited references overall strongly support this contention.

There are other factors which can reduce density[19] (free volume) and in some cases embrittle materials without affecting the low temperature transitions.[20] Annealing of polyethylene terephthalate below the T_g is one such case in which presumably molecular ordering (but not of the level expected with crystallization) can occur. This is a phenomena separate from antiplasticization and should not be confused with or used as evidence against the contention that antiplasticization is the result of polymer molecular motion reduction via reduction in secondary relaxation processes. After all, the diffusion data previously cited strongly implies reduction in polymer mobility.

TABLE I (Reference 3).
Poly(aryl ether) structures

Polymer no.	Polymer repeat unit	η_{sp}/c 100 ml/g	T_g°C	Water absorption wt. %
1	[structure: $-O-C_6H_4-C(CH_3)_2-C_6H_4-O-C_6H_4-SO_2-C_6H_4-]_n$	0.48	184	0.72
2	[structure: $-O-C_6H_4-C(CH_3)_2-C_6H_4-O-C_6H_4-O-C_6H_4-]_n$	0.50	122	0.13
3	[structure: $-O-C_6H_4-C(CH_3)_2-C_6H_4-O-C_6H_4-C_6H_4-]_n$	0.70	154	0.17
4	[structure: $-O-C_6H_4-C(CH_3)_2-C_6H_4-O-C_6H_4-C_6H_4-C_6H_4-]_n$	1.1	182	0.18
5	[structure: $-O-C_6H_2(CH_3)_2-C(CH_3)_2-C_6H_4(CH_3)-O-C_6H_4-SO_2-C_6H_4-]_n$	0.61	178	0.66

TABLE I (contd.)

Polymer no.	Polymer repeat unit	η_{sp}/c 100 ml/g	T_g °C	Water absorption wt. %
6		0.41	230	1.30
		0.55	175	0.32
8		0.87	262	1.36
9		0.50	190	0.67
10		0.51	172	0.52

4   STRUCTURAL VARIATIONS OF POLYARYLETHERS

a)   Experimental

All polymers were prepared as described elsewhere.[3] The mechanical loss and shear modulus–temperature data on the structures listed in Table I were obtained using a torsion pendulum similar to the original design of Nielsen.[21]. The samples were conditioned under the following procedures. Dry samples were placed in a vacuum oven at 100°C for ~ 12 h then stored in a vacuum dessicator prior to testing. Wet samples (equilibrium water solubility listed in Table I) were immersed in water until constant weight was attained. Samples were quickly mounted in the torsion pendulum and immediately quenched to liquid nitrogen temperature by heat exchanging dry nitrogen gas with liquid nitrogen. Dry N_2 gas, therefore, blanketed all specimens from −180°C to 0°C. Above that point electrically heated coils provided the temperature variation.

Data on polysulfone under conditions of dry, wet, and 50% relative humidity are shown in Figure 3. Note the effect of water on the magnitude of the −100°C

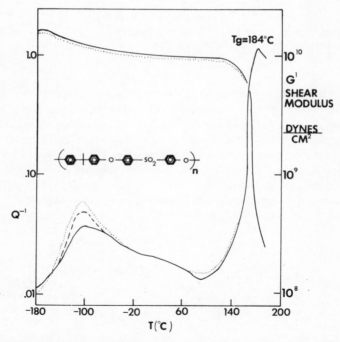

FIGURE 3   Shear modulus and mechanical loss versus temperature for bisphenol A based polysulfone. (————) 0%, (− − −) 50%, (· · · · · ·) 100% rel. humidity. (Reference 3).

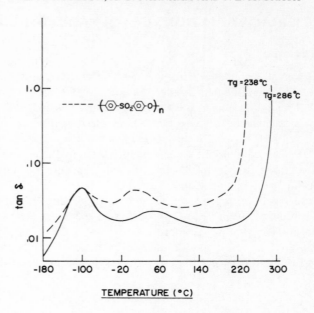

FIGURE 4   Mechanical loss data for bisphenol S polyether (– – –) and Astrel 360 (———), conditioned at 50% rel. humidity. (Reference 3).

transition as has been observed by other investigators. Bis S polyether and Astrel 360 (Figure 4) also exhibit a similar $-100°C$ transition wth another prominent relaxation at $\sim 0°C$. The origin of this relaxation has not been adequately elucidated by prior investigators and will not be speculated on in this paper.

The polyarylethers synthesized by the modified Ullman reaction[3] do not contain the sulfone moiety, and as will be shown by the data do not exhibit any appreciable change in dynamic mechanical characteristics in spite of the observation that they exhibit a pronounced secondary relaxation at $-100°C$. The Bisphenol A-oxydiphenyl polymer

data are shown in Figure 5. The $-100°C$ transition is prominent, but is unaffected by water sorption. Likewise the Bisphenol A-biphenyl polymer structure

(Figure 6) yields similar results. The data on the Bisphenol A-terphenyl polymer (Figure 7),

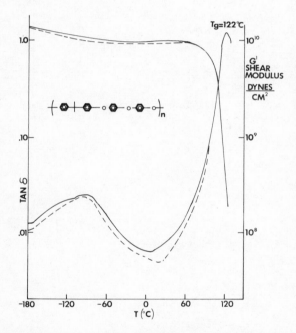

yields a transition at $-100°C$ however, its magnitude appears lower due to the higher loss at temperatures above this transition. The affect of water is judged minimal in this sample also.

FIGURE 5   Shear modulus and mechanical loss versus temperature data for oxydiphenyl—bisphenol A poly(aryl ether). (——) dry, (— — —) wet. (Reference 3).

For the sulfone containing polyarylethers, water exhibits a noticeable effect on the $-100°C$ transition. With dry dimethyl Bisphenol A polysulfone, no transition is observed at $-100°C$ (Figure 8) however, this transiton appears to have been shifted to $45°C$. Water addition, produces a transition at $-100°C$ with no

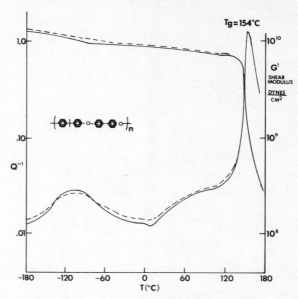

FIGURE 6   Shear modulus and mechanical loss versus temperature data for bisphenol A–biphenyl poly(aryl ether). (–   –   –) dry, (————) wet. (Reference 3).

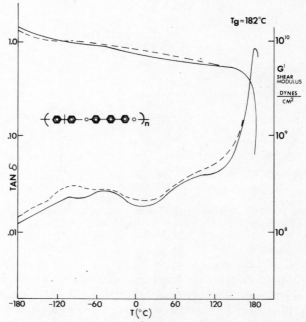

FIGURE 7   Shear modulus and mechanical loss versus temperature data for bisphenol A–terphenyl poly(aryl ether). (————) dry, (–   –   –) wet. (Reference 3).

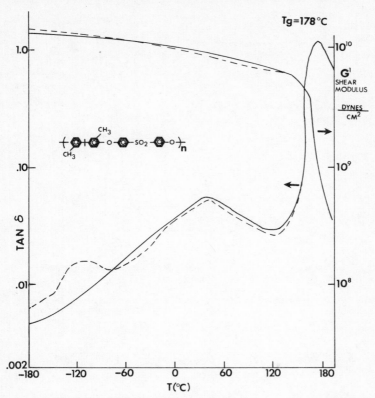

FIGURE 8   Shear modulus and mechanical loss versus temperature data for dimethyl bisphenol A based polysulfone. (————) dry, (— — — —) wet. (Reference 3).

effect on the 45°C transition. This result provides a very important clue to the origin of the typical −100°C transition and the role water plays on its magnitude.

With tetrasubstituted Bisphenol A based polysulfones similar results were noted as shown in Figures 9, 10 and 11, for tetramethyl Bisphenol A polysulfone, tetramethyl biphenyl polysulfone, and tetraisopropyl Bisphenol A polysulfone. All three samples appear to have a small transition in the vicinity of −100°C which appears increased in magnitude with water sorption.

The dynamic mechanical data for the dichloro substituted Bisphenol A polysulfone (Figure 12) and the Bisphenol A polysulfone with dichloro substitution on the diphenyl sulfone moiety (Figure 13) yields quite similar characteristics. These polymers are also equivalent to the dimethyl Bisphenol A polysulfone shown in Figure 8 and provide further evidence to support the hypothesis which will be discussed in the following section.

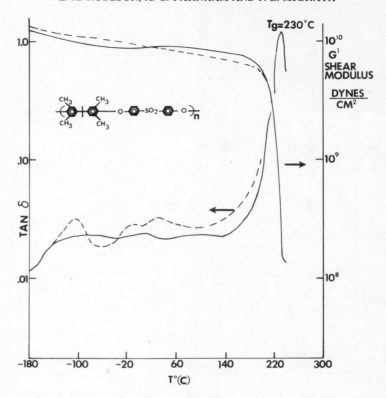

FIGURE 9   Shear modulus and mechanical loss versus temperature data for tetramethyl bisphenol A based polysulfone. (————) dry, (– – – –) 100% relative humidity. (Reference 3).

5   DISCUSSION OF EXPERIMENTAL RESULTS

The non-sulfone containing polyarylethers do not exhibit the effect of water sorption, i.e., the magnitude of the $-100°C$ loss peak is not increased as in all the sulfone containing poly aryl ethers. Dimethyl Bisphenol A polysulfone does show the appearance of the $-100°C$ transition with water sorption, which strongly implies the existence of a relaxation involving the water–SO_2 complex. These data therefore, strongly suggest that the $-100°C$ transition is the result of two separate relaxations in the case of wet sulfone containing polyarylethers, namely, (1); relaxation involving motion (i.e. rotation) in the aryl ether bond, (2); relaxation involving the sulfone–water complex. The data on dichloro substituted polysulfone (Figures 12 and 13) provide additional evidence to support this hypothesis.

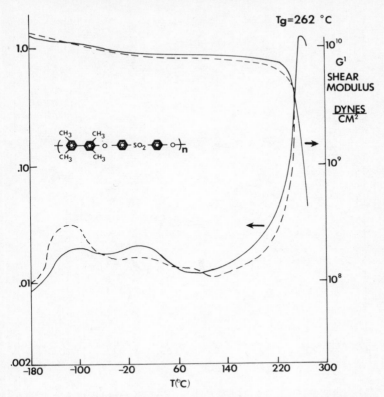

FIGURE 10   Shear modulus and mechanical loss versus temperature data for tetramethyl biphenol based polysulfone. (————) dry, (– – – –) 100% relative humidity. (Reference 3).

While this hypothesis is somewhat different than previous investigators, the literature data tend to support rather than reject the hypothesis presented here. Of course, the availability of polymer structures cited here have allowed these new explanations.

Kurz *et al.*[6] ascribed the −100°C transition (when dry) to motions involving the SO_2 as it was the logical site for water adsorption. They discounted the possibility that the aryl ether group could yield a −100°C transition when dry since

$$\left[\bigcirc - CH_2 - \bigcirc - SO_2 \right]_n$$

yielded a similar transition when dry. One might, however, expect the aryl—

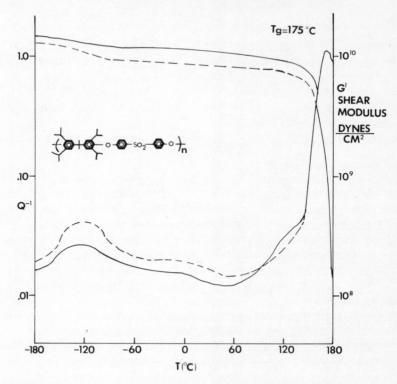

FIGURE 11   Shear modulus and mechanical loss versus temperature data for tetramethyl bisphenol A based polysulfone. (————) dry, (– – – –) wet. (Reference 3).

methylene group to exhibit similar characteristics to the aryl ether group. In fact, with poly(p-xylene)

$$\left[CH_2 - \bigcirc - CH_2 \right]_n$$

a large transition was noted at $-114°C$ compared to $-111°C$ for polysulfone by Chung and Sauers.[9] We believe this analogy offers sufficient reasons to discount the structural conclusions of Kurz et al.[6]

Allen, et al.[7] observed that the wet samples of various polyarylethers (dielectric loss data) exhibited a peak temperature independent of water content, but different than that of dry specimens. It was also noted that the activation energies for the wet samples, while substantially different than for dry specimens,

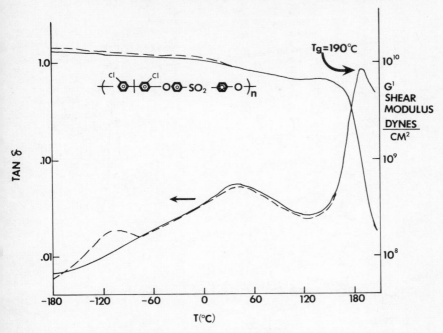

FIGURE 12   Dynamic mechanical data for dichloro substituted bisphenol A polysulfone.
(———) dry, (— — —) wet, 100% relative humidity.

was independent of water concentration. These two results would imply the
possibility of separate relaxations involved with wet vs. dry samples as is present
in our hypothesis. Unfortunately, Allen did not have sufficient structural variations available to arrive at a clear decision on the origin of the $-100°C$ relaxation
and therefore concluded in rather general terms that the absorbed water bonds
to polar groups of the polymer and takes part in the molecular process which
involves this relaxation.

The data on the tetra substituted Bisphenol A and biphenol based polysulfones (dry) are similar to PPO in that a broad relaxation occurs above liquid N_2
temperatures. Peaks appear to be resolvable around $-110°C$ and possibly $0°C$.
However, when wet, a prominent peak at $-100°C$ appears for these samples. The
origin of the dry peaks may be indeed different from that of the polyarylethers
with no aromatic ring substitution which could restrict motion (i.e. rotation) of
the aryl ether and Allen found that this relaxation in PPO was affected by water
sorption in dielectric loss measurements but not in mechanical loss measurements.
This does not detract from our conclusions here since dielectric loss is influenced
markedly by even low concentration of polar constituents (i.e. H_2O).

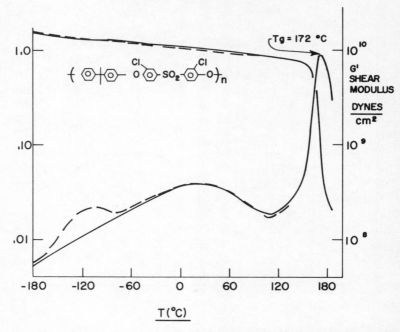

FIGURE 13   Dynamic mechanical data for Bisphenol A polysulfone with dichloro substitution on the diphenyl sulfone group. (————) dry, (− − −) wet, 100% relative humidity.

SULFONATED   POLYSULFONE

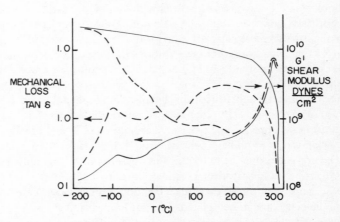

FIGURE 14   Dynamic mechanical data for sulfonated polysulfone as a function of water content. (————) dry, (− − −) 60% water.

As final supporting evidence, recent work by Noshay and Robeson[22] in these laboratories on sulfonated polysulfone reveals an extremely large relaxation at −100°C on highly sulfonated samples (both acidic and neutralized, i.e. −SO$_3$H and −SO$_3$Na) when equilibrated with water. However, the dry specimens show that the −100°C relaxation is lowered in magnitude and shifted to slightly higher temperatures when compared to the base polysulfone. This is illustrated in Figure 14 for a substitution of one −SO$_3$Na) group per repeat unit of polysulfone. Note that the glass transition temperature is markedly shifted to higher temperature by virtue of a highly polar group substitution.

The glass transition, T_g, data points out obvious structural features which can markedly affect the resultant value. The polar −SO$_2$ group is particularly effective in raising the T_g whereas the non-polar isopropylidene unit is not effective. Dimethyl and dichloro substitution have virtually no effect whereas tetramethyl substitution through steric hindrance raises the T_g significantly. Tetraisopropyl substitution lowers the T_g due to the flexibility of the side group. This trend has been observed in other investigations.[23-25] The highly polar −SO$_3$Na group is extremely effective in raising the T_g of the base polysulfone. The difference in T_g's between the dichloro substituted on the Bisphenol A and the dichloro substituted on the diphenyl sulfone can be rationalized by some branching at the 3, or 3′ position for the monomer,

Finally it is of interest to discuss the toughness characteristics of these materials. Polysulfone and all the non-sulfone aryl ethers cited (Table I: structures 1−4) all exhibit ductile characteristics under unnotched impact testing similar to Bisphenol A polycarbonate. Bisphenol S polyether and Astrel 360 likewise exhibit similar toughness. However, di substitution and tetra substitution (Table I: structures 5−10) exhibit brittle characteristics under unnotched impact testing. Structure 8 is somewhat borderline in this regard and structural similarity to PPO may provide an interesting analogy as PPO is ductile in the same testing. Water sorption does not change the relative brittle−ductile rating cited above. As the water −SO$_2$ complex relaxation can be considered a side group relaxation, this observation is in line with the generalization that side group relaxations are not effective in promoting ductility.

References

1. J. B. Rose, *Polymer [London]*, **15**, 456 (1974).
2. R. N. Johnson, A. G. Farnham, R. A. Clendinning, W. F. Hale and C. N. Merriam, *J. Polymer Sci. A-1*, **5**, 2399 (1967).
3. L.M. Robeson, A. G. Farnham, and J. E. McGrath, ACS *Polymer Pre-prints* **16**, 476, (1975). *Applied Polymer Symposia*. No. 26, 373 (1975).
4. J. Heijboer, "Modulus and Damping of Polymers in Relation to their Structure," lecture presented for the Plastics and Polymer Group of the Society of the Chemical Industry, London, Oct. 17, 1967.
5. M. Baccaredda, E. Butta, V. Frosini, and S. DePetris, *J. Polymer Sci. A-2* **5**, 1296 (1967).
6. J. E. Kurz, J. E. Woodbrey, and M. Ohta, *J. Polymer Sci. A-2* **8**, 1169 (1970).
7. G. Allen, J. McAinsh, and G. M. Jeffs, *Polymer [London]* **12**, 85 (1971).
8. L. M. Robeson and J. A. Faucher, *Polymer Letters* **7**, 35 (1969).
9. C. I. Chung and J. A. Sauer, *J. Polymer Sci. A-2*, **9**, 1097 (1971).
10. W. J. Jackson, Jr. and J. R. Caldwell, *Adv. Chem. Ser.* **48**, 185 (1965).
11. W. J. Jackson, Jr. And J. R. Caldwell, *J. Appl. Polymer Sci.* **11**, 211 (1967).
12. W. J. Jackson, Jr. and J. R. Caldwell, *J. Appl. Polymer Sci.* **11**, 227 (1967).
13. L. M. Robeson, *Polymer Engrg. Sci.* **9**, 277 (1969).
14. L. M. Robeson, A. Noshay, M. Matzner, and C. N. Merriam, *Angew. Makromol. Chem.* **29/30**, 47 (1973).
15. S. E. B. Petrie, R. S. Moore, and J. R. Flick, *J. Appl. Phys.* **43**, 4318 (1972).
16. N. Hata, R. Yamauchi, and J. Kumanotani, *J. Appl. Polymer Sci.* **17**, 2173 (1973).
17. M. G. Wyzgoski and G. S. Y. Yeh, *Polymer J.* **4**, 29 (1973).
18. N. Kinjo and T. Nakagawa, *Polymer J.* **4**, 143 (1973).
19. H. Bree, J. Heijboer, L. C. E. Struik, and A. G. M. Tak, *J. Polymer Sci. A-2* **12**, 1857 (1975).
20. P. H. Geil, *Ind. Eng. Chem., Prod. Res. Dev.* **14**, 59 (1975).
21. L. E. Nielsen, *Rev. Sci. Instr.* **22**, 690 (1951).
22. A. Noshay and L. M. Robeson, *ACS Polymer Preprints* **16**, 81 (1975).
23. R. F. Boyer, *Polymer Preprints* **13**, 1124 (1972).
24. R. F. Boyer, *Rubber Chem. Techn.* **36**, 1301 (1963).
25. A. Eisenberg, *ACS Polymer Preprints* **14**, 696 (1973).

DISCUSSION

R. L. Miller (*MMI*): Is there any evidence for crystallization in any of these polymers? Could the relaxation spectra be interpreted in terms of crystallization?

J. E. McGrath: With difficulty, several of the structures may be crystallized however Bisphenol A based polysulfone and substituted Bisphenol A based poly-sulfones are not crystallizable. During the method of preparation cited these materials were judged non-crystalline by virtue of the extremely low modulus above the T_g. With crystallinity levels of 1 to 2% a resultant modulus plateau above the T_g would be easily resolvable. We see no way one could interpret the relaxation in terms of crystallization (especially since crystallinity cannot be detected in most of the structures). Prior investigators cited here have disregarded any crystallinity explanations.

J. L. Work (*Armstrong Cork Co.*): Was there any tendency toward gelling in the preparation of these polymers?

J. E. McGrath: No, the system is normally highly difunctional and therefore produces only linear macromolecules. However, polymers derived from 3,3′-4,4′-tetrachlorodyshenyl sulfone may contain some branching.

R. F. Boyer (*Dow Chemical Co. USA*): I noticed that the 3/4 T_g rule seems to be violated in your polymers; the β-peak, which should appear at about 3/4 T_g, seems to be missing. Do you have any comment?

J. E. McGrath: The transition at $-100°C$ due to the aryl ether bond as well as the SO_2—H_2O relaxation has no dependence on the T_g. As one can vary the structures of polyaryl ethers (and likewise the T_g); one would not expect a relaxation associated with a specific group to be a function of the T_g. The 3/4 T_g rule for β peaks is generalization which we do not ascribe to. In our studies involving many polymer structures, we have not observed this relation-ship to be of great predictive benefit.

Index